THE CHARACTER OF RACES

THE CHARACTER OF RACES

AS INFLUENCED BY PHYSICAL ENVIRONMENT, NATURAL
SELECTION AND HISTORICAL DEVELOPMENT

BY

ELLSWORTH HUNTINGTON

Research Associate in Geography in Yale University

CHARLES SCRIBNER'S SONS
NEW YORK · LONDON
1924

COPYRIGHT, 1924, BY
CHARLES SCRIBNER'S SONS

Printed in the United States of America

CONTENTS

		PAGE
PREFACE	. . . .	vii

CHAPTER		
I.	RACIAL CHARACTER AND NATURAL SELECTION	1
II.	FIRST STEPS IN HUMAN CHARACTER	20
III.	THE EARLIEST GREAT MIGRATIONS	33
IV.	GLACIATION AND THE SUPREMACY OF EUROPE	47
V.	THE SUPPRESSION OF AMERICA	61
VI.	THE CLASSIFICATION OF RACES	73
VII.	THE ANOMALIES OF ABORIGINAL AMERICA	88
VIII.	THE ASIATICS WHO DWELL IN TENTS	112
IX.	JEWS, ARMENIANS, AND TURKS	129
X.	CYCLES OF CHINESE HISTORY	148
XI.	NORTH VERSUS SOUTH IN CHINA	158
XII.	THE SCOURGE OF FAMINE	170
XIII.	THE SELECTION OF THE CHINESE	184
XIV.	THE THREE GREAT RACES OF EUROPE	205
XV.	THE CHARACTER OF MODERN EUROPE	220
XVI.	THE CONTRAST BETWEEN GREEKS AND IRISH	236
XVII.	THE DISPERSAL OF THE NORTHMEN	252

CONTENTS

CHAPTER		PAGE
XVIII.	Warlike Normans and Peaceful Icelanders	264
XIX.	The Persistence of a Selected Inheritance	277
XX.	The Direct Effect of Environment on Character	286
XXI.	The Selection of Modern Americans	301
XXII.	The Racial Tendencies of Civilization	332
XXIII.	A Racial Test of Cities, Democracy, and Feminism	346
	List of References	374
	Index	379

ILLUSTRATIONS

FIG.		
1.	The Pressure Zone between Glaciation and Deserts	*Facing page* 48
2.	Generalized Zones of Primitive Migrations . . .	" " 76
3.	Head-form and Routes of Migration in Europe . .	" " 78
4.	Maps Showing the Correlation of Cultural Conditions with the Cephalic Index . . . . *Facing page* 80
5.	Birthplaces of High Officials in China	" " 162
6.	Percentage of Scientists among Eminent Europeans	*Between pages* 222 and 223
7.	Percentage of Eminent Europeans Engaged in Religious, Educational, and Philosophical Work . . *Between pages* 222 and 223
8.	Percentage of Eminent Europeans Engaged in Art and Literature	*Between pages* 222 and 223
9.	Percentage of Historians among Eminent Europeans	*Between pages* 222 and 223
10.	Percentage of Eminent Europeans Engaged in War and Politics	*Between pages* 230 and 231
11.	Distribution of Civilization in Europe .	" " "
12.	Eminent Europeans Born since 1600 A. D. per 10,000 of Estimated Population in 1800 A. D. . . . . . *Between pages* 230 and 231
13.	Distribution of Health in Europe . .	" " "
14.	Distribution of Climatic Energy in Europe . . . *Facing page* 232
15.	Approximate Variations of Rainfall in California during the Christian Era . . . . *Facing page* 232
16.	Deaths of Men Compared with Women at Various Ages in Iceland, Norway, and Switzerland, 1876–1915 . . *Facing page* 314

ILLUSTRATIONS

FIG.
17. Persons in *Who's Who* (1922-1923) Born in Each State per 1,000,000 of Mean White Population 1850-1880, Giving Double Weight to 1860 and 1870 . . . . . *Facing page* 314

18. Residents Included in *Who's Who* (1922-1923) per 1,000,000 Population in 1923 . . . . *Facing page* 316

19. Increase (or decrease) (1912 to 1922) in Proportion of Eminent Persons Born per 100,000 of Population . . . *Facing page* 316

TO
THE ONE TO WHOM I OWE MOST
MY MOTHER

PREFACE

Some nine years ago, in a book called *Civilization and Climate*, I set forth the thesis that the general distribution of civilization and progress depends largely upon climate. I also attempted to show that the influence of climate is strongly modified by migrations of peoples and cultures, by racial mixture, by inventions and discoveries such as iron tools and agriculture, and by ideas such as religion. After reasonable allowance had been made for all these factors, I still felt that certain highly significant features of the distribution of human progress remained unexplained. Why for example did such meteoric intellectual brilliancy occur in Greece? Why are such wonderful Hindu ruins located in Cambodia in Indo-China? Why has a high type of culture been able to persist for a thousand years in stormy Iceland? And why do we see in China the curious anomaly of a progressive South and a conservative North? After years of search I believe that at least a partial explanation is found in natural selection arising most frequently under the stress of over-population and migration. The principle involved is so simple and obvious that it seems almost incredible that it has been so largely ignored by students of geography, history, and sociology.

A study of famines in China was the step which disclosed the importance of this principle. Having long been puzzled over the fact that north China does not show the degree of progress that one would expect on the basis of either its climate, its institutions, or its past achievements, I was stimulated to further investigation by an article by Doctor Carl W. Bishop in *The Geographical Review* (vol. VII, 1922). Doctor Bishop points out most clearly the nature of the contrast between the "staid and conservative" North of China and the "alert and progressive" although semi-tropical South. It occurred to me that famines might have something to do with this. Accordingly, I wrote to an old friend, Reverend Robert E. Chandler, a missionary at

Tientsin, China, and asked his opinion as to the effect of famines in selecting certain physical or mental types for destruction, preservation, or migration. He not only sent his own answer, but procured a series of most interesting letters from other missionaries, including Doctor Arthur H. Smith, author of *Chinese Characteristics* and *Village Life in China;* Doctor F. F. Tucker, head of the American Hospital at Tehchow; Professor Guy W. Sarvis, of Nanking University; Reverend George D. Wilder, of Peking; and Reverend Henry Lieper, of Tientsin. All of these men have had personal experience of famines and of relief work. Later I visited China and interviewed still other missionaries, among whom I must mention Doctor William E. Souter, of the China International Famine Relief Commission; Reverend W. E. Macklin, of Nanking; Reverend R. A. Torrey, of Tsinan, who has lived among the famine-stricken people of Shantung and has thought deeply on the significance of what he has seen; Doctor J. H. Ingraham, of Peking; and Mr. E. W. Edwards, of the Peking Y. M. C. A., secretary of the China International Famine Relief Commission. All of these men, as well as others too numerous to mention, were extremely kind in giving me a wealth of information which seems to show that the process of natural selection through over-population, famine, and migration is taking place in China to-day at a rapid rate. Historical records prove that it has taken place with equal or greater rapidity during a large part of the last two thousand years or more. The case is so clear that the chapters on China form perhaps the most important section of this book.

Iceland furnishes another equally clear-cut case. The conditions there, however, are quite different from those of China, for in that cool northern island the main selection of the inhabitants took place once for all; the noteworthy fact is the extraordinary persistence of a high inheritance when once it had been segregated. The outward contrast between China and Iceland is so pronounced, while the similarity of the principles involved is so great, that in spite of the scanty population it has seemed worth while to treat Iceland almost as fully as China. The rest of the book is devoted to an attempt to apply the biological principle of natural selection to man's early history and distribution, and then more fully to specific historical instances. The number of such instances is so large and their intrinsic interest so great that it has been no easy matter to choose among them.

Three peculiarities of this book perhaps need explanation. First, natural selection is constantly emphasized, while other equally important factors are made subordinate, although great care is taken to mention them; second, only a few of the hundreds of fine examples afforded by history are here treated; and third, much space is devoted to a somewhat speculative account of man's early evolution and migrations.

The reason for all three peculiarities is the same: One cannot do everything in a single book. This book is the logical companion of *Civilization and Climate*. Each illustrates a great principle which has generally been overlooked in the study of history. Neither pretends to be complete, for where a new subject is first presented the mere limitations of space, as well as the dictates of psychology, make it necessary to concentrate upon a single theme. For a complete, well-rounded view of history one must combine these two books with many others which discuss such matters as geographic location and natural resources; human inventions, discoveries, and ideas; the influence of men of genius; the economic forces that bind mankind so closely; the growth and pressure of population; and the interplay of war, religion, intrigue, and human ambition. Some day the progress of science will give us a scientific history in which all these factors, as well as the factors of climate and natural selection, are co-ordinated and given due weight. As a help toward such a sketch, it has seemed to me wise to show how the principles of climatic change and natural selection probably apply to the earliest human development, as well as to outstanding examples from later periods. It has also seemed wise to concentrate on natural selection through physical causes, and to devote to other factors only enough space to convince the reader that I appreciate their great importance in modifying human character.

In this book, as in every other that I have written, I owe much to a group of kind friends, to all of whom I here tender most hearty thanks. The manuscript has been read and criticised in whole or in part by my colleagues Professors Irving Fisher, Richard S. Lull, and George G. McCurdy; by Professor Griffith Taylor, of Australia; and also by my wife and by Mr. R. V. Coleman, of the Yale University Press. Had it not been for Mr. Coleman's insistence I doubt whether this book would have been written. Five years ago he persuaded me to write on a topic which at first was tentatively expressed as *The Influence of Climate upon Mankind* and later as *The Effect of Physical Environment upon the Origin and Modification of Racial Character*. Professor Frederick W. Williams, of Yale University, and Professor Stephen S. Visher, of Indiana University, have not only read the manuscript but have devoted much time to providing material which has been the basis of long sections of the book.

I also wish gratefully to acknowledge my indebtedness to the following publishers and authors for permission to quote from their books: to Stanislaus Novakovsky, for use of a passage from his unpublished manuscript on "Hysteria Artica"; to E. P. Dutton & Company and E. H. Parker, for the use of passages from "China, Her History, Diplomacy, and Commerce: From the Earliest Times to the Present Day"; to Harper & Brothers and Sir Philip Gibbs, for the use of passages from "Now It Can Be Told" and "Adventures in Journalism"; to Hurst & Blackett, Ltd., and Havelock Ellis for the use of a passage from "A Study of British Genius"; to Oxford University Press for the use of passages from "Studies in History and Jurisprudence," by James Bryce; to Roland B. Dixon, for the use of passages from "The Racial History of Man"; to William McDougall for the use of passages from "Is America Safe for Democracy?" to Macmillan & Co., Ltd., of London, and The Macmillan Company of New York, for the use of passages from "Homer and History," by Walter Leaf; to Yale University Press for the use of a passage from the preface; by Charles M. Andrews, to "The Colonizing Activities of the English Puritans"; to Columbia Studies and Edwin

L. Clarke for the use of passages from "American Men of Letters: Their Nature and Nurture"; to Columbia Studies and Mabel P. Lee for the use of passages from "The Economic History of China, with Special Reference to Agriculture"; to F. W. Williams, for the use of passages from a manuscript entitled, "A History of China"; to S. S. Visher, for the use of "A Study of the Type of Place of Birth and Occupations of Fathers of Persons in Who's Who in America."

E. H.

YALE UNIVERSITY,
NEW HAVEN, CONN.,
August, 1924.

THE CHARACTER OF RACES

CHAPTER I

RACIAL CHARACTER AND NATURAL SELECTION

A CENTURY or two ago it was widely held that all races and even all individuals are equally endowed by nature. Physically there might indeed be enormous diversity, but differences in character and mentality were supposed to be due entirely to training. Then there arose a school of thinkers who violently combated this idea. They wrote books like *The Inequality of Races*, by Gobineau,* in order to convince the world that mentally as well as physically one race actually differs from another. The theory of evolution reinforced their conclusions; according to that theory it is almost inevitable not only that different races should be in different stages of mental as well as physical development, but that they should develop along different and divergent lines. Then came Mendel with his explanation of the mechanism by which individual traits are passed from parent to child, and Galton with his insistence on the importance of heredity. At last the idea of racial differences became so firmly established that a leading scientist could say: "Race has played a far larger part than either language or nationality in moulding the destinies of man." (Osborn, in *The Passing of the Great Race*, by Madison Grant.)

Then the problem of race left the realm of scientific discussion and flared up as an idea belonging to the people as a whole. In our own day scores of writers have declared or implied that racial inheritance is the most potent of all forces. It, and it alone, we are sometimes told, has determined the rise and fall of nations. It is even asserted that no great and permanent advance in civ-

---

* See list of references preceding the index.

ilization has ever been made except under the leadership of one
limited race, the Nordics. To-day the world is vibrating with
racial questions. The various races, or at least the people who
suppose that they belong to various races, are beginning to look
askance at one another as never before. There is a fierce cry for
the rights of racial minorities; there are rumors that great and
devastating racial wars are brewing for the next generation. If
open war is not threatened, there is even greater danger that the
highest racial values will be irrevocably swamped by those of
lower caliber.

The modernness of this whole idea of the importance of race
is well brought out by Lord Bryce in his lecture on *Race Sentiment as a Factor in History*. From a review of history he concludes that the mere consciousness of racial affinity has had almost no effect upon the contact of nation with nation. The
Greeks, to be sure, differentiated sharply between themselves
and the barbarians. Their cities fought side by side in an effort
to stay the advance of the Persians, and seem to have done this
largely because the enemy was of an alien race. But the Greeks
also fought fiercely against one another, and thereby contributed
greatly to their own decay. The Phœnicians likewise had a
strong racial feeling which caused them to inform the Persian
kings that while they were willing to fight their maritime rivals,
the Greeks, they would not serve against their kinsmen of Carthage. But perhaps it was mere community of language, or the
exigencies of commerce, which held the Phœnicians and their
colonies together. In the same way, but more markedly, the
ancient Jews showed an almost unequalled racial solidarity.
Nevertheless, the Israelites not only intermarried with neighboring Egyptians, Philistines, and Hittites, but adopted their gods.
The various tribes of ancient Israel never thought of helping one
another against the Syrian invaders. To-day religion is probably more potent than race in preserving Jewish unity.

Aside from the Greeks, Phœnicians, and Jews, Bryce finds no
case previous to the last century where people have waged war
or refrained from waging war because of racial motives. He may

have overlooked an agelong antipathy between the Iranian Persians and the Turanian Tartars, and likewise between Persians and Semites, and Semites and Tartars. These aversions appear in both ancient and modern history and appear to be neither religious nor geographic. Nevertheless, Bryce is not greatly in error when he says that people have fought for plunder, for land, for conquest, for religion, for commercial supremacy, but practically never because of racial unity or racial antipathy. Yet in our own day this motive is appealed to again and again. It did not, to be sure, prevent England from siding against the Germans in 1914, even though both nations suppose that they are more closely akin to one another than to the French. Nevertheless, the consciousness of racial kinships is abroad in the earth. It expresses itself in the so-called principle of nationality. In the last century this inspired Greeks, Italians, Poles, and Magyars to strive for political independence. Later it aroused Serbs, Roumans, Bulgars, and Armenians. In our day a wave of racial feeling has swept India, Egypt, and many other lands. As Bryce puts it, "Race consciousness sprang into life and became the core of Nationality." That this racial consciousness is fast pervading the world can scarcely be questioned. According to Clutton-Brock's illuminating suggestion in *The Atlantic Monthly*, many a man who knows that in his own self he has no special superiority over his fellows takes refuge in the thought that he belongs to a superior race. Thus the most incompetent Anglo-Saxon often looks down upon the most competent Italian, because the Italian belongs to the "Dago" race, or upon the greatest of Chinese savants because the wise man's skin is yellow and his eyes aslant.

Whether the present conception of racial differences is right or wrong, it seems destined to play a great part in the history of the next few generations, for it has become embedded in the world's equipment of ideas. If once a certain tenet is accepted by the mass of the people its verity or falsity makes little difference in its potency, as Le Bon well shows in his *Psychology of Peoples*. The mere fact that an idea has become the common property of the masses as well as of the leaders means that it cannot be eradicated for a long time. A new, and let us hope a truer, idea must first grow up in the minds of original thinkers, then it must spread among serious students. Next it must be popularized among the intelligent classes, and only then can it replace the old idea as the heritage of the whole people and be a main determinant of history.

If racial consciousness is playing and is likely to play so great a part in the world's progress, it is of the first importance to understand what racial character is, and how that character has come into existence. Is racial character something inherent, as Bryce seems to imply when he says, "After all, it is the natural racial divisions rather than political divisions that count everywhere outside western Europe"? (Letter in Huntington's *Civilization and Climate*.) Or is it the result of environment, as the same author suggests when he says in his lecture on *Race Sentiment*: "The importance of what may be called the racial constituent in national character has been much exaggerated. Something is due to it, but much more is due to the conditions, physical and economic and social, under which the nation has been developed. In the thought and imagination of every civilized people there is an unquestionable racial strain. But the habit of referring to this cause, practical aptitudes, such as the merits or defects visible in a nation's political life, has been pushed much too far."

But what is a race? Almost every intelligent person has some idea of five major races. They are variously described as: (1) white, Caucasian, or European; (2) yellow, Mongolian, or Asiatic; (3) red, Amerind, or American; (4) brown, or Malay; (5) black, Negro, or African. A glance at the faces of the people in any cosmopolitan city discloses many of the criteria on which this grouping of races is based. But it should be noted that in every case one or more geographical designations accompany the designation based on complexion. It has long been recognized, however, that other means of classification may be quite as important as those based on complexion and geographical location.

RACIAL CHARACTER AND NATURAL SELECTION

The white race of Europe is now commonly divided into three main types in which the form of the head, the stature, and the type of hair as well as the complexion are important criteria. The Nordics, as is now generally agreed, are a long-headed, tall, blue-eyed, fair-haired people, generally with curly or wavy hair. The Alpines, who predominate in the east and center of the continent, are broad-headed, of medium height, and more stocky than the Nordics. They are typically brown-eyed and brown-haired, and generally their hair is either straight or only slightly wavy. The Mediterraneans, who dwell largely in the south and southeast, are long-headed like the Nordics, but short of stature, black-eyed, and with straight black hair. Each of the three races is best developed in certain areas, although there is a vast amount of overlapping and intermixture.

Sometimes the division of mankind into races goes still farther and is based on still other criteria. Although the Aryan or Indo-European race is not really a race but a group of people speaking kindred languages, it occupies an extraordinarily large place in history and literature. Again, we often ascribe certain characteristics to the French as a whole or to the French of Provence, for example, and say that this is because the French are Latins by race. It is sometimes said that the people of the Faroe Islands are a race of hardy fishermen. The Turks are called a race of Turanian invaders despite the fact that Greek, Armenian, and other blood is probably as abundant among them as is that of the original Turkish stock, the Turkomans of central Asia. Evidently the word race is used in very various senses. It would be better to use the word stock instead of race if only it had an adjective corresponding to "racial." In the preceding paragraphs the criteria by which one race or stock is separated from another fall into three groups: first, inherited bodily characteristics, including the form of the head, stature, complexion, and the character of the hair; second, geographical conditions, including position on the earth's surface and occupation; and third, cultural conditions, such as language in the case of the Indo-Aryans and Latins, and religion in the case of the Turks.

Government might also be added, because there is a tendency to consider the people who live under a single government as allied in race. Yet after all is not mental and moral character more important than any other factor in differentiating race from race? One race holds aloof from another largely because each one, rightly or wrongly, believes that the other is somehow inferior or perhaps superior; and these terms generally mean of less or greater mental ability and energy. These mental differences are what we want to study.

In the strictest sense of the word, races should be defined only in terms of heredity. But certain geographical factors, such as climate, food, and occupations, have a distinct effect in changing racial characteristics. They cause people to grow up with certain habits; they also select certain types for preservation and eliminate others. Thus the fishing industry tends to eliminate people of a timid disposition. Such people may succeed as farmers, but not on the sea. Hence, in course of time, perhaps a very long time, each environment and each occupation tends to make its people slightly different from those in other environments and other occupations. Moreover, social conditions such as religion, language, government, education, and local customs have a great deal to do with determining a people's character. They play a part not only in controlling the training of children and youth, but in determining what types of people shall intermarry, what types shall have large families, and what types shall die out. Thus physical and social conditions may have a great effect upon human character both directly and through inheritance. Hence the character of specific groups of mankind depends on inheritance, physical environment, and social environment. Our task is to attempt to show how far these two types of environment select certain kinds of character for preservation or destruction and thus cause certain mental characteristics to become a permanent part of the racial inheritance.

From the biological standpoint we are quite certain that the inherent mental and physical differences between one race or stock and another are largely due to three chief causes: first, sudden mutations, or possibly small but progressive deviations from the standard type in a definite direction; second, racial mixture; and third, natural selection. No clear understanding of mutations and deviations has yet been attained. We merely know that they occur. Not only does brother differ from brother and sister from sister for no assignable reason, but sometimes certain groups of plants and animals, and presumably of men, show persistent new traits which apparently tend in a definite direction. As to how or why such mutations or deviations occur we have as yet no definite knowledge. On the other hand, we know a great deal about natural selection and racial mixture, for we can see them in operation. They are the two great processes to which we can thus far appeal for an explanation of the innate differences which distinguish species from species and race from race. Hence this book is an attempt to investigate the interrelations between migration, racial mixture, and natural selection. The purpose is to discover, if possible, how these three processes co-operate with mutations or deviations in giving rise to the character of races or of racial stocks. A parable and an example will make the matter clear.

Here is the parable: Now it came to pass that in a certain land there dwelt a farmer. His lands were narrow and his children many. When his sons began to take wives unto themselves, they were hard pressed to obtain food and raiment. So they consulted together and said: "What shall we do that we may live and find comfort?"

Pioneer was the eldest son and he was the first to speak. "I love to travel," quoth he, "and fain would have adventures in those far lands which God has made in such strange fashion. Men say that twenty days' hard journey beyond the mountains there is much good land. No people dwell there, or if there be any they are monstrous strange and wild. Thither will I go and have adventures and make a home for myself."

And his wife said: "Yea, that is good, I will go with you, even though my woman's heart misgiveth me. But I am strong of body, and we are set on high adventure. Our children shall not be slaves of poverty as we have been."

Then Cityman, the second son, began to speak. "Why play

the fool and stick your neck within the noose of a land where all is uncertainty and danger? Is not the city a better place for him that would get much from life? Thither will I go and amass great wealth, and there shall I live in luxury. Always my wife demandeth silks, satins, fine jewels, and costly furniture. She would see theaters and great people, and she wearieth my ears with her cackle of princes and great dames, and dances and dinners. To the city will I go and fill her with that for which her soul doth crave."

Now the third son was slow and timid, and was espoused to a maid who loved the kine and the pigs and the making of golden butter. To him and to her there was no sound so sweet as the clucking of a hen who findeth a paltry worm for her chicks. When Plowman, for that was his name, had talked well with his betrothed, he came to his father and said: "The maid and I have no love for new lands, nor yet for cities. It is enough for us that we go once a month to market. We would stay here with you, and keep this good old home of ours. Let my brothers go, and I will buy their inheritance."

There was still a fourth son of that father, and Seeker-for-Truth was his name. He was a quiet, studious lad, brave yet shy. His mother cherished him because he understood more quickly than his brothers and could be trusted farther. "Stay here, my son," she said. "Our hearts have need of you." But sadly he answered: "Nay, I have read all books that here are found, and have become the School Master's master. Yet have I only set lip to the goblet of knowledge. Deep must I drink. Then must I find the fountain whence my draft was drawn; from its depths I must raise new truth to fill the goblet for them that follow me athirst. I must fare forth alone, to farther, stranger realms than any that men yet have seen save in their dreams."

And it came to pass as each one had said. Beyond the mountains in a land of plenty lay the new-made home of Pioneer. And other pioneers settled around him, and they had many sons and daughters. Some of the sons and daughters inherited a richer dower than others, for they had double measure of that high spirit which brought their brave fathers and braver mothers out to the wilderness. These in their turn pressed onward to new lands, and when their children were of age, the pioneers among a race of pioneers moved on once more. Thus the farthest of the good lands beyond the mountains were peopled by a new race—a race in which the spirit of achievement was bred in the bone. In their very blood ran the thirst for adventure and progress, and they became the greatest nation of their time.

With Cityman it fared far differently. He helped to build a city full of wonderment, but "Eat, drink, and be merry" was the watchword of his house. Few were his children: they were trained in luxury; and they espoused the sons and daughters of other city men brought up in equal selfishness. Had they and their children and their children's children been left alone for a score of generations that city would have perished from the face of the earth, for the vigor of its sons and daughters faded away. The city lived only because new people came from the old homes on the farms.

But Plowman's life was of another kind. Poor he was, indeed, and ignorant. Not even that which would have brought him wealth seemed in his eyes desirable if it demanded aught that was new in thought or deed. "Nay," said his wife who loved the clucking hens, "why change? We know the old and we can meet it, but we neither know nor want that which is new." For "Nay" was ever the word that she loved. Now Plowman and his wife had sons and daughters. So too did the neighboring plowmen and their wives. And if by chance among their children or grandchildren unto the twentieth generation there arose a pioneer, a city man, or a seeker for truth, that son or daughter fared forth to other lands, until at last the plowmen were all plowmen.

Worst was the fate of Seeker-for-Truth. A score of years he wandered with never a thought of home. At length he found another seeker and her he wed, but she had long been wandering and her youth was gone. Two sons she bore him, but there were no other seekers in those parts. So one of those two sons went to his death unmated. The other wed a pioneer, but with their son that race of seekers ended, and its genius was forever lost.

This is the parable of the farmer and his sons. It is also the parable of history—the parable of races. Let us put the parable into the language of history and see its interpretation. We shall begin with an example which illustrates the main principles as they apply to pioneers. Examples of the other types will come later. The details of the present example are only dimly known, but the general outlines are unmistakable. I insert it here not as something well proved, but as an interesting example which follows out the spirit of our parable—some day this same example may be rewritten with far more detail and exactness. In later chapters we shall cite many other examples less picturesque, perhaps, but well-known even in their minute details.

One of the world's most famous ruins is Angkor Wat. In the dense forests of Cambodia in French Indo-China, some two hundred miles northwest of Saigon, scores of wonderful ruins stand near a desolate lake embowered in tropical vegetation. Once Angkor Wat and its neighbors must have been great cities of marvelous beauty. To-day the ruined temples excite the profound admiration of every thoughtful visitor. The numerous huge structures with their lofty walls, carefully carved statues, and vast size indicate that they were built by a people not only of great patience and industry, but of high artistic capacity and real originality. No ordinary minds could plan such great and complicated buildings. Nor could any one man have designed all parts of each of the greater structures, with their innumerable groups of figures carved in high relief. One scene alone contains a thousand figures of warriors, charioteers, and all the actors in a mighty battle. The builders must have had in their midst many men of high artistic genius. They must have been assisted by many others who were skilled in the arts of quarrying and transporting stone, of carving it true to the line, and of planning the work of thousands of laborers. Likewise they must have had at their command large numbers of people who could be supported by the work of others while they labored year after year on the temples and palaces. In other words, the builders of Angkor Wat, Angkor Tom, and the other ancient temples must have been a people among whom there were at least a few great geniuses and many men of much more than the average ability and energy.

One of the strange facts about the art of Cambodia is that it begins suddenly without precursors and comes to an end with almost equal suddenness. In Central America, where the ruins left by the Mayas have many qualities like those of the Cambodian ruins, one finds a long series of earlier structures gradually leading up to the most beautiful of those that now persist. In Cambodia the most beautiful ruin is the oldest. It is a small temple of surpassing delicacy and great originality, and was probably built in the seventh or eighth century after Christ. Older ruins may perhaps be discovered when the damp recesses of the dense forest are finally explored, but at present one must go fifteen hundred miles to India to find a type of art like that of Cambodia. But even there one does not find the direct predecessors of the Cambodian ruins, for Angkor and its neighbors are unique in many respects. The builders evidently derived the foundations of their knowledge from India, but they evolved new ideas and a new type of art which are distinctly their own.

For four centuries or more they built according to this new style. Generation by generation the size of the buildings increased, but true beauty and delicacy declined and originality diminished. In Cambodia, as in modern China, the later ruins tend toward a grandiose and highly ornate style less beautiful than the original delicate simplicity. Yet even the latest ruins inspire profound admiration not only by their size but by their genuine beauty. Only men of high ability could have planned and executed them.

Before the last temple was finished some catastrophe, or perchance merely a little revolution or the death of a great leader or architect, put an end to the work at Angkor Wat, so that the last of the main temples was left unfinished. Thus the great artistic age of Cambodia came to an end after enduring about four hundred years. Only in a few rare instances does one find so sudden an outburst of art, or one which rises so promptly to such brilliancy and ends so abruptly without leaving any real successors either in its own region or elsewhere, so far as we are aware.

Little by little the outlines of the history of Cambodia are taking form. Its builders were the Khmers, a part of the Brahman race, those Children of the Sun who appear to have been the last main body of invaders to pour into India before the Christian era. According to an unverified tradition, during the fourth century before Christ a certain prince of Delhi strove to make his father divide the kingdom with him. Failing that, he moved eastward with a great band of followers. This band, it is said, became the ancestors of the Khmers, who finally reached Cambodia. How true this story is we do not know, but this much we may accept with considerable certainty: The Brahmans came from somewhere in central Asia, migrating down into India across the western end of the great mass of central mountains or through the deserts of Afghanistan and its neighbors. Such a migration, as we shall presently see, is a strenuous process and involves great natural selection. If the men migrate alone there is almost invariably much fighting as they press into new regions. The most inefficient, cowardly, and dull are left at home or soon fall by the way. As the migrants press farther into new lands, the ones who are less brave, less resourceful, and less vigorous are killed in battle or die of disease. Thus such a migration weakens the community from which it starts, but produces at the end a small body of highly picked and very competent people, the cream of the group from which it fared forth.

If the whole community migrates—men, women, and children —as appears to have happened in most of the ancient migrations and presumably in the case of the Brahmans, the process of selection takes place even more intensely. Imagine what it is like for women with little children to move with their husbands into new and perhaps hostile lands across mountains, rivers, and deserts; in summer or winter; in storm and flood; and at any and every period of their lives regardless of whether they are fit or not. Under such circumstances the death-rate almost invariably becomes enormous: only the most vigorous and intelligent mothers can survive, only those who have the most endurance, initiative, and spirit of adventure; and among the children only those who are exceptionally bright, vigorous, and attractive. As a rule the personal attractiveness of the child, which is merely another name for its degree of mental alertness and physical perfection, counts greatly; for the attractive child is carried by the adults when it is tired, fed when it is hungry, given the best place to sleep, and in almost every way given advantages and chances for preservation which are denied to the stupid, crying, unattractive child. And the same is true of the women, for those who are most attractive, which in general means those who are not only most perfect physically but most alert mentally, are the ones whom the men are most willing to help; and they are likewise the ones who do not give up and die or fall by the wayside and allow themselves to become the captives of the enemy. Thus a hard migration of a whole community means that the process of natural selection is at work with great vigor to weed out all who are weak and to preserve a remnant who are especially capable. Among the Brahmans we have no record of this, but in later periods we shall see the intimate details of the selection which is even now altering the character of races.

It was presumably due in considerable measure to this stringent process of natural selection that the Brahmans who reached India were able to conquer the earlier inhabitants and impose themselves as a ruling class which later became a caste. To this may also have been due much of the unusual genius which enabled the people of India to evolve two new religions, Buddhism and Hinduism, to frame a new social system, to devise a noteworthy type of art and architecture, and to write a series of great poems which cause Sanskrit literature to rank among the greatest of those developed among primitive people.

The Khmers appear to have been an offshoot of these competent Brahmans in the days when the original ability of that people was still active. By what route the Khmers travelled, just when they left the old home in northern India, how many different waves of migration there were, and how many generations or centuries each set of migrants spent upon the road we do not know. Nor do we know the causes of the migration, the peoples with whom the migrants fought, and the number of migrants who were killed by the hard conditions of their journeyings. But we do know that they must have travelled at least two thousand five hundred miles, and that in their journey they must have encountered enormous physical difficulties. Moreover, unless they came by sea, which is highly improbable in view of the inland location of the ruins, they must have passed through many regions that seem to our eyes more favorable for settlement than the place where they finally came to rest. First their way led across the fertile plains of the Ganges for eight hundred or perhaps a thousand miles. On their way they must have encountered many inhabitants, for India then appears to have been fairly well peopled. And almost certainly they had to fight again and again. And it is almost equally certain that many men, women, and children—the weaker ones—must have died of disease and hardship, or been taken prisoners or killed by enemies in that first stage of the journey. Perhaps the migrants tried to settle, just as modern migrants in western China have tried again and again, but each time were forced to move on, either because they chose a poor place, or because of the hostility of their neighbors.

Beyond the Ganges plain a more difficult country faced the migrants. The great Brahmaputra River had to be crossed, and one may well imagine that it delayed the wanderers many years, and cost them many lives, because that same delay presumably exposed them to the prolonged enmity of hostile tribes along the river's bank. Beyond the Brahmaputra the wet plains with their floods and fevers give place to a maze of great mountains, an extremely difficult tract in which to wander, especially among unfriendly people. That the people of the mountains were unfriendly can scarcely be doubted, for that is practically always the case under such conditions. Beyond the mountains lies the fertile and pleasant plain of Upper Burmah. There, if anywhere, one would expect a migrant people to settle. Perhaps they did settle there for generations, although as to this we have no evidence, but ultimately something drove them on. It may have been their own restlessness, curiosity, and desire for a better land, or the advent of hard times and famine, or the push of new invaders, or a revolution among the former plains-dwellers whom they had perhaps conquered. Or possibly the Khmers were never able to conquer the Burman plain, and had to keep to the rough mountains that surround it.

At any rate the Khmers moved on, and thereby inevitably doomed themselves to new hardships, more deaths, and a still more stringent process of natural selection. Thus they crossed the great Irrawadi River and the Salwin in its enormous gorge amid a maze of the wildest mountains. Then they either came down into the fertile plain of the Menam River in the best part of Siam, or kept near the headwaters of that river among difficult mountain ranges. But even if they reached the plain, something still drove them on. So they traversed vast somber tropical forests, pushed their way through the jungle, climbed the tree-clad hills, endured the fevers of the swampy lowlands, and finally reached Cambodia. A hazardous journey of two thousand five hundred miles or more lay behind them; decades, generations, or even centuries had been spent among all sorts of hardships and dangers, and an enormous percentage of all who had been born on that long hard journey had almost certainly died before they became old enough to be the parents of the next generation. Natural selection, then, had presumably had an opportunity to do its utmost. At the beginning the Khmers appear to have been an especially adventurous and active portion of the Brahman race which had already been highly selected, and which had been making notable contributions to human progress. At the end the selected Brahmans had presumably been still more highly selected. Is it then surprising that they achieved great things?

Just why the Khmers finally settled in Cambodia we can only conjecture, but probably it was because there they finally mastered the original population to such an extent that though few in numbers they could impose their will upon their subjects. From time to time new bodies of migrants reached Cambodia, the most noticeable migration being in the fifth century A. D., when the Khmers as a nation rose into prominence under S'rutavarman. When their wanderings came to an end, the energy and ability which had hitherto been devoted to the task of preserving life and migrating into fresh fields were presumably turned to governing a relatively inert tropical people, subduing the tropical forest, and building up the framework of civilization. Then came an era of architecture, and ultimately a period when the rulers established the brilliant court which at a much later date is vividly described by a member of a Chinese embassy. But that was near the end of the thirteenth century, long after the period of constructive progress had come to an end. In the work of building up their own peculiar type of civilization the genius of these highly selected Khmers turned especially to architecture. For their main ideas they naturally looked back to India, with which they may then have communicated by sea rather than by land. But they were by no means mere copyists: the process of natural selection had apparently compelled the mating of pioneer with pioneer for generations—unusually competent men with unusually competent women—and had left little opportunity for degeneration through the mating of the competent with the incompetent. This may be the main reason why the small body of Khmers who dominated the native Cambodians were exceptionally gifted. Their gifts expressed themselves in a sudden and brilliant burst of architecture and sculpture, and, by inference, in a form of government which was able to utilize vast numbers of the natives as workmen and yet keep the remainder employed on the land in such a way that there was wealth enough to maintain both the common people and the luxurious rulers for several hundred years.

Just why the Khmers fell from their high position we do not

know. The immediate occasion of their fall seems to have been an incursion of a Siamese people into Cambodia from the northwest in the latter part of the twelfth century. At about the same time the Khmers were at war with the people to the east of them. A period of decline then set in and culminated in an uprising whereby the common people of Siam, the Thais, were freed from the yoke of the Khmers to whom they had long been subject. The royal race was thus expelled from Siam, probably at the very end of the thirteenth century. This general period, it should be noted, was a time of world-wide physical disasters, such as floods, droughts, famine, and pestilence. For this reason, presumably, it was also a time of almost innumerable migrations, rebellions, and wars. At the end of the disturbed period the Khmers were finally ousted even from their Cambodian capital, Angkor Tom, and disappear from history in the fifteenth century.

The vigor of the Khmers had presumably been sapped by centuries of life in Cambodia. The climate with its malaria and other diseases may have been one factor in weakening them; luxury and idleness may have been another; it is also highly probable that little by little, in spite of strict prohibitions, there had been more or less intermarriage with the native population. Thus the original high inheritance may have become diluted, and perhaps at last was so weak that even if the climate had been good, and diseases and luxury had been absent, the ruling Khmers might not have maintained the innate ability, energy, and originality which their great temples prove beyond question that they possessed in earlier times.

We have traced the story of the flowering, fruiting, and decay of civilization and of racial character among these old Khmers because it is typical of what has happened hundreds of times. Later we shall see the same thing illustrated in Greece and in the barbarian invasions of Europe. We shall see that China, Norway, Iceland, New England, and California, in one way or another, are repeating the story of Cambodia. In western Europe during and after the barbarian invasions, in Egypt and Mesopotamia time after time, in India, Java, and many other regions a similar history can be traced. Always the essential elements seem to be the same. First, in any given race there are sure to be differences of physique and of character. Second, some cause, perhaps war or famine, starts a migration. Third, practically every migration is more or less selective. All types of people are not equally likely to migrate; the pioneer type migrates farther than others. In general, the longer and harder the migration the more highly selected are the survivors and the more likely are they to give rise to a race which more or less permanently inherits the characteristics which have been most important in causing survival. The survivors are usually of a high type, because mental as well as physical strength seems to be a potent selective factor.

Perhaps the most striking element in the problem becomes evident when the migration reaches its end, and the migrants at length become so well established that they are free from the stress involved in mastering the new environment. Then the energies and abilities which have hitherto been devoted to the struggle for existence are released. At once there seems to be a sudden impulse toward creative effort. The thing that is created may be a new form of government, a new social system, a revival of science. Often the creative spirit expresses itself in architecture. Or perhaps we should say that almost always architecture is one of the forms which it ultimately assumes, and architectural structures are more likely than any others to survive to tell the story to future generations.

Now comes the last and saddest part of the story. For some reason every selected race seems to degenerate. Ordinarily the growth of luxury and idleness are given as the causes. Sometimes the cause may be the poor climate into which the migrants have penetrated. But back of these the laws of biology indicate that almost invariably there may be other, deeper causes. One of those causes is the deterioration which is sure to take place in any selected stock if active selection ceases to be effective. In a herd of horses the ability to run invariably declines from generation to generation, unless the slower colts are removed and only the best runners are allowed to breed. Every animal which is not of a pure strain has so mixed an inheritance that bad combinations of qualities as well as good are sure to be produced. The same is true of man. Another important cause of racial deterioration is the restriction of the birth-rate which almost invariably becomes active among the upper classes when overpopulation begins to create a condition of economic stress. A third cause is the mixture of the competent selected stocks with the incompetent, unselected stocks. Such intermingling may occur through mixture with a conquered race, through the introduction of slaves, through unrestricted immigration, through the breaking down of the barriers between different classes or castes, or in various other ways. If man is subject to the same laws of inheritance as animals—and we are practically certain that this is the case both physically and mentally—each step in such dilution means a decrease in ability. If the pioneers among pioneers, to return to our parable, are selected in each generation, it may perhaps be possible to build up a pure strain which will ultimately breed true for generation after generation. In Iceland, as we shall see, this seems almost to have happened, but certainly it is very rare. If the pioneers are selected only in a single generation, and still more, if the selected pioneers intermarry with people who are not selected, deterioration is inevitable. The processes of selection and degeneration, and their relation to other great processes in the development of racial character form the principal theme of this book.

CHAPTER II

FIRST STEPS IN HUMAN CHARACTER

In the preceding chapter we seem to have been dealing with one of the great generalizations of history. Migrations, or changes of environment even without migration, seem to cause natural selection. The character of races is thereby altered. Migration also causes the mixture of races, and thus changes racial character still further. Any other cause, such as overpopulation or famine, which kills off a part of the population at an unusually rapid rate, may act in the same way. In other words, the biological laws which apply to animals also apply to man. The races of mankind are plastic. They are varying on a small scale before our eyes. They appear to have varied on a much larger scale in the past. Let us apply the principles of variation and selection to man's history from the earliest times and see what we can learn.

In thus applying the principles of biology to human races we shall begin in the dim past with the origin of man. There we shall avowedly deal with theories rather than facts. We shall build up a theoretical history of the evolution of the human species. Here and there we have bits of solid fact on which to base our conclusions, but they do not give certainty. Nevertheless it is worth while to begin at the beginning. When we get to the period where written records are available we shall find that the actual facts as to the evolution of character among races which are now playing their part in the world are almost innumerable—far more abundant than we can use in this book.

It is generally agreed that early man originated somewhere in Asia. Formerly it was supposed that he came from the warm, tropical parts of the continent. Little by little this view has given place to the idea that man's early home was in what are now the central deserts and plateaus, the vast region between Mesopotamia and the Caspian Sea on the west, and eastern

Tibet and Mongolia on the east. There is abundant evidence in archæology and history that the greatest of all human movements have been from the central parts of Asia outward. One great stream of migrants presumably went by devious routes southwest into Africa; others went west to Europe; India has almost always been entered from the northwest, though some minor streams came from the northeast. Australia appears to have been peopled by primitive men who went down the Malay peninsula and across the islands. China has again and again been overwhelmed by people invading from the north and northwest, while the original inhabitants of North America are now almost universally agreed to have come in one or more streams from Asia by way of Alaska, and thence to have penetrated to South America. Moreover, man's chief domestic plants and animals came from this same general region and its borders. The horse is a noteworthy example. What is true of him appears to be equally true of the sheep, camel, ox, yak, dog, goat, hen, duck, and other animals. It is likewise true of wheat, rice, barley, and many other highly valuable plants. Although America has supplied a few things like corn, potatoes, and tobacco, and Europe may be the original source of some others such as oats and rye, the contributions of all the other continents to man's domestic plants and animals are only a fraction of those which presumably came from the central parts of Asia. Thus the evidence that man originated somewhere in what is now the plateau region extending from Mesopotamia and Persia to west China is strong.

This conclusion as to where man originated is supported by another line of evidence in the form of man's relation to climate. Matthew, of the American Museum of Natural History, in his book on *Evolution and Climate,* seems to be right in saying that man is not of tropical origin. This common assumption, he thinks, is only partly true.

Its general acceptance is perhaps due, among other reasons, to the supposed relation between loss of hair on the body and the wearing of clothes, the first being regarded as an earlier specialization in an environment of tropical forests, the second as a secondary adaptation resulting from migration to a cold climate. But here, it seems to me, we are putting the cart before the horse. We may more reasonably regard the loss of hair in the human species as a result of wearing clothes and conditioned by this habit, rather than attribute it to any climatic conditions. This view is supported by several points in which the loss of hair in man is differentiated from the partial or complete loss of hair common in tropical animals, the following two being most clearly significant.

(1) It is accompanied by an exceptional and progressive delicacy of skin, quite unsuited to travel in tropical forests. I do not know of any thin-haired or hairless tropical animal whose skin is not more or less thickened for protection against chafing, the attacks of insects, etc.

(2) The loss is most complete on the back and abdomen. The arms and the legs and, in the male, the chest, retain hair much more persistently. This is just what would naturally happen if the loss of hair were due to the wearing of clothes,—at first and for a long, time, a skin thrown over the shoulders and tied around the waist. But if the loss of hair were conditioned by climate it should, as it invariably does among animals, disappear first on the under side of the body and the limbs and be retained longest on the back and shoulders.

According to this view the comparative absence of hair even among tropical savages, who wear little or no clothing, means that the hair on man's body was lost in the earliest human times. It must have disappeared while man was still near his original habitat, and before the ancestors of the present tropical races branched off from the other races.

This conclusion as to the origin of man in a mildly warm rather than a tropical climate is supported to a considerable degree by recent studies of the climates to which various races are best adapted. Numerous studies in the United States, England, France, Italy, and elsewhere indicate that for all European races the optimum or most favorable conditions of climate are essentially the same. The best conditions are a temperature which on the average for day and night does not rise above 70° F. in summer, or fall below 40° F. in winter. The extremes may of course be greater. Physical health and activity are at their highest when the outdoor temperature for day and night averages about 64° F., while mental activity seems to be greatest at a lower temperature, possibly as low as 40° F. among people who heat their houses. A fairly high degree of humidity appears also to be favorable at moderately low temperatures, but not at high temperatures. Variability and storminess seem also to be important. It is significant that not only do fair, broad-headed Mongoloid Finns and dark long-headed Mediterranean Sicilians have the best health at the seasons when their respective countries approach most nearly to this ideal climate, but the same is true of the Japanese. Even the negroes of the southern United States show approximately the same response to climate. Their optimum temperature on the basis of the death-rate, and of the degree of activity with which they work does not appear to average above 70° F. In other words, while it is higher than the optimum for the white race, it is distinctly lower than the mean temperature of 80° F., or more in which their ancestors lived for thousands or tens of thousands of years, and in which they themselves still live for much of the year. While all races have not yet been investigated in this respect, many facts suggest that for practically the whole human race a temperature that averages between 60° F. and 70° F. is better than one which averages above 70°. This adaptation to moderate rather than great warmth seems to be so universal that it may be a primitive trait belonging to the whole human race. If this is so, it would suggest that man did not take the greatest steps in his evolution in a tropical climate such as that of the peninsula of India or Africa south of the Sahara, but in a somewhat cooler climate. Hence our final conclusion agrees with the following statement of Matthew, except that in pointing out the exact place of man's origin he has perhaps been a little more definite than the facts yet warrant.

> It will not be questioned that the higher races of man are adapted to a cool temperate climate, and to an environment rather of open grassy plains than of dense moist forests. In such conditions they reached their highest physical, mental, and social attainments. In the tropical and especially in the moist tropical environment, the physique is poor, the death rate is high, it is difficult to work vigorously or continuously, and especial and unusual precautions are necessary for protection from diseases and

enemies against which no natural immunity exists, and which are absent from the colder and drier environment.

This lack of adaptation to tropical climate is also true, although to a less degree, of the lower races of man. Although from prolonged residence in tropical climates they have acquired a partial immunity from the environment so unfavorable to the newcomer, yet it is by no means complete. The most thoroughly acclimated race—the negro—reaches his highest physical development not in the great equatorial forests but in the drier and cooler highlands of eastern Africa; and when transported to the temperate United States, the West Coast negro yet finds the environment a more favorable one than that to which his ancestors have been endeavoring for thousands of years to accustom themselves. In tropical South America, the Indians, as Bates long ago remarked, seem very imperfectly acclimatized and suffer severely from the hot moist weather; much more than the negroes, whose adaptation to tropical climate has been a much longer one.

In view of the data obtainable from historical record, from tradition, from the present geographical distribution of higher and lower races of men, from the physical and physiological adaptation of all and especially of the higher races, it seems fair to conclude that the center of dispersal of mankind in prehistoric times was central Asia north of the great Himalayan ranges, and that when by progressive aridity that region became desert it was transferred to the regions bordering it to the east, south, and west. We may further assume that the environment in which man primarily evolved was not a moist or tropical climate, but a temperate and more or less arid one, progressively cold and dry during the course of his evolution. In this region and under these conditions, the race first attained a dominance which enabled it to spread out in successive waves or migration to the most remote parts of the earth.

Having come to a tentative conclusion as to the place where early man evolved and the climatic conditions of that evolution, let us now apply these conclusions to the problem of how and why man's anthropoid ancestors became human. The change from beast to man is generally supposed to be connected with man's descent from the trees. That descent was presumably accompanied by a change from a stooping, four-handed gait to one adapted to the erect position. Such a change involves many alterations in physical structure and in character. Modern research indicates that the most primitive human types such as the ape-man of Java, go back to the Pliocene period of geology, while his semi-human ape-like precursors presumably lived during the preceding Miocene. Therefore we must examine these geological ages perhaps one or two million years ago, and see what kind of environment surrounded our primeval prehuman ancestors.

During the Miocene period the climate of the earth seems to have been distinctly warmer and milder than at present. At that time, to judge from the fossils, the great deserts which now extend from the Sahara to Mongolia did not exist, or at least were small and were not located as at present. In place of most of them there was presumably a country covered with a fairly dense forest, not positively tropical but fairly warm and moist, and without such severe winters as now characterize large parts of the desert area. Moreover, many of the greatest mountain ranges had not yet been formed, and moisture-laden winds were therefore able to blow in freely from the oceans. A great inland sea extended from the Mediterranean eastward to Persia, and possibly to China, with a branch running northward from the Caspian and Ural region to the Arctic Ocean. This in itself doubtless had much to do with keeping the climate mild, but its effect may have been increased by other conditions such as an abundance of carbon dioxide in the air and a scarcity of the cyclonic type of storms.

The relief of the land throughout most of central Asia in those early Miocene days was probably gentle almost everywhere. Where now the Himalayas raise their great crests 20,000 feet into the air, there was an ocean in which were deposited limestones that are now found on the mountain-tops. Elsewhere there were either rolling plains or gently sloping, low mountains such as those whose remnants are now seen on the plateaus of Tibet and Tian Shan. As one climbs toward the plateaus one sees to-day a curious topography. The precarious track follows tortuous steep-sided gorges, very young, as the physiographer counts age. This indicates that the streams which cut them have been at work only a little while. Many canyons are so narrow as to be impassable, and one cannot float down the rivers because they are so young that they have not yet had time to smooth out their falls and rapids. That is why the Brahmaputra has never been followed from source to mouth. Even on the Indus, although the river's whole course has been mapped, there are many places where the gorge is so narrow that one cannot follow the stream, but must clamber over difficult mountains. At the heads of these streams, however, one comes upon a country of a wholly different aspect, a country of old hills smoothly rounded. In Tian Shan a gently rolling old upland plain, diversified with beautiful green hills, sometimes forms a lofty plateau almost surrounded by the well-nigh vertical cliffs of young gorges.

In India I once ascended the splendid canyon of the Shyok River, a tributary of the Indus. For many days we were hemmed in by steep cliffs on either side of a broad, rocky flood-plain. We longed in vain to see what lay on either side. Then one day a final ascent of a minor valley brought us out into a different world. We had reached the great plateau and soon were traversing its gentle slopes and broad plains. Low rounded mountains replaced the steep cliffs and sharply serrated peaks of the plateau border. It was easy to travel anywhere. One night we made camp at an altitude of over 16,000 feet. Next day we expected to cross the Karakorum Pass, which is about 18,300 feet high, and marks the divide between India and China. That night the cook gave us a warning:

"To-morrow we are going to cross a pass where the air is very poisonous. I advise you not to eat much, and especially not to eat any meat."

"Oh, nothing is the matter with the air," I answered. "It is merely because the pass is high."

"No," said the cook, "it's not high. You wait and see."

I insisted that it was as high as Chang La, an equally lofty pass with a long, steep, toilsome ascent of about 6,000 feet on each side, which we had crossed two weeks before. He was unpersuaded and clinched his argument by saying:

"When we crossed Chang La we climbed steeply and with great difficulty. Here we scarcely climb at all. We call a thing high when you have to climb to it. We don't climb here, and so we think the air is poison."

The old plateau whose gentle relief made the cook think it was low represents approximately the sort of topography, although not necessarily the place, in which primitive man is supposed to have originated. But in those remote times the plateaus presumably stood far lower than now. Nor were there any such young, high mountain ranges as those of the Himalayas, Kwen Lun, Tian Shan, and Hindu Kush. The climate was probably moist and mild, and vast regions that are now high plateaus or low desert basins were presumably covered with forests whose remnants are found as fossils. It was probably in these forests that the anthropoid ancestors of man had been gradually, but probably very slowly, evolving more and more power of brain. There or somewhere else they had taken the fullest advantage of the fact that their feet can be used as hands. Perhaps they had gone so far as to use sticks or even stones as missiles, although stones may have been relatively inaccessible, and hence unimportant, since most of the earth's surface was presumably deeply covered with soil and vegetation.

At this stage of geological history there began a series of slow changes in the earth's crust. First there was a general uplifting of the whole region of central Asia so that the inland sea finally disappeared, while the shores of the Indian and Pacific Oceans receded. At the same time mountain ranges began to be slowly upheaved, thus shutting out the moist winds from the surrounding oceans. These conditions and perhaps others gradually led to the aridity which is now so extreme in central Asia.

Thus far we have been rehearsing either actual facts or conclusions which are widely if not almost universally accepted. Now, in attempting to show the effect of these facts on human development, we must venture into more speculative ground. Long before there was anything which could be called real aridity in those old Miocene days, the climate probably became dry enough, and the dry season long enough to have a pronounced effect on early man. As Elliot, Barrell, and others have shown, one of the first results of the incipient dryness must have been to make the forests thinner so that the trees were more or less isolated instead of forming a dense, close cover. Such conditions would oblige the anthropoid apes to come down to the ground in order to pass from tree to tree. As the forests grew thinner, patches of grassland or savanna must have appeared with only a few trees scattered here and there and with what are known as gallery forests in strips along the streams. Then the anthropoids must have been put to still more stress. Formerly the forests had been their refuge. Carnivora larger and fiercer than those of to-day prowled about, but the apes could take refuge in the trees. There, too, they found a large part of their food. But now it was not possible to swing from tree to tree except in the gallery forests. There the more timid or less brainy types, or those less able to walk, were forced to remain, but found their habitat constantly diminishing. Those with quicker wits and more adventurous dispositions and with the greatest ability to stand upright presumably ventured to walk from tree to tree. But at such times they were relatively defenceless against the carnivora unless they took tools in their hands in the form of sticks that could be used as clubs, or stones that could be hurled. In such a critical period mental activity along almost every line must have been peculiarly helpful.

As time went on and the aridity became greater, the trees in the open savannas must have disappeared almost entirely, giving place to broad grasslands, the home of numerous animals such as the horse, sheep, and ox. These, together with wild grain growing among the grass, would furnish food to man in place of the fruits and nuts of his former forest home. Thus when primitive man left the trees he would tend to become an eater of cereals and meat instead of fruits and nuts.

As the mountains gradually rose higher, and the sea receded farther, the increasing dryness of the climate must ultimately have almost destroyed even the gallery forests along the watercourses, for these are now rare. This would mean the final extinction of the apes which for generation after generation may have persisted where trees were still abundant. It is not impossible that the anthropoids which were not brainy enough to take to the ground, or were not physically capable of doing so, may have survived for hundreds of thousands of years after their line of descent separated from that of the progenitors of man. But ultimately all perished except possibly certain groups which more or less unconsciously migrated southward and seaward in the early stages of the great change, and hence remained in the forests. Although the process of separating the erect, small-jawed, two-handed, big-headed, brainy human type from its bent, heavy-jawed, four-handed, small-headed, and unintelligent relatives was slow, it was inexorable and irrevocable. As the environment gradually killed off the animals which could not walk erect, which did not know how to use their hands for holding tools, and which were not able to take refuge in caves, make themselves shelters, or protect themselves by fire or otherwise, the human race may be supposed to have grown up, and at some indefinite stage to have become man instead of beast.

Although man's descent from the trees probably depended mainly upon mental ability there are other phases to be considered. When the anthropoids began to leave the trees and walk upright, as Barrell points out, there took place during many generations "a physical transformation shown in changed foot structure, changed ratios of the limb-lengths, a changed profile of the backbone, a shortening of the jaw and a changed dentition."

As man stood more and more erect, he probably reaped certain great advantages in the freedom of his hands and head. As long as he had to help support himself by touching his hands or his knuckles on the ground, and still more while he used his hands as well as his feet to help him among the trees, his hands were useless for the purpose of carrying things any great distance. Even though he found and shaped a good club, he could not carry it with him. At almost the first hint of danger he had to drop it and use his hands in climbing. Nor could he pick up a handful of good round stones and have them ready to throw at the enemy. Until his hands were free, the kind of brain that could devise and fashion a rude bag of vines, such as Elliot loves to imagine, would be terribly handicapped in finding means of expressing itself. But once the hands were permanently free, the clever brain would enable its owner to make a bag, fill it with stones, and hang it around his neck. Moreover, so long as his hands were essential as organs of locomotion, they could acquire only a moderate degree of the delicacy of touch and the ability to pick up and manipulate small objects which is one of man's strongest characteristics. In other words, although certain types among the anthropoids may have had the mental capacity to use tools, this apparently gave them relatively little special advantage and hence was only a minor factor in self-preservation and in the preservation of the species until the front limbs ceased to be needed for locomotion.

Another point which is well made by Wetzel and Elliot is that up to the time when man walked erect the development of his skull and brain was hampered by the heavy muscles of his jaws and neck. Until his hands came freely into use and were aided by tools, the teeth were the main means of rending, tearing, and even carrying food. Hence the jaw had to be very strong and large, and required powerful muscles. Moreover, such a jaw and its use for heavy and violent work involve a great strain on the neck, and the muscles that support the head must be correspondingly large. This is still more the case when the head hangs forward as it does in the apes, instead of resting on the top of a bony column as it does in men so long as they keep away from desks. It requires far less exertion to hold one's head erect than to hold it bent forward. Because of the strain of the muscles of jaw and neck, the sutures in the upper parts of the skull of the anthropoid close early in life, whereas in man they do not close for many years, not till about the age of forty. Moreover, since the sides of the skull are not clamped by strong muscles, the human skull can expand backward, forward, and sidewise, as well as upward, which is the only free direction in lower forms. Perhaps some reflection of this plasticity of the human skull is found in the observation of Vern that at Cambridge, England, the students whom he measured showed an increase in breadth of skull during their university course. Between the ages of nineteen and twenty-three their cephalic index rose from 77.9 to 79.2. Of course mere size does not mean that a brain is necessarily of high caliber, for the largest brain ever measured was that of an idiot. Nevertheless, it is a fact that as we go upward in the scale of evolution, both in men and animals, the size of the brain in proportion to the body is on the whole closely correlated with the degree of intelligence.

As to the mentality of the ape-man who came down out of the trees, it is hard to form any definite estimate. We have already indicated some of his qualities, and we may well conclude this chapter with a quotation from Elliot which suggests certain higher attributes:

> Most anthropologists recognize that a strictly scientific treatment must not omit reference to those moral and spiritual instincts which are beyond and above, though inextricably connected with, the brain. It seems to us that especially at this critical period the Pliocene precursor of man must have been exceedingly inquisitive. He was more or less in safety so long as he remained an arboreal animal; but when he left the trees or only resorted to them for safety and shelter at night, this involved all sorts of moral qualities. He had to be excessively wary, and yet bold and courageous; enterprising, and yet patient; the maintenance of his young must have required of him very hard work and an extraordinary amount of affection.
>
> All these qualities involve an unusual character. The moral quality of the Pliocene ancestor was surely beyond that of the best of the animals, even though these do possess embryonic states of both virtues and vices. It will be seen, therefore, that at the critical moment, three lines of development happened to coincide. For a whole geological period, perhaps, the power of standing, of running, of using hands and fingers had been slowly perfected, involving, as we tried to show, a thorough-going modification in the ground-plan of the body. At the same time, increase of brain had been accompanied by a development of intelligence; eye, ears, finger and thumb were being more and more directly controlled by the mind. Then, also, frequent crises required of the male heroism and self-sacrifice, and of the mother incessant daily devotion.
>
> But at some particular moment the accumulated knowledge that he had gathered of all the evil things in a very dangerous world, and his power of enjoyment of what was good in it, suddenly changed into a knowledge

of Good and Evil. The change was not more sudden than the first stroke of a complicated piece of machinery set going for the first time, nor of the first abrupt explosion of a volcano which has been extinct for centuries. But so far as man was concerned, it transformed everything; from being one of the other animals, he became *in posse* master of all.

CHAPTER III

THE EARLIEST GREAT MIGRATIONS

It is not by accident that the chief early development of mankind took place in Eurasia, and that it occurred during a glacial period. These facts are in accord with two great principles which are rapidly assuming a fundamental place in geology. One principle is that the world contains a few great centers of evolution. These are located in the interiors of the continents in medium or high latitudes. The Asiatic center far overshadows all the others combined, while that of North America comes next. Central Asia, as has been well shown by Osborn and Matthew, and more recently by Andrews's expedition for the American Museum at New York, was the original home of many great families of mammals such as the horse, tapir, rhinoceros, deer, dog, sheep, pig, and giraffe. A few like the camel apparently originated in North America.

The main reason why evolution is so rapid in the interior of large continents is that those are the places where the physical environment is most variable. In a warm, moist plain in low latitudes an animal never experiences any great extremes of either heat or cold. Very rarely does it suffer from severe drought, and its food-supply is relatively abundant and constant. In such an environment the animals are not likely to acquire any great specialization. Even if they are not sluggish, they are usually primitive. A dry rugged interior in middle latitudes produces quite the contrary effect. The extremes of heat and cold cause weak non-resistant animals to perish. The variations in rainfall from one year to another are relatively great. Since the number of animals increases rapidly when several good years come in succession, the excess of population leads to an unusually fierce struggle to get food or avoid being eaten. Hence among certain species only those animals survive that are especially fleet of foot, keen of sense, quick to take alarm, or able to hide and thus avoid their enemies. So too in other species the animals which are able to go unusually long distances in search of food and water, to endure long periods of fasting or thirst, or to persevere to an unusual degree in the pursuit of prey are the only ones that can survive. If the region is rugged, still other specializations, such as the ability to climb steep, rocky slopes, are a great help in enabling an animal to live and reproduce its kind. Hence in the Asiatic center there has for many ages been a tendency to kill off the unspecialized kinds of animals and to evolve types with a higher and higher degree of adaptation to a severe and variable environment.

From this there follows another important geological generalization which also applies to man. The dry interiors of the continents, especially of Asia, have been the great centers of migration. This is not only because new types originate there, but because such an interior, especially if it lies at a fairly high latitude, suffers far greater changes from one geological period to another than does any other part of the world. Sometimes the inner parts of continents have warm, moist, monotonous climates such as we have already described at the beginning of the Miocene before the early anthropoids left the trees. At other times they have a desert climate like that of central Asia to-day. When the climate is mild, the number of animals of a given species becomes large; when it is severe, the animals either die or migrate. In low latitudes like the Amazon or Congo basins there is little reason to think that during any part of geological time the climate has been greatly different from that of to-day. In Illinois and central Russia, however, even in recent geological times, we have had the contrast between a glacial climate accompanied by a thick sheet of ice on the one hand, and a climate milder than that of to-day on the other. In central Asia the corresponding changes have been almost equally great, and probably even more important in promoting evolution because they have never completely extinguished all life. During or immediately after the glacial epochs huge lakes hundreds of miles

THE EARLIEST GREAT MIGRATIONS

long filled basins that are now deserts, while the surrounding country was covered with grass and forests. At the other extreme the country became a desert even more terrible than that of to-day. The central parts of continents have also endured great changes due to the uplifting of mountains and the isolation of the interior from the oceans. The upheaval of the Himalayas, Kwen Lun, Tian Shan, and other ranges in Asia, and of the Sierra Nevada and Rockies in the United States, have in themselves been sufficient to produce deserts where the climate had previously been mild, moist, and oceanic.

A single illustration will show how all these changes in climate and relief not only select certain types of animals for preservation and kill others, but how they drive the more primitive types away from the centers of evolution. In central Asia the Andrews expedition of the American Museum of Natural History found the remains of a very primitive horse, which appears to be not far removed from the type whence all other horses have sprung. Suppose that when such a species is most numerous and most widely spread, perhaps with its center in Mongolia but extending over a considerable area, the climate is such that broad grassy plains extend far and wide. Now suppose that greater aridity ensues, becoming evident first in a relatively limited central area and then spreading outward. The number of horses in the desiccated central area must diminish, for there is not food enough to support so many as formerly.

This diminution takes place in two ways: first, the death-rate becomes high, for many horses are undernourished, and many mares cannot supply milk for their foals. Second, as the horses wander irregularly in search of food a certain number who happen to be on the borders of the dry area stray into better regions and never come back. The horses that remain permanently in the dry area are likely to be a peculiar type. Perhaps they may be ones whose toes have become elongated, thus giving the animals greater agility and fleetness, and hence making them better able to cover long distances in traveling back and forth between the scattered water-holes and the places where the scanty grasses are found. Thus in the dry area a new species may arise, while on the outskirts the old species persists. If the dry conditions continue long enough and become extreme enough, the old species may be pushed thousands of miles from its original home. Then the climate may moderate once more, but the old species will not necessarily come back into the old home. That region may be so completely occupied by the new and more highly specialized species that there will be no room for the other.

Again there comes a dry period. A still more specialized type of horse is developed, perhaps the one-toed type, succeeding the more primitive three-toed form. It pushes out its predecessor and this predecessor pushes the primitive species still farther. The net result is that among mammals the most primitive forms are usually found either as ancient fossils in old strata near the place where the family originated, or as later fossils in modern strata or even as living species in regions far from their original home. The places where the primitive types are most likely to be found still living are areas of refuge such as tropical forests, isolated peninsulas, islands, or other inaccessible regions protected perhaps by mountains. There they are able to persist either because they have not been subjected to an extreme environment greatly different from that in which they originated, or because new types have not been able to drive them out.

All this applies to man quite as much as to animals. In fact, the growth of man's intelligence seems to be a wonderful example of the way in which evolution centers in regions of physical extremes and is accelerated by climatic variations. The whole of human history falls within an exceptional period in the earth's physical condition. Ever since the time when the apes first came down out of the trees, the continents have been unusually large, the mountains unusually high, the climate unusually severe, and the variations of climate unusually great. It is difficult to disabuse ourselves of the idea that the conditions with which we are familiar are those which prevail most widely. Most of the readers of this book are doubtless residents of regions where the hills and plains are covered with vegetation, and where the grass if left to itself, takes the form of turf. But that is not the usual condition throughout the earth as a whole. Most of the eight hundred million people of Asia have never seen real turf. Nor have most of those of Africa and South America. The same is true of large parts of North America, Australia, and even southern Europe. It is probably no exaggeration to say that to nearly three-fourths of the sixteen or seventeen hundred million people in the world turfy grass does not seem normal. They are like a Kurd and an Armenian with whom I was once riding in Asia Minor. We came to a beautiful, green, grassy plain at the end of a small lake. Having seen nothing really green for two or three months, for it was August, and having seen no good turf for two or three years, I was delighted.

"See how beautiful that is," I said. "In my country it is like that everywhere." But the impression on my companions was not what I intended; "How unhealthy your country must be!" was their answer. To them a place where the grass is green in midsummer meant standing water, mosquitoes, and malaria.

Just as we who are to-day the world's leaders live in a relatively unusual and restricted environment, so the human race throughout its whole history has lived in a decidedly unusual environment. Throughout geological times by far the most common condition has resembled that which prevailed in early Miocene times before the primitive ape-men came down from the trees. At that time, it will be remembered, the lands were relatively small and low, arms of the sea penetrated far inland, most of the mountain ranges were rounded, gentle, and low, and the climate was so mild that corals flourished far to the north, great forests presumably throve in central Asia, and beech-trees, oaks, planes, poplars, limes, magnolias, holly, ivy, and grapevines grew in Greenland, Iceland, and Spitzbergen.

Then began the great change which we have pictured, the uplifting of the lands, the upheaval of vast mountain chains, and the progressive cooling and drying of the atmosphere. At first the change was probably very slow and gradual, imperceptible in the life of any individual. Perhaps it was never rapid enough so that even if modern man had been there he would have been conscious of it except on the basis of records kept from one generation to another. But ultimately the change culminated in one of the most perturbed periods the world has ever seen, the Pleistocene glacial period. During at least four epochs, and perhaps in several earlier epochs whose record is lost or not yet known, there were great climatic pulsations which culminated in vast continental ice-sheets. To-day we are still living in an unusual period—the last part of an ice age. It is by no means usual for Greenland to be shrouded in ice or for the whole Antarctic continent to be covered with an ice-sheet. Our great extremes of temperature from summer to winter, as well as our cold oceans, are quite different from the conditions which have ordinarily prevailed. Thus from the standpoint of environment one of the outstanding facts in organic evolution is that during the period of strictly *human* development the physical environment has been unusual; the greatest advances appear to have been made when the environment has been most extreme. The last step in the development of the mental capacities in which man now rejoices was taken during the last ice age, whose waning power is still seen in the glaciers not only of Greenland and Antarctica, but even of Norway and the Alps.

The preceding outline of biological and geological principles prepares the way for a study of the kind of events that must have happened to early man during the various ice ages. Here, as in the last chapter, our story must consist largely of inferences rather than of historical facts. Only at long intervals have we even the faintest trace of man at this time. In the whole vast continent of Asia there has not yet been found a single authentic trace of man before the last glacial epoch. Nevertheless, on the basis of certain generally accepted principles of ecology and paleontology we can sketch a rough outline of certain steps in man's upward progress. The sketch is like some of the stories of the Bible whose main object is to illustrate broad principles rather than supply an exact account of what actually happened.

THE EARLIEST GREAT MIGRATIONS 39

Its value is that it enables us to view man in perspective, it brings out the close similarity between man and other animals, and it gives a consistent background against which to project the definite facts that will engage our attention during the later stages of human development.

In the preceding chapter we carried our primitive ancestors to the point where they had permanently assumed the erect attitude and no longer depended upon the forests either for shelter or food. They were living, presumably, somewhere between the Caspian Sea and western China, in a region of steppes and savannas, but not of deserts. Perhaps they had already spread over most of this region. The limits to which they expanded were possibly set by the range of the grass-eating animals which probably formed their main source of food. In Europe the earliest human remains are almost invariably associated with animals such as the bison, elephant, mammoth, and rhinoceros. The continent of Eurasia as a whole, and most of the individual mountain ranges stood much higher than a million or so years previously, when the first apes began to walk unsteadily on two feet from tree to tree in the open glades. But the mountains were apparently not so high as they became during the period of perhaps half a million years which has since elapsed.

During those half-million years human evolution has been chiefly guided by two sets of facts, namely, the mutations that have occurred in man, especially in his skull and brain, and migration and natural selection under the influence of great pulsations of climate. Before the beginning of the first glacial period central Asia must have reached a condition almost as dry as now. That could scarcely happen without causing early man to spread widely from his ancestral home. Either then or in some later interglacial period he reached at least four main types of environment, according to whether he traveled south or north, west or east. Those that went southward may be represented by *Pithecanthropus erectus*, the ape-man of Java, whom some authorities believe to be preglacial. Probably the modern and much mixed descendants of some group of southward-moving migrants may be recognized in the primitive Pygmies of southern Asia, central Africa, and the East Indies, and in the Australoid and Negroid peoples of the torrid zone in general. In their most primitive forms such people live largely in inaccessible tropical forests where they are protected from later arrivals by the denseness of the trees and the general unhealthfulness of the surroundings.

Those of the early migrants who went south would doubtless have congratulated themselves on their good fortune, had they known and understood what happened to the others. But their seeming good fortune was their undoing. By getting into a tropical climate they almost inevitably condemned themselves either to stagnation or to an evolution slower than that of their kindred who stayed in the old home or went in other directions. One reason for this was that their new environment did not differ greatly from that in which their ancestors lived before leaving the trees. In fact their southward way brought many of them into the forest environment of the apes. The profusion of trees, insects, and beasts of prey in the torrid zone encourages an arboreal life. Among the Pygmies there has been a considerable development in this direction. Those little people build their houses high among the branches, they are wonderfully adept at climbing, and the trees are to them a refuge, much as they are to the apes. Moreover, in a tropical environment it is easier to get a living than on the steppes and savannas. Less activity and more patience in lying in wait are required in hunting. There is no pronounced change of seasons to oblige people to adopt one method in winter and another in summer, and thus make them adaptable. There is less likelihood of prolonged periods without food, so that the hoarding instinct, and the power of self-control in order to make the food last through a period of privation are not fostered so much as farther poleward. This is the more true because tropical man can find vegetable food more abundantly than can the man in the cooler grasslands, while at the same time the tropical food is far more perishable than the other. Fruits and even roots are by no means easy to store for any length of time, whereas the seeds of grasslike grains such as wheat and barley spoil much less quickly. Here, too, the temperature and humidity make a great difference, for even the great tropical grains, rice, millet, and corn, which will keep perfectly during a winter in central Asia when the thermometer goes down to zero, are relatively hard to store in a warm climate which averages near 80° F. Moreover, in the cooler climate, even during its warm season, there are not so many insects, rodents, and other creatures as in tropical lands, nor so many moulds, fungi, and the like to destroy the food that man attempts to save for times of scarcity.

Added to all this is the fact that even the more backward races seem to have the best health and greatest energy in a climate with an average temperature of 70° F. or less rather than 80°. The fact that the temperature is constantly too high seems not only to weaken the body somewhat so that it is more easily a prey to disease, but to make people do things in the way that requires least effort. Great activity in a hot, moist region raises the bodily temperature too high—so high that the cooling mechanism of the sweat glands and lungs cannot keep the temperature down to normal. Thus the active people are frequently afflicted with what is really a slight fever, harmless perhaps in itself, or in a single occurrence, but tending toward weakness in the long run. The man who can get a living with the least effort, that is, without heating himself or without going into places where he will be stung by noxious insects, has a real advantage in warm, moist countries, and his kind is likely to survive where the active kind die out. Hence from a score of standpoints the primitive men who followed the seemingly favorable climate southward from central Asia, and who finally migrated into a tropical environment were at a disadvantage compared with those who unwittingly remained in the lands where life was hard.

Let us go back now to central Asia, and inquire what presumably happened to the primitive men who spread westward and eastward during early times. How far they actually migrated during the period when the Asiatic mountains were rising, and the climate of central Asia was growing drier, we do not yet know. If the so-called Foxhall and Red Crag flints and the "coprolite jaw" found in England are of preglacial age, as some scholars believe, preglacial man must have spread into Africa and Europe as well as Asia. He could get to Europe only by way of North Africa. At Suez there was a broad connection between Africa and Asia, whereas Europe appears to have been cut off from Asia by a continuous sea, which extended from the Black Sea to the Caspian and thence to the Arctic Ocean by way of the great Ural Gulf. On the other hand, Africa and Europe were united at Gibraltar, and perhaps at Sicily, so that men and animals could migrate freely. It is not here necessary, however, to know just when man spread to various parts of the world. We are concerned with the effect of increasing aridity in central Asia and of corresponding climatic changes in other parts of the world, no matter whether these occurred before the first glaciation or during an interglacial or post-glacial epoch. The same general sequence of climatic events occurred repeatedly, and in each case the general effect upon man must have been similar.

As central Asia grew drier there must always have been on each border of the desert a belt of moister climate, in some places wide, in others narrow. Its inner portion consisted of grasslands and tree-studded savannas, like those of the supposed first home wherein man came down from the trees. Beyond this lay forests. As the dry area expanded the savanna belt retreated and pushed back the forest. The men who for generation after generation slowly migrated eastward or westward with the savanna belt probably did not appreciably change their type of habitat. Since they encountered few new types of vegetation and animals, and since the climatic conditions to which they were exposed were closely similar to those under which their ancestors had lived, there is little reason to think that the evolution of new types went on with any great activity. In China the northern part of the country, namely Mongolia

THE EARLIEST GREAT MIGRATIONS 43

and western Manchuria, is not forested, and would not be covered densely with trees even if man were not there. The fact that the rain comes only in the summer, and is often delayed until June or even July, makes it impossible for forests to prevail, except locally. In north Africa the treeless region is well known to reach as far as the Atlantic. Hence both the eastward and the westward migrants may have almost reached the ocean without suffering any great change in environment or in character.

This does not mean that evolution came to a standstill. Among the eastward and westward migrants and likewise among those who went south or north, mutant types may have arisen at any time. Moreover, the processes of sexual selection, so far as it has any importance, and whatever advance there may be through the struggle for existence as described by Darwin, must have gone on at all times. Each set of people may likewise have met certain new conditions due to movements of the earth's crust or to the presence of plants and animals hitherto unknown, or new combinations of mountains, plains, valleys, rivers, lakes, and oceans. In each migrant group there may have arisen men of unusual ability who made new inventions such as better traps, more effective bows, more shapely flints, and more convenient clubs, huts, and other devices. The people who were clever enough to adopt these devices would be the ones who would be most likely to survive. Thus human evolution presumably would have made progress even if the climate had remained constant. But all the causes of progress or change mentioned in this paragraph would perhaps in the long run act almost as strongly upon people who migrated in one direction as in another. What we are here concerned with is the conditions which would produce noteworthy differences in the rate of progress among migrants to different parts of the world.

The primitive people who went north from the dry central region of Asia or who went north from Africa into Europe, must have encountered conditions more severe than those which confronted the migrants to the south, east, or west. With the lower temperature and longer winters the need of clothing, shelter, and fire would increase. Thus skill in procuring these necessities and in making the tools with which they are prepared would be at a premium. Moreover, among people in cooler climates the low temperature encourages activity: it leads a man to pursue the chase with vigor; it makes him enjoy the task of gathering wood for a fire, or of cutting a tree with a stone axe; it puts a premium on continued and repeated physical activity. There would likewise be greater need for forethought in order that fuel, clothing, and a place for shelter might at all times be available in cold and snowy weather. Any great movement toward the north would also bring the migrants into the forest. Whether the forest environment is as stimulating as that of the savannas need not now concern us. It is different, and therefore those of the wanderers who were most adaptable, and best able to cope with new circumstances would have an advantage. In many ways those who entered the forests to the south must have had the same advantage, but the northern forest is more difficult because food is harder to find. Still more important is the fact that in the northern forest the hunter who has wandered far from home may freeze to death at night unless careful precautions are taken, whereas in the southern forest primitive man could anywhere climb a tree and remain in comparative comfort and safety without clothing, shelter, or fire. Thus among the northern migrants the selection of the most intelligent types for preservation probably went on faster and more effectively than among those who wandered in other directions.

But how about the remnant who remained in the original home in spite of the steadily increasing aridity? Generation by generation their struggle must have become harder. Like all the other men of that ancient time they were presumably hunters who eked out their supplies of meat with whatever vegetable food they could lay hands on, but as yet agriculture was not even a dream. Judging by the present deserts, just as we have judged the other environments by what happens in them to-day, the people who remained in the central dry area of Asia must

THE EARLIEST GREAT MIGRATIONS

have been subjected to the severest type of natural selection. As the mountains rose higher throughout the long ages they had to endure greater extremes than formerly of both heat and cold. Game must have been scarcer than formerly; longer and harder journeys were required in pursuit of it. Greater endurance than ever was required not only in this respect, but in the ordeal of hunger and thirst which must have been experienced more often than at any previous time. In many ways the factors that acted to select certain types for preservation were the same as those farther north, except that aridity and vast uninhabitable areas, with their many severe trials, took the place of prolonged cold and vast forests. The process of weeding out the human beings who fell below a certain level must have gone on rapidly. But an environment which in moderation tends to weed out the less competent may be so extreme that it weeds out the more alert types, leaving those that are able to survive through sluggish endurance rather than through activity and alertness, as we shall see fully in China. From the standpoint of the evolution of human intelligence the environment of the deserts and of the northlands was good in so far as it put a premium on activity and alertness; it was bad when it became so extreme that it drove out the active types, or gave the main advantage to those which were able to survive only by means of passivity, semi-hibernation, or lack of a keen nervous system.

We have dwelt on these early migrations and their presumable effect upon human character because the same process must have been repeated time after time. With the growth of human knowledge and the building up of the vast mechanism of civilization which is passed on from generation to generation, the response of man to any given environment changes somewhat, but the principles by which certain variations in human ability are selected for preservation or destruction remain constant. In the first great migrations those who went to the tropical regions subjected themselves unknowingly to conditions which presumably tended toward stagnation or even toward retrogression, for moderate activity was often more profitable than great activity, while the abundance of resources and lack of the exigencies of the seasons tended to give the stupid almost as good a chance of survival as the intelligent. Among those who migrated east or west, there was probably no great selection of one type rather than another because there was no marked change in environment. They progressed to the extent that other causes determined, but in those respects were presumably little better off than the rest of mankind. But those who went to the north, either in Asia or from Africa to Europe, and those who remained in the dry primeval habitat, provided the environment did not become too extreme, seem to have been subjected to a series of conditions which placed a maximum survival value upon the qualities of mental and physical alertness, upon the ability to persist in the chase or in the search for water, upon the kind of mentality that takes thought for the morrow and that can provide itself with tools, clothing, and shelter, and that can utilize that great weapon and comforter, fire. It is, then, in central and north central Asia or possibly in the north Atlantic parts of Europe and the Asiatic area draining into the north Pacific, that we should expect the greatest development of human ability during the period of the first dispersion of mankind.

CHAPTER IV

GLACIATION AND THE SUPREMACY OF EUROPE

THE next great step in the evolution of races was the coming of the first glacial epoch. This means much more than the mere advance of an ice-sheet, for that is only one of many results all springing from common causes. Those causes include in the first place the gradual elevation of the land and the uplifting of mountain chains, together with a possible decrease in the amount of carbon dioxide and water vapor in the air. But in addition there seem to have been other causes which acted more suddenly, and were capable of repeatedly reversing themselves and of varying irregularly in periods which had a length of tens of thousands of years for the greater cycles, or of only scores or hundreds of years for the minor cycles. Irregular upward or downward movements of the lands are almost certainly included among these causes, as is intermittent veiling of the sun by vast quantities of volcanic dust.

Still more important, as it seems to me, are variations in solar activity, as explained in *Climatic Changes* and *Earth and Sun*. The climatic cycles of the glacial period seem on a large scale to be almost identical with the small cycles whose extremes are now observed at times of sun-spot maxima and minima. One of the main characteristics of the present cycles is an increase in the number and intensity of cyclonic storms at times of many sun-spots, and especially a tendency for the paths of the storms to lie in relatively high latitudes. The frequency and intensity of the storms cause an unusually large amount of warm air to be carried aloft in their centers. How this happens is familiar to every one who has observed the way in which a warm wind from a southerly quarter, especially in winter, is apt to precede a storm, whereas after the storm there comes a strong cold wind from the north and west. The cold air actually blows under the warm air and lifts it up. Curiously enough, although it has now been demonstrated beyond question that the earth's surface is cooler at times of many sun-spots than at times of few, the sun seems to give out more heat at times of many sun-spots. This anomaly seems to find its explanation in the upward movement of vast quantities of warm air in the centers of the storms.

On this basis we interpret a glacial epoch as a time which may begin with a certain amount of uplifting of the lands, and an unusual degree of volcanic activity, but which is mainly characterized by increased storminess and strong winds, especially in high latitudes. The actual sunshine may be warmer than usual, but the air on the average is unusually cool because vast amounts of warm air are drawn away from low latitudes and carried aloft in the centers of storms. For a considerable period, presumably several thousand years, the accumulation of snow continues in certain especially stormy areas, which are usually but not always highlands. The chief areas were Canada on both sides of Hudson Bay and Scandinavia, as appears in Fig. 1, but there were numerous minor centers such as Scotland, the Alps, and the Canadian Rockies. In course of time the depth of the snow increased to hundreds of feet.

Because a snow-field is almost invariably cold, each large region where snow remained permanently must have become an area of high atmospheric pressure. Therefore strong winds presumably blew outward as they do on the borders of the ice-sheets of Antarctica and Greenland, where the gales often reach velocities of a hundred miles an hour. This would tend to blow the snow continually outward toward the edges of the snow-fields. At the same time the high atmospheric pressure would force the storms to seek lower latitudes than formerly, and most of them would sweep around the equatorward border of the main snow-fields. At some stage the snow would become so deep that it would gradually be converted into ice, and would begin to flow outward and thereby add to the ice-covered area. Thus at the height of each of the four glacial epochs much of north-

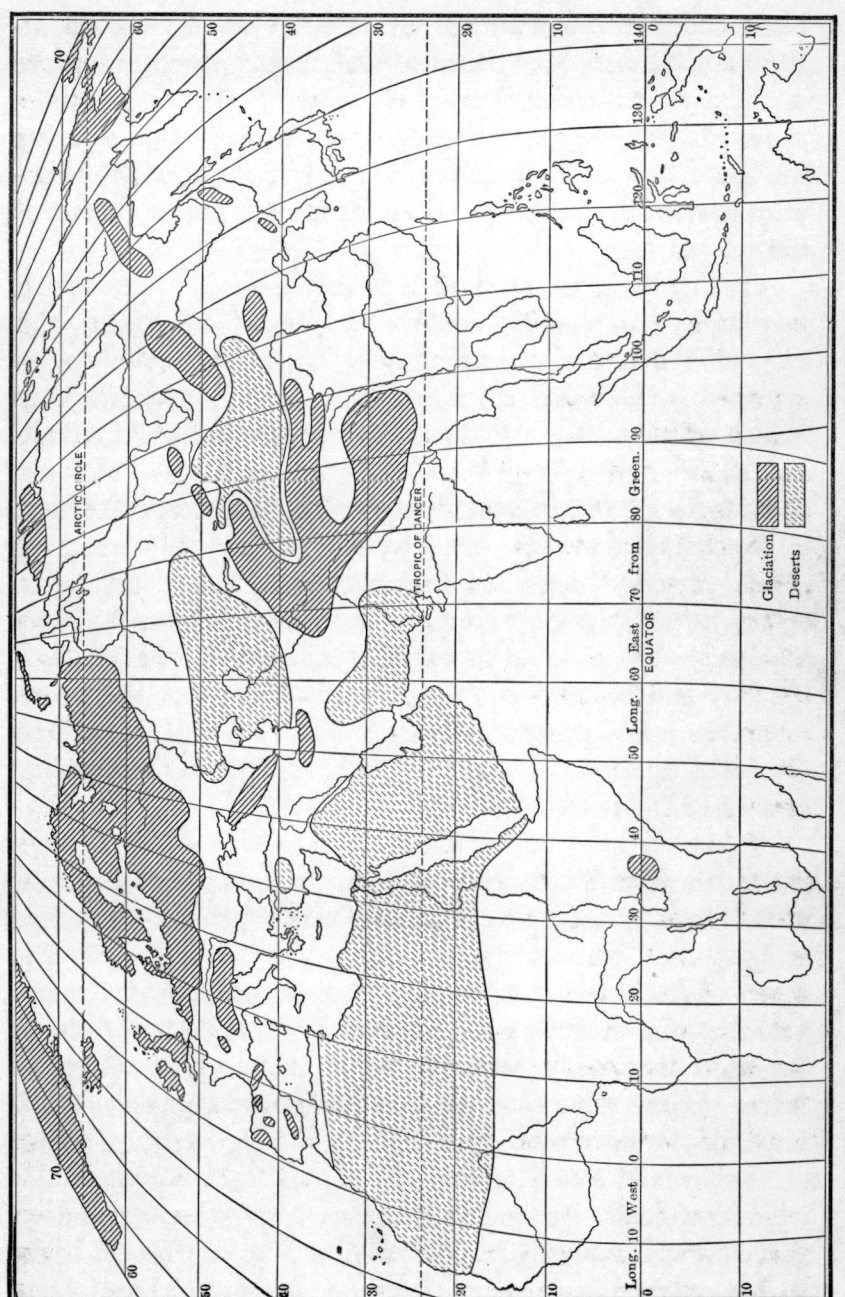

FIG. 1. THE PRESSURE ZONE BETWEEN GLACIATION AND DESERTS.

western Europe was shrouded with ice, while much of the region north of the Pyrenees and Alps probably had a climate as severe as that of Lapland to-day. Farther south in the Mediterranean region and north Africa, and extending eastward to central Asia and northern India, there was apparently a broad belt of cyclonic storms and highly variable weather, with abundant rain at practically all seasons. Still farther south the great deserts of Sahara and Arabia were apparently greatly narrowed, and were far less dry and much more habitable than at present.

Let us now see what probably happened to the primitive men who had become established in different parts of the Old World when the glacial period came on. In the tropical regions the coming and going of such a period would make relatively little difference. The general temperature would be lowered somewhat, enough perhaps to make the climate quite pleasant. For example, the vast area within ten or fifteen degrees of the equator may have had an average temperature not far from 70° F. instead of above 80° F., and there was probably more variability than now from one day to another. In a general way the climate probably tended to reproduce many of the conditions which prevailed farther north, in the supposed primeval home of man before his descent from the trees. A mild, pleasant climate of this sort would improve the health of the tropical people, and make them more active. It would permit a larger proportion of the children to grow up, and thus would tend to make the population more dense. The amount of food available for human consumption would presumably not be altered greatly, and the methods of getting it would not have to be changed. The problem of getting a living would be easier than formerly because people would be more active and vigorous, but the problem might also be made more difficult if a lower death-rate appreciably increased the density of population. Such an increase might have an important effect in putting a premium upon all the qualities which enable a people to make a living, but perhaps the extra effort required because of the denser population would be balanced by the greater energy.

Even if the climatic amelioration in tropical lands during a glacial epoch were a real factor in raising the general level of the people, as may well have been the case, it would apparently introduce few new factors and few new types of natural selection. Not even the most severe glacial period would apparently cause the equatorial regions to have any great contrast of seasons in excess of the regular periods of more or less rain which prevail at present. In the genuine equatorial regions, with which alone we are now concerned, the mode of life, so far as I can see, would remain practically constant no matter how many glacial epochs came and went. The universal evolutionary factors such as the struggle for food, and possibly mutations, would still pursue their courses. Migrations, likewise, would sometimes introduce new races, but the partial isolation of the tropical regions by deserts on either side, their relative impenetrability, the danger to newcomers from disease, and the relatively unchanging character of the environment, all appear to have combined to make migrations less important near the equator than farther north. Thus while tropical man may not have been stationary in the march of evolution, he appears never to have been subjected to any such active evolution as has been the case in higher latitudes and more variable climates. In this respect man merely repeats the experience of family after family of mammals and other animals, whose most primitive types dwell in the habitats where changes have been least numerous, and where the environment has demanded the least adaptation to special circumstances.

The contrast between the action of tropical and non-tropical environments, whether through natural selection or through the stimulation of mutations, seems to be one of the most important causes of differences in racial character. It appears to be a biological law that a tropical environment, because of its uniformity, tends to perpetuate primitive, unspecialized forms. Since man split off from the apes his specialization has been in the size, complexity, and functioning of the brain. Other specializations, such as changes of complexion, stature, and hair, have been of minor importance. In equatorial regions the mental type of specialization has apparently been slow, largely because there have been no really great changes throughout man's history, not even during the severest glacial epochs. That, presumably, is one of the chief reasons why it is so difficult to impose upon equatorial people anything more than the outer husk of northern government, northern religion, northern ideals, and northern culture. Of course many tropical or semitropical people are of highly mixed ancestry and contain elements in which mental specialization has gone so far that they can assimilate other forms of culture; but that is a different matter. Moreover, when various races have first migrated to low latitudes, the new environment may in certain respects tend to select for preservation the individuals with certain specialized characteristics. It is not impossible that if a few million of the most competent Europeans were to be set down in a tropical country where they were fully isolated from the present inhabitants, they might ultimately develop into a new race of extremely high type. After their numbers had been cut down remorselessly by disease for generations, the survivors might be those who possess great mental ability, adaptability, and self-control, and are at the same time able to take life easily but seriously, to work wisely, rest freely, and protect themselves assiduously not only from the ordinary tropical diseases but from whatever more insidious dangers there may be in the climate. But this does not apply to the present people of the torrid zone. They are largely descended from stocks which apparently experienced their main mental evolution long ago, and have almost stagnated ever since.

In high latitudes the effect of the glacial period must have been very different from its effect near the equator. The first thing that the inhabitants noticed when a glacial epoch began to come on in central Europe, for example, may have been that they experienced more rain, more snow, and longer winters than were indicated by the traditions of their fathers. More probably, they became aware that whereas the traditions told of hunting Merck's rhinoceros, the lion, the urus, or the hairless elephant, they themselves much more frequently pursued the woolly rhinoceros, the lynx, the bison, or the hairy mammoth. The types of trees likewise changed. The tulip-tree, beech, maple, elm, and their congeners gave place to larch, hemlock, birch, and pine. Just such transformations are recorded by fossils in central Europe.

As time went on the conditions became still worse. The hunting must have been so bad that many a day the wife scolded her husband for bringing home nothing, and the children cried from hunger. Many of the children and an unduly large number of the older people doubtless perished each winter. An intense selective process thus went on whereby only those with the greatest endurance and the greatest skill could survive. But their skill was of little use, for their environment left them few resources and little opportunity, energy, and spirit for devising anything new, or for even the most primitive art. Moreover, when things reached this stage, or perhaps earlier, the cleverest of the primitive families and those in whom initiative, the power of leadership, and the love of adventure were most developed presumably migrated. The wisest of all would deliberately choose their direction of migration, and go toward warmer lands, more like those described in the oral traditions of their fathers. Among the less competent who remained in the old environment, the hardships attendant on the long, cold, snowy winters must at some point have begun to cause degeneration, for in the harshest climates the mere power to endure is one of the chief characteristics. Finally, in all northwestern Europe, the few hardy families who still survived must have perished or else followed their wiser and more adventurous companions who had had the wit and knowledge to move in the right direction before the glacial period caused them to deteriorate.

This brings us to what I believe to be another highly important step in understanding the evolution of racial character. In northern Asia, as well as in northern Europe, the approach of an ice age would cause three things to happen. First, some of the inhabitants, presumably the more adventurous and intelligent, would migrate southward into milder regions. Second, a large percentage of the population, though not a large number as we count population, would be exterminated from generation to generation. Third, the remnant which survived would go through a process of regressive selection, whereby the survivors would be those in whom passive qualities of resistance to hunger and discomfort were most highly developed. The nervous, active types who lead the march of human progress would be at a disadvantage compared with those of a more phlegmatic constitution, as we shall see more fully in studying the people who went to America and China. This phlegmatic type is now represented by the Lapps, Eskimos, Chukjis, Tierra del Fuegans, Samoyedes, and Alaska Indians. In ancient Europe it was perhaps represented by the late Mousterian people whose retrograding culture may be inferred from the flints and other implements left by them during the period just before the last glacial epoch, the so-called Wurm advance of the ice.

At this point we must emphasize the distinction between Europe and Asia, a distinction which seems to have been one of the most vital factors in human evolution. The distinction is that in Europe it was the cold and ice of *glacial* epochs which caused the greatest weeding out of human types and produced the conditions that most fostered migrations and mutations. In Asia it was the drought and deserts of *interglacial* epochs that presumably produced the most intensive human evolution. During the height of each glacial epoch most of Europe down to southern England, northern France, central Germany, and central Russia, was shrouded with ice and snow; another large ice-field deployed from the Alps; and ice from the Pyrenees and Carpathians spread far down the valleys. The rest of the continent north of the Pyrenees, Alps, and Carpathians must have been extremely stormy and disagreeable, and was probably covered with snow well into the summer. Hence with the recurrence of each glaciation absolutely all the inhabitants were driven out or exterminated north of latitude $50°$. It is not to be supposed that they were killed suddenly, but merely that they gradually diminished from generation to generation because of migration or an excessive death-rate especially among the children. South of this in the tier of countries including France, Austria, Hungary, Rumania, and southern Russia, only a handful of survivors can have remained. They must have been a highly selected remnant, selected because of their ability to withstand cold, hunger, long winters, and privation. During the long process of exterminating, repressing, or driving out the people of Europe, a considerable number of the more energetic, intelligent, hardy, and adventurous men, with or without their families, must have drifted southward and eastward into the Mediterranean peninsulas, north Africa, and southwestern Asia. These were presumably the people of greatest ability and initiative. Even among them the great majority probably died, for only the strong survive in any primitive migration. The women and children suffer especially from all sorts of hardships, such as lack of food and shelter, while many of the men are killed in the fighting which is almost inevitable when migrants enter a new land already populated. Whole tribes and even whole racial groups may have been practically exterminated. In other words here we have one of those inexorable processes of selection which weed out those who are weak in body or mind, and preserve those who are strong, resourceful, energetic, and adaptable. The shrouding of Europe by ice in four successive epochs carried out such a selection time and again. It presumably drove the selected remnant into southern Europe and into the neighboring parts of north Africa and western Asia. During at least the first three ice ages the Ural Gulf probably prevented migration to Asia except by way of Asia Minor or the Caucasus. Thus the countries from Persia, Asia Minor, and Mesopotamia around by Egypt to Morocco, together with the southern peninsulas of Europe itself, formed a belt into which must have migrated the selected remnants of the Europeans of greatest physical strength, mental ability, and initiative who survived the migrations due to the advancing ice. There the migrants must have encountered some of the most vigorous races of Asia. Thus there presumably arose not only a complex mingling of races but a severe problem of adjustment whereby many of the former inhabitants were killed or driven out. In the parts of Europe well beyond the ice-sheet and its uninhabitable climate, and in the neighboring parts of Africa and Asia we have the very antithesis of the tropical regions. A semicircle of selected people who have been through radical changes of environment and who have made long migrations, are mingled with other races. They have experienced almost every condition that leads to rapid evolution. They surround an empty continent, ready to penetrate it and expose themselves again to the evolutionary processes of migration and expansion as soon as the ice and the climate permit.

Turn now to Asia. There the precipitation appears never to have been great enough to cause any extensive ice-sheet. One of the facts that geology has only lately realized is that snowy precipitation is far more important than low temperature in causing glaciers. The coldest part of Asia, namely the region around Yakutsk, shows no hint of glaciation in the lowlands, either now or in the past. In the mountains farther south, to be sure, and likewise in the Taimir peninsula to the northwest, and the Anadyr peninsula to the east near Bering Strait, there was glaciation, for in these places the temperature is higher and the precipitation greater than in the region from Yakutsk northward. If by chance these far northern regions contained any inhabitants in preglacial or interglacial times, they were of course driven out of the glaciated portions. Elsewhere, even in the coldest parts they may possibly have survived, although there is no evidence as to this either one way or the other. If people can live where the January temperature averages 60° below zero F., as they do in north central Siberia, they could perhaps survive a temperature 10° or 15° degrees lower, although assuredly on a very low plane and in very small numbers. The point that we would make, however, is that the climatic change in northern Asia was not nearly so great as in Europe so far as human habitability is concerned. In northwestern Europe places like Denmark, that now have some of the finest climates in the world, were transformed into snowy regions covered with an ice-sheet hundreds or thousands of feet thick. In Siberia the change was merely from a dismally cold, repressive climate to one that was still more dismal and repressive. In Europe a large population, that is, large for primitive hunters, presumably grew up during interglacial times. It was then subjected to an exceedingly severe selective process whereby the most competent people were gradually sorted out and pushed into an outlying ring where the climate was comparatively favorable. In northern Asia, provided there were any people there before the last glacial advance, a similar process presumably took place, but it must have been relatively ineffective in promoting evolution. This is partly because the interglacial climate was not nearly so good as that of Europe, and the population must have been correspondingly smaller, and partly because the change in habitability was not nearly so great as in Europe.

In this discussion of the effect of glacial epochs upon racial character we have left the deserts until last. In contrast to the relatively mild changes in the latitudes south of the great desert belt and to the great and repressive changes to the north, the deserts themselves must have experienced a great and most favorable change. That is, the change was favorable from the point of view of mere comfort, though not perhaps of evolution. While northern Europe was being slowly gripped and choked by the ice, the deserts were growing more and more habitable. Storms began to sweep over them, at first in small numbers and only at certain seasons, but later in large numbers and throughout a far longer season. The barren sands, gravels, and clays became clothed with grass; the places that had been grassy began to support forests; the salt lakes expanded; new lakes were formed; wild animals, especially those that live in the grasslands, must have increased enormously. All these conditions made life easy. The abundant food, the freedom from the constant menace of famine, the stimulating climate, and the presence of new areas ready for occupation must have caused a rapid growth of population. Such a growth would not exert any

GLACIATION AND THE SUPREMACY OF EUROPE 57

special selective influence, for even the stupid and weak would have plenty of opportunity. But it would allow freedom for man's inventive genius to express itself, and for new ideas and institutions to expand and bear fruit.

The only great drawback in such a time of prosperity and expansion would be the influx of invaders from the northern regions that were growing cold and icy. On the margins of the dry areas there must have been constant conflict, conquest, and racial mixture. At the same time the push from these northern regions must have co-operated with the attraction of what had formerly been the deserts which were now becoming more habitable. Thus the people dwelling between the cold northern environment and the desert would tend to move desertward. This movement must have been greatest on the northern side of the great desert belt, that is, in north Africa and central Asia. Farther south the Sahara and Arabian Deserts, although more habitable than now, were presumably not so habitable as those to the north because less fully under the influence of cyclonic storms. In the northern belt far more than the southern we find traces of expanded lakes such as those which filled the basins of the Dead Sea, the Sea of Aral, and Lop Nor in western China. Moreover, as we have just seen, the southward migrations due to the cold climate of the far north must have been far less important in Asia, which was merely cold, than in Europe with its ice-sheet. Hence during the approach and culmination of an ice age the inhabitants of the Mediterranean region and north Africa must have been terribly squeezed, as it were, between the desert on the south and the European migrants on the north. Accordingly great migrations may have taken place from that region eastward into the improving deserts of central Asia, and thus back toward man's ancestral home. Such movements may have conflicted with the southward movements from the cold regions in the north of Asia.

Now reverse the process. Let us see what must have happened when the glacial climate began to relax. Here we must remember that in all probability the atmospheric conditions which induced glaciation passed away considerably before the ice-sheet finally melted. This is important because it means a tremendous stress in certain areas. First, however, let us briefly record what would happen in other areas. In tropical regions the passing away of a glacial epoch, like its approach, would have relatively little effect. So far as it influenced the equatorial people it probably tended to reduce them to a lower plane of progress. In addition to this it may have driven migrants from the deserts into low latitudes. Of course such migrations would come mainly from the deserts north of the equator, but also from those to the south, provided mankind had penetrated to all parts of the world.

The amelioration of the glacial climate in high latitudes would cause an effect the opposite of that in low latitudes. Siberia, for example, must have become more habitable, although there, as we have already seen, the change was apparently of only moderate proportions. Where the environment had been extremely repressive it became moderately repressive. This would permit some migration from central Asia northward and northeastward. But in Siberia, as in the tropical countries, the effect upon human progress was probably slight.

On the other hand, consider what must have happened in the desert belt from north Africa to Mongolia; and likewise in the neighboring regions of Europe on the one hand, and China, Chosen, and Japan on the other. As soon as the storminess which appears to have been one of the main causes of glaciation began to decline, the interior of Asia and of north Africa apparently began to return toward its desert condition. The relatively large and progressive population which had presumably become established there must have been put under great stress, because there were too many people in proportion to the supply of food. Large numbers doubtless perished because of recurrent famines due to drought. Others presumably migrated. Where would these migrations lead them? A few, as we have just seen, would migrate northward, but the northern lands would still be much less able to support people than they are in our own day.

GLACIATION AND THE SUPREMACY OF EUROPE 59

Southward the lines of migration would be blocked partly by the great plateaus and mountains from Tibet and the Himalayas westward, partly by deserts such as those of Arabia and the Sahara. Some migrants would break through into India and Persia, as has happened repeatedly during historic times. A migration might follow the lofty western side of Arabia thus skirting the desert and finding a way across into Africa at the southwestern corner of Arabia where during much of the glacial period Africa and Arabia were connected. But the greatest tendency would be to migrate eastward into China and adjacent countries such as Chosen and even Japan, and westward into the relatively well-watered border-lands of Asia such as the Caucasus, Mesopotamia, Asia Minor, Syria, and likewise Egypt and northern Africa. Meanwhile much of Europe would still be shrouded in ice and could not receive many new migrants. The result would be that in western Asia and north Africa there would be one of the most tremendous processes of selection and racial mixture that can well be imagined. Tribe after tribe would move in from the desert. On the way, or before they started, they would have lost a large share of their weaker, less intelligent members. They would come into conflict with the picked remnant of the former inhabitants of Europe. Thus there would take place a mixture of races probably unparalleled elsewhere, and the races thus mixed would be the selected remnant from Asia on the one hand and Europe on the other. Small wonder then that in the region from Persia or Transcaspia westward into north Africa we find the homelands of most of the world's more progressive races. In such a region, according to the commonly accepted principles of biology, evolution is bound to be more rapid than elsewhere, for there we have the maximum environmental change, the maximum migration, and the maximum mixture of different types.

In view of all these circumstances it is not surprising that the world's greatest early civilizations arose in the region which extends from Persia through Mesopotamia and Syria to north Africa, with a similar but lesser development in northern India and China. Moreover, from this same region, as the ice retreated from Europe, came the tribes which peopled that continent after the ice had passed away. Some moved northward from the Mediterranean regions, others westward by the route through Asia Minor or north of the Caspian Sea. As they poured into the new continent renewed by its long rest under ice and storms, the various streams met and fought, and still another strenuous conflict and great mixing of races took place. Thus here, even more than in western Asia and in north Africa, we should look for the development of highly specialized races.

Our survey of the biological effect of glacial periods seems to lead to the conclusion that as each glacial epoch passed away, the people in a great belt from China to Gibraltar were left with higher abilities than had been possessed by their predecessors. Since the effect of four successive glaciations was concentrated in virtually the same area, the evolutionary effect of glaciation must have been multiplied fourfold. Hence as the last ice-sheet waned, the highest abilities were naturally concentrated in China, north India, and especially southwestern Asia, northern Africa, and the neighboring Mediterranean areas where migrations not only had produced their greatest selective effect, but where there had been little or none of the repressive selection which seems to take place in high and low latitudes. Thus, seemingly, when man had evolved to a certain point, it was only to be expected that agriculture and what we know as civilization should evolve in exactly the countries where we first find traces of them. It was equally logical and inevitable that in Europe, as the waning ice opened a virgin field of settlement, there should be concentrated the selected outpourings of the progressive belt of countries where civilization was beginning to grow up. Such migrants would naturally be among the most competent of all the races that the world has yet seen. If they found themselves in a superlatively healthful and invigorating climate, as they apparently did in Greece and Rome, and later in the North Sea region, it would indeed be strange if they failed to dominate the world.

CHAPTER V

THE SUPPRESSION OF AMERICA

Just as Europe has presumably been a constant gainer, America appears till recently to have been a constant loser through the vicissitudes of climate. Many fascinating books have advanced the idea that America was peopled by migrants from the Old World to the New in low latitudes. Some have gone so far as to allow an Egyptian culture to migrate to Central America by way of the hypothetical, a continent of Atlantis, which supposedly foundered a few thousand years ago. Others, less extreme in their theories, suggest that the germs of the relatively high civilizations in Central America and Mexico must have come from the Old World by ships from the Mediterranean regions, or else from Indo-China across the Pacific. As time goes on, however, these theories, fascinating as they are, seem to find less and less support. A stray boat or two may perhaps have been driven across the Atlantic by the trade-winds, and may have brought certain new phases of culture which the most intelligent aboriginal Americans, already well on the road toward civilization, were able to adopt. Beyond this, however, it seems probable that the culture of America before the days of Columbus was of purely native origin, except in so far as it was brought by the original immigrants. Those immigrants are generally agreed to have come originally from Asia by way of Siberia and Alaska. Whether they came before or after the last ice age has not yet been definitely settled. Parts of human skeletons which are sometimes interpreted as of interglacial or preglacial age have indeed been found in Florida, California, and elsewhere. Even as I write, the newspapers report a Pliocene skull near Los Angeles. But the fact of any such great age is by no means established as yet, and the supposedly ancient bones, except perhaps in this latest case, appear to be of surprisingly modern types.

In spite of this, it seems probable that man came to America before the last glacial advance, and perhaps earlier. One reason for this view is that men with heads of practically every shape are found in both North and South America. Such great variety seems to indicate a series of successive migrations from Asia. Unless a long period is available it is scarcely probable that so many migrations could have penetrated to the far corners of the continents. If the aborigines came to America after the last glaciation, the ancestors of the first migrant group can scarcely have lived north of latitude 60° in Asia until the ice had completed at least half of its present retreat. That would be perhaps twelve or fifteen thousand years ago. It does not seem as if that were enough time for all that has since happened. Moreover, if these various migrations have all taken place during so recent a period, most of the migrants must have crossed at least fifty-six miles of open ocean at Bering Strait, or a much greater stretch farther south. The strait is now so important a barrier that only a handful of American Eskimos are found on the Asiatic side, and practically no Asiatics on the American side. In other words, no appreciable migration from one continent to the other is now going on, and none from Asia has occurred so recently that any Asiatic language has taken root on the American side. Bering Strait is a real barrier. In its present state, it might prevent migrations for thousands of years. In earlier times, however, there were long periods when the land stood higher than now, and people could pass freely from Asia to Alaska.

After crossing into Alaska the migrants of each successive wave presumably spread widely in both Americas, driving the preceding migrants before them, and being themselves driven out by their successors. The migrations occurred so long ago that fifty-five different linguistic stocks have been developed north of Mexico, and many others farther south. Few of these stocks have less than three or four dialects, and most are divided into numerous dialects, or even languages more distinct than French and Italian. A language does not grow up in a day—nor even in a century. Again, it must be remembered that it requires a long period for a hunting race to develop the arts of agriculture, architecture, and writing, and bring to fruition a civilization like that of the Mayas in Central America. More than two thousand years ago the written characters of the Mayas' language had reached such perfection that records could be carved upon stone in hieroglyphics which appear to represent syllables rather than words, a system more advanced than that of either ancient Egypt or modern China. Moreover, the Maya calendar was extraordinarily perfect, more perfect than any ever known until our Gregorian calendar was devised a few hundred years ago. Such a development of language, of astronomy, and of accurate records presumably requires thousands of years of preliminary progress before people begin to leave written records inscribed on permanent materials such as metal and stone.

Without going into further details we shall adopt the working hypothesis that the American Indians are a mixed people, the result of several invasions from Asia by way of Alaska. Further reasons for this will appear in a later chapter. We shall assume that at least part of the earlier invasions took place during interglacial times. If we put aside for the moment the possibility that some of the Asiatics came to America by sea—a possibility which we shall consider later—we may be quite sure that the others could scarcely have come during an ice age, for ice then covered the peninsulas on both sides of Bering Strait. Even now the climate of northern Siberia and central Alaska is so severe that we can scarcely believe that primitive people with scanty protection against cold and storms would migrate thither except under strong compulsion. Hence, unless the earliest human inhabitants date back no more than ten or fifteen thousand years, they were presumably pushed out from central Asia during relatively mild interglacial times, perhaps by reason of the expansion of the Asiatic deserts even beyond their present limits. Thus we may think of America as having been peopled by wave after wave of migrants in the same way as Europe, Africa, and Australia. The first of these may have been pressed northeastward by the increased aridity of the central Asiatic deserts during an interglacial epoch, and may have finally trickled over into Alaska during times when the deserts were perhaps larger than now, and the climate of the far north more hospitable. When the change toward a glacial climate ensued, the impulse toward migration presumably relaxed because the deserts became more habitable and the northern regions more inhospitable. Ultimately the increasing severity of the climate presumably drove the inhabitants of the colder regions southward in both continents. This process of driving people out of the north must have been especially active in North America, where ice covered practically all of Canada. Some of the races who are thus supposed to have come in interglacial or even preglacial times, may have disappeared entirely, leaving no traces except perhaps in fossils such as the skull found near Los Angeles. Others who came in the last interglacial epoch may at a later date have been reinforced by post-glacial migrants, and may have been among the ancestors of part of the modern American Indians.

The point that concerns us here is not the number or date of the Asiatic migrations to America, but the fact that no matter when they occurred, practically all who migrated by land must presumably have spent many generations in an extremely rigorous climate. Even in the mildest interglacial epoch this was probably true to a considerable degree, and during all other periods for the last hundred thousand years the climate was presumably well-nigh as severe as now, and much of the time more so. A long migration of perhaps fifteen hundred miles in the coldest part of Siberia, followed by another twenty-five hundred miles, more or less, through the coldest part of Canada under the cruel compulsion of an impending ice age, must have been a terrible experience. The part of Siberia through which the migrants must have passed is the coldest portion of the earth. Even in November and March practically the whole region averages below 10° F.; in January the entire region, including a large part of Alaska, averages lower than $-$ 10° F., while certain parts average 50° F. below zero.

It must be remembered that the vast majority of migrations take place very slowly. A rapid movement like that of the Huns under Genghiz Khan attracts wide attention, but such movements probably do not account for one per cent of the migrations even during historic times. Among primitive people the tendency is for migrations to be even slower than among the more advanced people who are helped by their superior knowledge and by their widespread command of the forces of nature. This is especially the case when wholly new lands are being penetrated, and still more when the migrants do not know what lies before them. Most of all must it be true when the migrants are being forced against their will into regions less hospitable than those where they formerly dwelt. It took the British settlers more than two centuries—at least seven generations—to advance from one side of the United States to the other. Yet they were the most progressive and best-equipped people of their day, or of any day except our own. Moreover, they were taking possession of one of the most favored lands in the whole world. It may have taken the primitive ancestors of the American aborigines many times seven generations to advance a similar distance, for they had no iron tools, no beasts of burden, no compass, no highly developed arts, and no means of making permanent descriptions and maps of the lands that they were about to penetrate. Still more would this be true when such primitive people were being forced into lands which became more inhospitable as they advanced. It is almost certain that no human race would migrate into such regions except as the more favorable regions were already occupied. If ever such people decided consciously to migrate and thereby better their condition, they would surely move backward into warmer lands, not farther north into those that were almost inexpressibly inhospitable. Hence we believe that the aboriginal migrations from Asia northeastward to America, and especially the first one, must have taken many generations, possibly thousands of years.

In *The Red Man's Continent* I have advanced the idea that such a climate may be one reason for the stolid character of the

Indians. It produces such stress that food is often scanty, and starvation a frequent visitor, especially in winter when game is hard to get. Here is what the famous explorer Wrangel tells of the strain among the Yukagir in northeastern Siberia even during the summer hunting season:

> Summer had arrived but the reindeer had not yet come. It is difficult to imagine the extent to which hunger prevails among the aborigines, whose existence depends primarily on chance. Even in the middle of summer people often begin to feed on the bark of trees and on skins which have hitherto served them as bedding and clothing. A reindeer caught by chance is divided among all, and is eaten up in the full sense of the word. All, even the internal organs and the crushed bones, are used, because one has to have something to fill the empty, tormenting stomach.
>
> During our stay here the coming of the reindeer was the only subject of conversation. Finally, on September 12, to the joy and relief of the aborigines, there appeared on the right bank of the river a countless herd of reindeer, which covered all the hills along the shore. Everyone came to life. From all sides the aborigines pressed toward the river in the hope of ending their sufferings by a successful hunt. Glad anticipation shone in all faces, and everything promised abundant game. But to the horror of all, the grievous word suddenly spread that the reindeer were leaving. And in truth we saw that the entire herd, no doubt frightened by the crowd of hunters, left the banks and disappeared in the mountains. Desperation replaced joyful hope. One's heart was torn by the sight of the people suddenly deprived of all means of maintaining their miserable existence. The universal grief and desperation were heartrending. Women and children moaned, wringing their hands, others fell to the ground and with howls dug into the snow as if preparing graves for themselves. Fathers of families remained silent, immovable, gazing at the hills where their last hope had disappeared. (Translation by S. Novakovsky.)

After such a summer when the herds of reindeer have been scarce, the winter is almost unbearably miserable. The extremely low temperature and the shortness of the days make hunting and fishing well-nigh impossible. There is little to do except wait in misery until the pangs of hunger and the monotony of inactivity cause the nerves to break, or till death claims its victims. Under such conditions two types of natural selection appear to take place. First, the weakest in both mind and body, especially those in whom the power of passive endurance is not highly developed, die at a rapid rate. Second, a certain number of the bolder, more adventurous, and more intelligent spirits break away from their fellows and struggle back toward a better environment. Most of them perish, but some reinforce the people in their old homes, while the loss of all weakens the people of the north.

In the colder Arctic regions the process of natural selection goes on apace among white men as well as aborigines. Among the whites, as I have explained in *The Red Man's Continent*, it is especially likely to drive away or kill people of a relatively active, nervous, or mentally high-strung type. Every spring the first boats returning to civilization from Alaska carry an unduly large proportion of men who have lost their minds because they have endured too many dark, cold winters. His companions say of such a man: "The north has got him." Almost every Alaskan recognizes the danger.

Thus the strain of life in the far north tends to eliminate the very type which is most likely to start some new idea and thereby cause progress. If such people do not break down mentally, they migrate to warmer climates. Only those survive whose nerves are in perfect condition, and who are able to bear the strain of prolonged cold, darkness, and monotony. The man of action, the one who has ideas, generally leaves Alaska because he is the one to whom the strain of inaction is least endurable during the unduly long winter. For women the Arctic life is far harder than for men, especially when they have children. Thus the reproduction of the active type of minds is greatly handicapped.

Another side of the Arctic problem must not be overlooked. The explorer Stefansson in those fascinating books, *The Friendly Arctic* and *The Northward Course of Empire*, has strongly protested against the common idea of the uninhabitability of Arctic regions. He believes that the view presented above does not fairly represent the reality. According to him, the Arctic regions are among the pleasantest in the world. People's health is good, the cold is not benumbing, and the long dark nights are not a disadvantage. In fact the Eskimos like the winter better than the summer, and the same is true of many white men. It is only a question of adapting oneself to the Arctic conditions, and learning how to use them. It is, as he points out, necessary to adopt a wholly new mode of life. One must learn how to clothe oneself, how to avoid the suffering due to the condensation and freezing of the moisture given out by the body, and accumulated in the clothing, how to build Arctic houses, and how to find an occupation which keeps one busy and interested through the winter. Nervous disorders, he holds, are quite unnecessary. They occur largely because people fear them and do not live in the way to forestall them. But the people whom Stefansson cites to prove his point are those who have chosen to stay in the Arctic because they enjoy the winters. In other words they are the selected remnant of a vastly larger group, the majority of whom have gone back to warmer climates. Few women and almost no children are included in the remnant. Thus while Stefansson may be right in thinking that the Arctic regions will some day contain enough people so that herds of reindeer and ovibos, as he prefers to call the musk-ox, will help to feed the vast industrial population of middle latitudes, his conclusions do not really conflict with those stated above. In fact he shows most strongly, although unconsciously, that a stringent selection is necessary in order to find people adapted to life in the far north. To find women and children who can stand it is far harder than to find men in the prime of life.

But how about the Indians? What type of natural selection takes place or has taken place among them? One answer to this is found in an unpublished work on the nervous disease known as Arctic or Siberian hysteria, which the Polish geographer, Novakovsky, has kindly put at my disposal.* This disease is a common form of disability in the north. It often comes upon people very suddenly, and causes them to do all sorts of strange things. For example, in one case a man was in a boat when suddenly he was seized with hysteria and would have thrown himself overboard had it not been for the efforts of his companions.

*An abbreviated form of this work appeared in *Ecology* for April, 1924.

In many cases it leads people to imitate others against their will and to do what they are ordered regardless of the consequences. For example, Jochelson, who spent many years in northeastern Asia as a political exile, tells how he was in a village while the people were fishing. His Cossack attendant, knowing of the curious nervous ailment which prevails there, took a live fish between his teeth and ran off with it up the hill. An old woman immediately did the same thing, and ran after him, panting but apparently unable either to stop or to let go of the fish. She kept crying out "Enough, enough," meaning that the Cossack should stop, but he persisted until he came back to the river. The old woman ran into the river, and would have been drowned had she not been helped. Only then did she let go of the fish.

This curious disease is closely connected with a tendency toward suicide. For instance, among the Chukchees it is not uncommon for a man deliberately to decide that the struggle of life is too hard. He calmly announces his purpose to his friends, who try to dissuade him. If he persists, they finally aid him in his purpose. Preparations for a festival and for the funeral rites are made by all concerned. After two or three weeks, at the end of a great feast in which the man who is to die is fed the very best in vast amounts, he is put to death by his friends in such manner as he himself may select.

Another curious feature of Arctic hysteria is that while men are subject to it, it especially affects women. Moreover, it has its worst effect at times of pregnancy and when the reproductive functions are most active. Whether it causes disturbances in reproduction or is caused by them matters little for our present purpose. The important point is that it tends strongly to reduce the number of children. It afflicts the people who are nervously or mentally active, and who are potentially the best material for human progress. Thus this type tends to be weeded out rapidly.

Siberian hysteria is peculiar in its geographical distribution, its seasonal distribution, and its distribution among different types of people. In one form or another the disease is found in almost all Arctic regions. In Greenland, for instance, Whitney, in his book on *Hunting with the Eskimo,* tells of a woman who began to rave during the night, and ran out into three feet of snow and partly froze herself. Such hysteria appears to be somewhat like the insanity which afflicts the white man in Alaska. But it is most highly developed in Siberia, where it assumes its peculiar form. Moreover, and this is a point which Novakovsky makes very clear, the hysteria is most abundant in the very coldest part of the country. "The whole of northern and part of southern Siberia is a region where the people suffer from nervous diseases more than in any other known regions of the world." So says Miss Chaplicka, a Polish scientist who personally conducted some interesting investigations among the aborigines of Siberia. Her conclusion is borne out by many other investigators on the basis of great amounts of evidence. As the "pole of cold" near Verkhoyansk, in the province of Yakutsk, is approached, the ravages of the disease become greatest. Often it is epidemic, and person after person, especially among the women, falls into convulsions and indulges in all sorts of strange and dangerous antics. Its results are apparently seen in the frequency of idiots and dullards.

Another noteworthy feature of Arctic hysteria is that it is worst in the winter and spring. In other words, it seems to be brought on by the long cold winter. How much the low temperature has to do with it and how much the long nights is not clear. It should be borne in mind that in these far northern regions all parts of the year bring a great deal of nervous strain.

Another interesting fact is that Arctic hysteria is least common among the races which have lived longest in northern Siberia. According to Novakovsky the Chukchees and the Koryaks are relatively free from the disease, and they are the most ancient inhabitants of northern Siberia. On the contrary, among the Yakuts, Lamuts, and Tungus the disease is more prevalent and severe. These people are relatively newcomers. In the same way, although Russians do not suffer from hysteria in their own country, they are afflicted with the disease when they migrate to the coldest parts of Siberia. Thus it appears that if a race stays long enough in the severest Siberian environment, it gradually becomes acclimatized, but that happens only, it would seem, through the weeding out of the types that are nervously and mentally most active.

In our own day some of the best observers maintain that the Indians as a whole still show the characteristics of a northern race. In Central America, for instance, it is generally recognized that Indians do not stand the heat and moisture of the lowlands as well as negroes. According to Brinton, one of the greatest authorities on the subject: "The American Indians cannot bear the heat of the tropics even as well as the European, not to speak of the African race. They perspire little, their skin becomes hot, and they are easily prostrated by exertion in an elevated temperature. They are peculiarly subject to diseases of hot climates, as hepatic disorders, showing none of the immunity of the African. Furthermore, the finest physical specimens of the race are found in the colder regions of the temperate zones, the Pampas and Patagonian Indians in the south, the Iroquois and Algonquins in the north; whereas, in the tropics they are generally undersized, short-lived, of inferior muscular force, and with slight tolerance of disease." "No one," says the famous naturalist Bates, "could live among the Indians of the Upper Amazon without being struck with their constitutional dislike to heat. The impression forced itself upon my mind that the Indian lives as a stranger or immigrant in these hot regions."

The importance of all this in appraising the character of the Indians is obvious. On their way from Asia to America almost every stream of migrants, no matter what its original racial characteristics, probably passed through the regions where Arctic hysteria is most prevalent, and where life often depends on sheer passive endurance rather than on energy, activity, or inventiveness. There for perhaps thousands of years they may have been subjected to a selective process which weeded out those who were of the more nervous, active types, the kind who lead, who invent, and who are largely responsible for human progress. Thus the American aborigines, whatever their race, may have acquired a certain uniformly dull, passive quality. They did not, to be sure, lose all capacity for improvement, as their achievements abundantly prove. Moreover, among them as among any race new mutants may have arisen, and may yet arise, possessing qualities surpassing those of the original Asiatic stock. Nevertheless when once a certain quality has been acquired by a species or race, the laws of heredity make it very difficult to get rid of that quality. So it would seem that the Arctic environment may have stamped upon the people of America a certain lack of originality, a certain tendency toward stoicism, a great ability to endure privation; and may have weeded out much, though not all, of the alertness, curiosity, and inventive faculty which are so essential to progress. This may perhaps help to explain some of the qualities which distinguish the American race, if race it is, from the races of Europe and Asia.

CHAPTER VI

THE CLASSIFICATION OF RACES

Up to this point we have been following the methods of geologists and ecologists. The foundations upon which we have built have consisted of two parts: first, certain generally accepted principles as to the effect of environment upon natural selection and perhaps upon mutations; and second, certain well-established facts and widely accepted conclusions as to the environmental changes through which primitive man must have passed if he originated somewhere in central Asia in Pliocene times. On this basis we have attempted to discover how the environment, and especially how the glacial period must have influenced the evolution of human character. We have found that this analysis seems to shed important light upon the origin of civilization and upon the present qualities of various races in Asia, Europe, North America, and the other continents. We might have gone on to show that in the most extreme deserts, and especially in the hotter deserts such as those of Arabia, the Sahara, and those south of the equator, the stress of life is so great that a repressive evolution seems to occur. There, as in extremely cold regions, the conditions under which the environment acts as a stimulant are passed, and retrogression begins.

We might also have shown that the low estate of the aboriginal people of south Africa, the southern part of South America, and especially Australia may be partly connected with the fact that the more primitive types of humanity have, as a rule, been the ones that were driven into these remote regions. We might also show that the low condition of the aborigines is apparently correlated with the fact that they were subjected first to the repressive or retrogressive environment of the equator and then to the almost equally unfavorable environment of the hot deserts which lack the stimulating qualities of the more northern deserts where there are strong changes of seasons. This is pre-eminently true of Australia, where primitive man is at his lowest. A people subjected to two such environments could scarcely be expected to develop greatly, even if they were originally gifted, which was apparently not the case, or if they were endowed with all sorts of natural and artificial advantages such as good domestic plants and animals, and helpful iron tools, the wheel, and the art of agriculture, which again was not the case. Thus in a broad way, the general distribution of racial character seems to be closely in harmony with what we should expect from the principles of paleontology and ecology, and from the geological facts as to early man and his environment.

It will now be of interest to approach the problem from a wholly different standpoint, namely, that of the anthropologist. Perhaps the most widely accepted generalization of anthropology is that the degree of kinship among races can be read in the form of their heads better than in any other way. This is because environmental circumstances appear to have less influence upon the form of the skull than upon that of any other part of the body, or than upon any mental trait or external habit. Some anthropologists, to be sure, question this generalization. Boas, for example, holds that a new environment may cause the form of the skull to change a little. He believes that his measurements of immigrants and their children in New York City indicate that when broad-headed Jews from Poland come to New York the children born in this country have heads not quite so broad as those of their parents or of their brothers and sisters born in Poland. On the other hand, the children of long-headed Sicilian immigrants, when born in New York, appear to be less long-headed than their parents or than their brothers and sisters born in Sicily. Whether Boas made sufficient allowance for the fact that the head-form of children and young people continues to change until complete maturity is attained is not yet certain. Even if he is right and there is an actual change in head-form when people migrate to a new environment, the change appears to be slight, and the children of both the broad-headed and the long-headed parents would be grouped in the same general categories as their parents. Moreover, there is no knowing how greatly a few unrecognized cases of racial mixture in the new environment may have influenced the results. Again, the change in head-form, if change there really is, appears to occur immediately after the arrival of the immigrants, and is probably a sudden mutation rather than a progressive tendency. Hence the facts thus far available do not seem to disturb the anthropologists in their general conclusion that the shape of the head is the best measure of the relation between one race and another.

Without committing ourselves one way or the other as to the degree to which environment can alter the form of the human body, let us see whither we are led if we accept the general view of anthropologists as to the importance of head-form as a criterion of race. Curiously enough the first broad application of this criterion to the world as a whole was made by a geographer, Griffith Taylor of Australia. His conclusions published in *The Geographical Review* in 1919 and 1921 were so contrary to generally accepted ideas, and were also inevitably based on data containing so many gaps, that they were received most sceptically. In 1923, however, in his book on *The Racial History of Man*, Roland B. Dixon of Harvard came to essentially the same conclusion. The generalization which both men make is this: The form of the head is the most permanent and distinctive of racial traits. The most primitive heads are long, narrow, and low, with small brain capacity. As man has evolved, his head has tended first to lengthen from back to front; then to become higher, and finally broader. Thus there has been a series of steps toward a round head. Such a head is biologically the highest and most specialized, because it can hold the largest brain in proportion to its surface and weight. If the people of the world are analyzed according to the form of their heads, we find the same peculiar distribution in each continent. Even in Australia this is shown imperfectly, especially if we count the Pacific islands with the continent. The general nature of the distribution is that the most primitive types with the narrowest heads are found in the margins of each continent, or in areas that are protected by mountains, by tropical forests, or in some other way. Toward the center of the continent and in the most easily accessible of the other portions, the breadth of the heads increases. Hence the central part of each continent is regularly occupied by broad-headed people of relatively recent origin.

This condition has been sketched by Taylor in a map which is reproduced as Figure 2. This map, as Taylor most carefully insists, is merely a rough generalization. Flaws can be picked in it right and left, as they can in every map which attempts to express an extremely complex relationship in a few simple strokes. Nevertheless, it is of high value because it presents graphically the main conclusion to which the premises of the anthropologists seem inevitably to lead; and which the more detailed maps of Dixon abundantly support. It suggests that narrow-headed people first spread over the world. They might have thought that the world was theirs if they could have known their own distribution. But even as they were advancing, new races with larger heads were developing in central Asia, and soon began to spread into the other continents and to displace their narrow-headed predecessors. And so the process has gone on. The most narrow-headed people live largely in tropical regions and on the edges of the continents, the tips of peninsulas, the places that could easily be defended against invaders and those that their successors have not wanted, or could not easily get into because of mountains, seas, deserts, or other barriers. Look at Figure 3, also from Taylor, and see how clearly the process is illustrated. The most long-headed people, as appears from the dark shading, occupy the mountainous peninsulas of Scandinavia, and Siberia, and the protected and rugged western portions of Great Britain together with Ireland. They likewise occur along the southern coast of Italy and in rugged northern Greece where they presumably represent a primitive race which we shall later discuss under the name Pelasgians or Ægæans. In north Africa it is the desert rather than mountains or seas that has protected the long-heads.

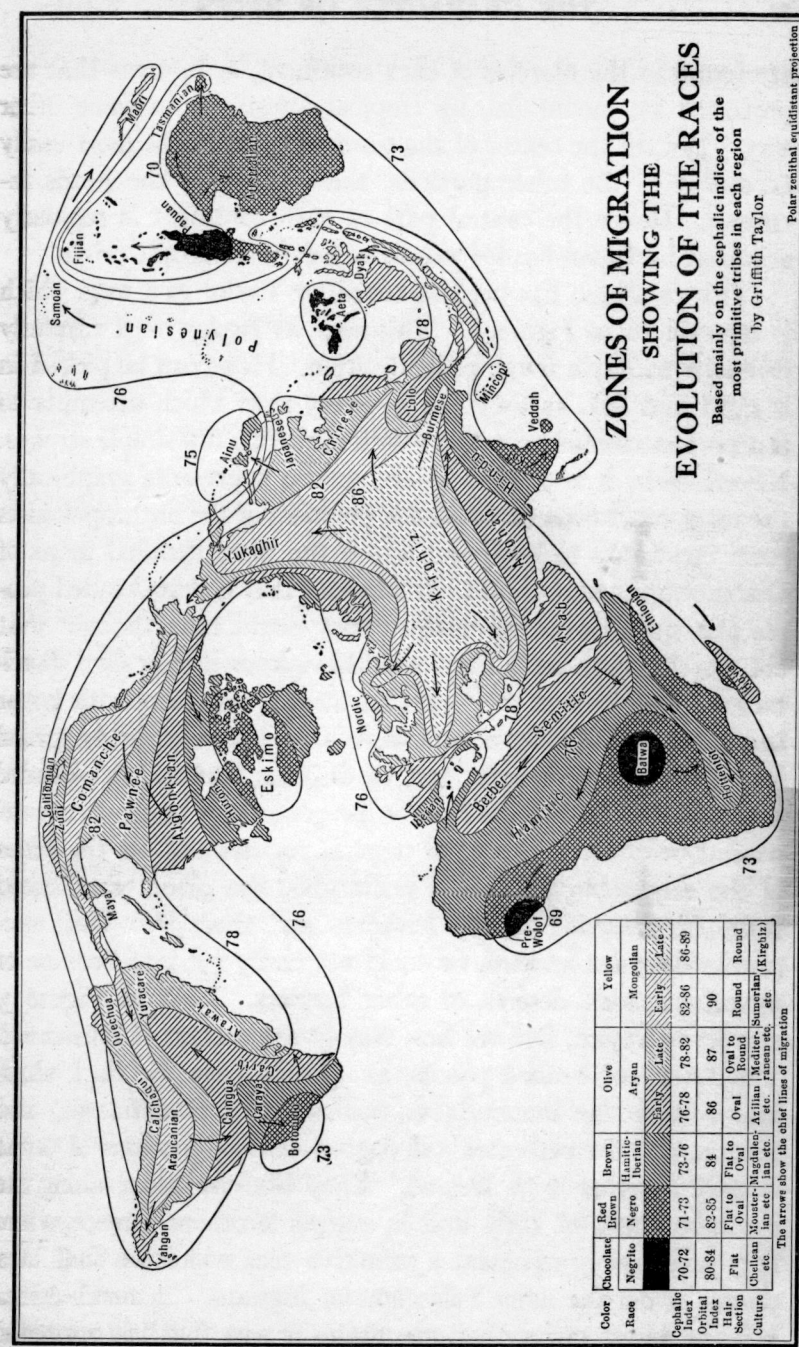

FIG. 2. GENERALIZED ZONES OF PRIMITIVE MIGRATIONS. (AFTER TAYLOR.)

THE CLASSIFICATION OF RACES

In the central parts of each continent (Figure 2) the broad-headed people are steadily pressing forward. Central Europe has suffered a great change in this respect even in our own day, for broad-headed Poles, Czechs, Russians, and others have been pressing into Germany. Figure 3 shows clearly how the people from Asia with a cephalic index above 80 have thrust themselves among the long-heads like a great wedge. The migration of the English, Scotch, and Irish to America, as well as of the early Spanish and French, was a movement of relatively long-headed people. If it occurred in prehistoric times the anthropologists would probably interpret it as a thrusting out of the long-heads by the round-heads. In America the first wave of long-heads has been followed by a wave of round-heads from eastern and central Europe. These round-heads, with their high birth-rate, are taking possession of the northeastern United States and driving the original long-heads westward. These later phases of migration do not appear in Figure 2, which deals only with the aborigines of America, Australia, and Africa. They merely indicate that the old process of replacing long-heads by round-heads is still as active as ever. One can almost feel their activity if one studies Figure 3 in detail and notes the arrows indicating lines of migration. All this is not pleasing to us who are Nordics, for our heads, though long and high, lack breadth. The south Germans, Rumanians, Turks, and Chinese approach nearer to the supposedly ultimate broad-headed type than we do. We hate to admit that potentially they may be the better people, but both Taylor and Dixon agree that our present seeming racial superiority is only an accident. But, as Taylor puts it, in many features the Mongolian is at least on the same level as the white race. The Mongols of central Asia and of the American Cordillera share with the Alpine (early Mongol) folk of central Europe the honor of possessing the highest cephalic index. They are of course farthest removed from the Negro and Negrito in this important respect. The same order obtains as regards orbital and aural indices and in the cross-section of the hair. As regards cranial capacity the white and yellow races are ranked together by Duckworth, though Clapham states that the average weight of the Chinese brain is greater.

The present superiority of the Nordics is due to environment, according to Taylor, and to racial mixture, according to Dixon. I believe that both these factors are important, and that a third factor, natural selection, works with them, no one of the three being any less essential than the others. Back of these lies still a fourth factor—the mutations or progressive deviations which may be of supreme importance; but as to whose causes we know so little. Unless this factor intervenes to produce other results, it seems probable that if the round-headed races should have an environment as good as ours, if they should make as fortunate mixtures with other races, and undergo as rigid selection, their rounder heads and greater brain capacity may win the day. The hope of the other races perhaps lies in so utilizing racial mixture and natural selection that the leaders will retain their environmental advantages and at the same time gain that type of physique—whatever it may be—which leads to greatest progress.

It must not be supposed that the form of the head is the only way of detecting the relation of one race to another. Other physical conditions such as the cross-section of the hair, the proportion of the limbs, complexion, and stature must likewise be considered. The most primitive hair is flat in cross-section, and therefore is kinky like that of the negroes, or even knots itself into "pepper-corns." Higher than this in the evolutionary scale comes hair that is oval in cross-section, and which therefore is wavy or curly as among Nordics. Highest of all is the kind which is perfectly round in cross-section, and which therefore remains like that of the Chinese, straight. As to the proportions of the limbs, the lowest types of men such as the Pygmies have short legs, long arms, and long bodies, but these features are variable and have thus far been of comparatively little use in determining the relationship of one race to another. The same is true of stature and complexion, for both of these conditions appear to be greatly influenced by environment. The Nordics, for example, may contain an appreciable share of the blood of a bleached

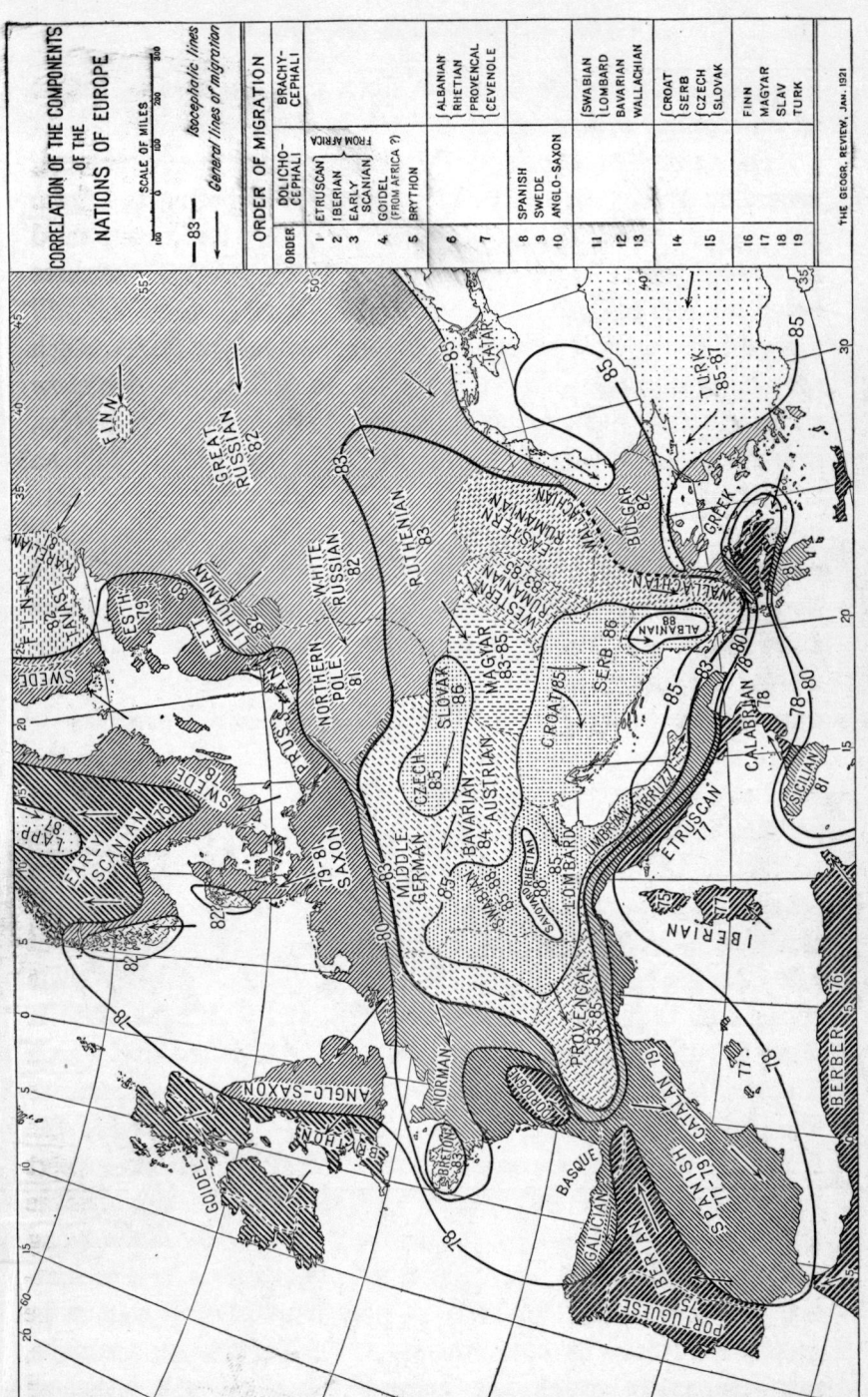

FIG. 3. HEAD-FORM AND ROUTES OF MIGRATION IN EUROPE. (AFTER TAYLOR.)

negroid race. Nevertheless, all these physical factors need to be used in connection with the shape of the head in determining how the various races have originated and how they are connected with one another. One of the merits of Taylor's presentation is that he takes account of all these factors and has even attempted to make allowance for the degree to which each may be modified by environment.

Still another set of factors may contain most important information as to the relationships of races. These are the cultural factors such as language, customs, implements, and the like. Taylor has boldly attempted to correlate some ten of these with the shape of the head.

The results are shown in Figure 4. In each of the maps there shown, the area where a certain cultural feature is known among primitive people is shaded or else indicated by a letter. Taylor has added lines like those of Figures 2 and 3, showing the approximate limits within which the cephalic index generally attains a certain level. For example, in the first map it appears that deformation of the head in infancy prevails almost entirely among broad-headed people whose cephalic index ranges from 80 to 84, but not among those with the broadest heads where the index rises above 84. In the same way totemism is found among people whose index is generally less than 77. The rest of the maps are almost self-explanatory. The couvade, it will be remembered, is the curious custom whereby at the time when a child is born the father takes to his bed, or at least pretends to suffer pangs like those of the mother. In the case of kin-words the languages of the unforked type use the same word for the kin of both parents; just as we say "uncle" for either a father's or a mother's brother. In the forked type, however, a mother's brother and a father's brother are called by different words; and so with all degrees of relationship. The Levirate is the custom whereby a man marries the widow of his deceased brother, as among the ancient Hebrews.

These maps of Taylor's have been criticised as being too highly generalized. It seems to me, however, that such broad generalization is necessary as the first step in any highly complex problem. Taylor has surveyed a vast number of facts; he has had the wisdom to gather together related data from all parts of the world, and to express them graphically in maps. He has thus caused ethnology and anthropology to work hand in hand with geography. It seems to me that in doing this he has opened the way for a vast amount of more detailed and accurate studies. Many of the details of his work will doubtless be modified, but if their main outlines stand, they will some day be recognized as a great contribution to human knowledge. The primary reason for this belief is that Taylor for the first time advances a broad anthropological hypothesis which is in harmony with the facts of modern geography, and also with the facts proved by geology as to the evolution of the great families of mammals. I am surprised at the way in which Taylor's conclusions accord in the main with those to which we have been led in previous chapters on the basis of paleontology, ecology, and geology. Another reason for believing that Taylor's main thesis is correct is that Willis in his remarkable study of *Age and Area* has demonstrated with mathematical precision that, other things being equal, the oldest families and genera of plants are also the most widely distributed. In plants, just as in animals, an ancient type may be missing over broad central areas, but its age is evidenced by the fact that it is found in outlying regions. What this means, according to the interpretation of Willis, is that the old forms each started in some central area. From this they spread outward in many directions, suffering slight modifications, but in general retaining the ancestral character. The original homes of plants, that is, the places where new species are now arising, are very largely areas like mountain-tops where extreme conditions prevail. In such areas new types originate and often drive out the old. They may even cause the older species or families to disappear over areas of almost continental extent. Yet the fact that the old forms are found scattered here and there almost certainly means that the intervening areas, so far as their environment does not forbid, were at some time occupied by these old forms.

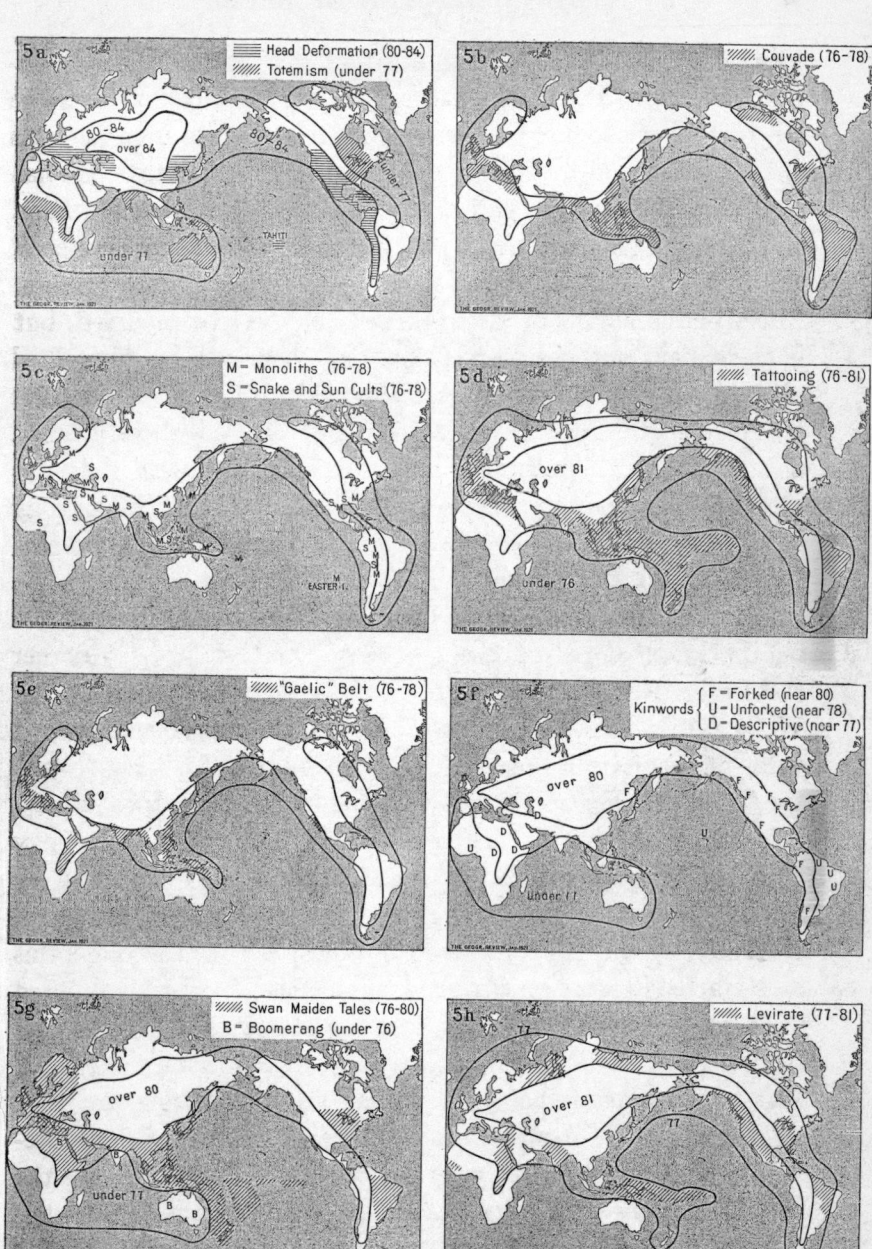

FIG. 4. MAPS SHOWING THE CORRELATION OF CULTURAL CONDITIONS WITH THE CEPHALIC INDEX. (AFTER TAYLOR.)

The old forms must have spread from the common centre along certain paths where now they have disappeared, as is proved by almost innumerable cases such as New Zealand where the facts as to the geographic distribution of plants are so clear as to leave little room for question. Hence *age* is roughly measured by *area*. This generalization, as Willis has shown, applies to living forms of animals as well as to plants. It is the same generalization which the geologists have made in respect to extinct animals, and which Taylor has made in respect to man. The agreement of these three lines of evidence is what we should expect in view of the increasing evidence that nature's main laws apply equally to plants, animals, and man.

Another important confirmation of Taylor's hypothesis is that Dixon has independently come to the same generalization, although his methods and temperament are absolutely different from those of Taylor, and many of his facts are drawn from different sources. Thus it would seem that by the present form of his head, and to a less degree by his other physical characteristics and even by traces of certain elements of primitive culture, man shows that like most of the great families of mammals, he originated in central Asia, and migrated outward in successive waves, largely under the influence of great changes of climate.

The book in which Dixon discusses the great anthropological generalization as to the significance of head-form is a remarkable example of what can be done by taking a single guiding principle and working it out with absolute logic. It will pay us to discuss it more fully. This does not mean acceptance or rejection of its conclusions, for the facts are not yet sufficient for that. Assuming that head-form is the main criterion of racial relationships, Dixon concludes that five main measurements of the skull are sufficient, or at least are all that can be used in the present incomplete state of anthropology. He shows, however, that other measurements agree so well with these that it is fairly certain that we cannot be led far astray if we rigidly rely on the five which he has chosen.

The five measurements are: (1) the breadth of the skull from ear to ear; (2) its length from the forehead to the most projecting part of the occiput at the back; (3) its height from the base to the top of the head; (4) the width of the nose at the widest part as measured on the skull, and (5) the height of the nose. These five measurements give rise to three indexes, each of which is arbitrarily divided into three groups according to whether it is large, medium, or small. The first index is the ordinary cranial or cephalic index, that is, the ratio of the breadth of the head to the length. This shows whether a head, when looked at from above, appears more or less oval. An elongated oval head is dolichocephalic or long-headed, and is normally accompanied by a long face. A head with a medium cephalic index is moderately oval and is called mesocephalic, while a head with a high index is brachycephalic, or broad-headed, and is normally accompanied by a broad face.

The second or altitudinal index is the ratio between the length of the head and the height. A high index means a high forehead and a head that approaches the circular form when viewed from the side without regarding the chin. A head with a high index is called hypsicephalic, one with a medium index orthocephalic, and a low head chamacephalic. Other things being equal, the high head contains a larger brain than the low.

The third index is the ratio between the breadth and height of the nose. A narrow nose is leptorrhine, a medium nose mesorrhine, and a broad nose platyrrhine. The three indexes, together with the degree of prognathism or projection of the jaws, determine the shape of the face.

With three indexes and three divisions under each index, there may be skulls of twenty-seven different types. Thus one skull may be relatively long, high, and broad-nosed. Two others may be equally long and high, but one may be medium-nosed and the other narrow-nosed. Still others may all be low and narrow-nosed, but one may be long, another short, and a third medium in length. Among the twenty-seven types eight contain no medium indexes of any sort. These Dixon regards as typical of eight fundamental races from which all others are derived by the mixture of race with race. Whether he is right in this assumption is open to question, but he at least shows not only that the other nineteen types can all be derived from the eight extreme types by various mixtures, but also that the assumption of eight types, which have migrated widely, fits a vast body of historical and archeological facts, and is not contradicted by any similar body of facts. Moreover, it explains many things which have hitherto been unexplained. Hence, while Dixon's eight types or primeval races may not be final, they at least suggest a helpful means of determining the degree of relation between one race and another. Since his method of using cranial indexes is new, he has been obliged to study the original data for all the individual skulls whose measurements have been published, and for many that are unpublished. The amount of work is colossal, but the results are highly suggestive.

Dixon names and describes his eight main races or types as follows:

Name	Probable Original Home	Shape of Head	Shape of Face	Shape of Nose	Degree of Prognathism	Skull Capacity
Proto-Australoid	Southeast coast of Asia	Long Low	Medium broad	Broad	Moderate	Small
Proto-Negroid	Northern and western Africa	Long High	Medium broad	Broad	Moderate	Small
Mediterranean	Eastern Mediterranean and Black Sea region	Long Low	Narrow	Narrow	None	Large
Caspian	Caspian region	Long High	Narrow	Narrow	None	Large
Mongoloid	Mongolia	Round Low	Broad	Broad	Moderate	Medium
Palæ-Alpine	Eastern central Asia	Round High	Broad	Broad	Moderate	Medium
Ural	South Ural (?)	Round Low	Medium	Narrow	None	(Largest)
Alpine	Western central Asia	Round High	Medium	Narrow	None	Largest

Having assumed that these are the fundamental races, the next thing is to discover the racial composition of any given set of people. If we are Nordics, Dixon's analysis indicates that we are chiefly of Caspian and Mediterranean origin, with some Proto-Negroid admixture. The Proto-Negroids are not the Negroes of to-day, but a far more primitive race whose origin is placed somewhere in northern or western Africa. Dixon believes that part of this race migrated into northwestern Europe. There they were bleached by the climate so that a race which was originally black-haired became blond. Later the same regions were invaded by two other races. The Mediterranean race, coming perhaps from far western Asia, probably encircled the Mediterranean, and then moved northward. In its purest form it is found in Sicily, but farther north it too became somewhat blond. The third race was the Caspian which came from somewhere near the sea of that name and migrated across Russia to the Baltic. It hurts the racial pride of the fair, lordly Nordic to think that he has in his veins the same blood as the humble Negro. It also hurts his pride to think that the round-headed Ural and Alpine races, such people as the Kalmucks and Chinese, have brains that are relatively larger and perhaps potentially more competent than ours. Nevertheless, if Dixon's premises are accepted, most of his conclusions are logical.

Having established the criteria of these eight types, Dixon assumes that all the intermediate types are derived from them by intermixtures, which is perfectly possible. It is not our purpose to discuss the validity of the types nor of the assumption as to the results of their intermixture. It is enough to say that Dixon's method produces results which agree to an extraordinary degree with the known facts of history. They also clear up a number of hitherto insoluble racial problems, as will be illustrated later in the case of South Africa. They introduce, to be sure, certain new problems which are highly perplexing, but many of these become clear when studied in the light of geography and ecology. Hence we accept his method, not as final, but as the best means yet devised of testing racial kinships.

In order to discover the real racial antecedents of any group of people Dixon goes back to the individual skulls, entirely discarding the use of averages. He classifies the skulls according to his twenty-seven types, and then analyzes the mixed types into the eight primary types. In doing this one of the most curious results is that the cranial types among women are found

THE CLASSIFICATION OF RACES

to vary distinctly less than among men. In case after case the cranial measurements of women indicate more primitive types than those of their husbands. Often the men show a preponderance of types which are known to have migrated into a region relatively recently, while the women display a larger proportion of types which migrated earlier. Of course at the immediate time of a migration this is natural, for men often make invasions leaving their women behind. They slaughter the men in the places where they settle and take the women as wives. But after such a migration is over, one would expect that among the children, the girls and the boys would show an equal inheritance of head-form from their fathers and their mothers. But such is not the case. For many generations after the invasion the men still show more predominantly the head-form of the invaders, and the women that of the earlier inhabitants. Thus women are conservative in head-form as well as in character. For the anthropologist this conservatism is most fortunate, for it presumably preserves indications of earlier racial relationships. Moreover, so far as it goes it suggests that Dixon's general method is right, even though his eight races may never be firmly established.

A single example will illustrate the curious way in which the history of races can be read in the form of their heads. In South Africa there are now three races, the Bantus, Hottentots, and Bushmen, while a fourth race, the so-called Strandloopers, became extinct during the early colonial period. Anthropologists have been much puzzled as to how these various people were related to one another. Language and customs give only a few important clues, but a comparison of the skulls brings out what seems to be a reasonable racial history. For instance, among the Strandloopers half the women were Mongoloid, a quarter Proto-Australoid, and a sixth Proto-Negroid. Among the Strandlooper men, however, the Mongoloid and Negroid elements are not quite so strong as among the women. This seems to indicate that the first people in South Africa were of the primitive Mongoloid and Negroid types and that the Australoids came later.

In all four of the South African races there is a considerable proportion of Mongoloid blood. This is largest in the female Strandloopers and next in the male Strandloopers. Then come the male Bushmen and the female Hottentots with approximately the same amount, while the male Hottentots and the Zulus have the least. All four of the races likewise contain a considerable percentage of Proto-Negroid blood, but this is relatively slight in the Strandloopers, increases in the Bushmen, is still greater in the Hottentots, and greatest in the Zulus. This would indicate that the Negroid blood was the result of an invasion later than that of the Mongoloid. As to the Proto-Australoid strain, it is about equal in men and women among the Strandloopers, rises to a large percentage among the male Bushmen and female Hottentots, and is most noteworthy among the male Hottentots and male Kaffirs, the latter being a branch of the Bantus.

On this basis, according to Dixon:

A logical theory of the racial history of the whole of South Africa may be outlined as follows: The oldest population is represented by the Strandlooper, who at a very early period, perhaps already in Palæolithic times, extended widely over the plateaus of the southern portion of the continent and reached northward, perhaps as far as the region of the Great Lakes. Primarily Mongoloid in type, they were blended with a considerable Proto-Austroloid factor and a lesser element of the Proto-Negroid type. Later a southward movement of peoples, mainly of Proto-Negroid and Proto-Australoid types, who had absorbed a small proportion of Pigmy, Palæ-Alpine folk, took place, and, passing through Rhodesia, forced their way into the region of the southern plateaus. Of the older Strandlooper population they absorbed a part, driving the remainder to the coast, where they survived down to the sixteenth century. From the mingling of the immigrants and the Strandloopers the Bushmen developed, and these occupied all of the better lands. Again a new drift, this time of Bantu-like peoples, forced their way over the same route from the northward, and established themselves in the region south of the Zambesi. These folk, who were pastoral, cattle-keeping people, took the better lands, driving the Bushmen into the less favorable localities, and in time absorbed a certain proportion of them, and thus developed the Hottentot, whose language, although possessing several clicks taken over from the Bushmen, nevertheless shows strong Hamitic relationships. Lastly, probably about the fourteenth or fifteenth century, a last thrust of Bantu peoples occurred, bringing into the region the warlike Zulu, Kaffir, and so forth. Like their predecessors, they drove the earlier occupants from much of the better lands, forcing the Hottentots west and north toward the Atlantic coast, and the Bushmen into the Kalahari and other desert sections. Just as the earlier groups of invaders, these latest Bantu conquerors mingled somewhat with the older peoples, so that when their progress was stopped by European occupation, they had already absorbed a small Mongoloid element, which thus differentiates them from the Bantu tribes farther north. If this suggested theory is correct, the Strandlooper, Bushman, and Hottentot represent three successive stages in the racial history of this part of Africa, the fourth and last stage of which was put an end to by the European colonization. That the Hottentot represented a very old Bushman-Bantu mixture has been often suggested by others; the present theory carries this a step farther, and derives the Bushmen themselves from the still older Strandloopers by a similar process.

We have given this example at length because it illustrates the remarkable way in which a mathematical analysis of the form of the head seems to bring out racial relationships. In later chapters we shall apply Dixon's results to other parts of the world, including our own, and shall see what they indicate as to the relation between inheritance and environment in the determination of racial character.

CHAPTER VII

THE ANOMALIES OF ABORIGINAL AMERICA

THE distribution of racial character and of progress in America before the coming of the white man seems at first sight to follow quite different laws from those which govern the corresponding distribution in Europe. The greatest progress was not found in the regions with the most stimulating environment, nor was it connected with the races whose head-form allies them with the most advanced races of Europe. In discussing this matter I shall in part follow *The Red Man's Continent*, where the culture of the red men is considered in its relation to physical environment; but I shall add many new points of view suggested by Taylor and Dixon and by the earlier parts of this book.

To-day the distribution of progress in America, as in Europe, is almost identical with that of health and climatic energy. In pre-Columbian times one would expect a similar distribution, for the climate appears to have differed little if any from that of to-day. Yet the most advanced American aborigines lived in regions which rank relatively low in both health and climate. In the tropical plateaus where dwelt the Aztecs and Incas the climate though cool is so monotonous that it is unstimulating. On the other hand in the bracing climate of the northern United States east of the Rocky Mountains, the American Indians were mere savages, living largely by hunting, while southern and central California were inhabited by almost the lowest of the aborigines. The decay of the high civilizations which preceded the relatively degenerate Aztecs and Incas of the time of Columbus may have been hastened by a change of climate, but climatic conditions certainly do not explain why the people of the northeastern United States and of the Pacific coast stood so far behind those same Aztecs and Incas, even in the days of their decline. The contrast between Europe and America may briefly

THE ANOMALIES OF ABORIGINAL AMERICA 89

be summed up thus. In practically all parts of the Old World the degree of progress throughout the historic period appears to have been closely in harmony with the environment, provided due allowance be made for other known factors, such as migrations, famines, the discovery of America, the Crusades, and the Reformation. In modern America the same is true. Why, then, were the conditions of pre-Columbian America so different?

From the standpoint of racial inheritance the anomaly of America may be as great as from that of environment. According to Dixon:

> Singly or together . . . [the Caspian and Mediterranean] types stand pre-eminent in the history of the Old World, yet, strangely enough, in the New they can boast little in the way of achievement. In South America the Caspian type appears in any importance only among the wretched and fast-dying tribes of the [Alikaluf in the] extreme south, while in the northern continent it apparently forms the dominant element only in the Eskimo. It is hard to conceive of a greater contrast to the forceful, conquering, intellectual peoples of this type in the Old World, than the timid and simple Eskimo. Is it not impossible that the one should be of common ancestry with the other? Have our criteria and analysis in this instance told the truth? It must be confessed that explanation of the failure of the Caspian peoples in America to live up to the achievements of their supposed kindred in Asia and Europe is difficult, although for the Eskimo a case can perhaps be made. For if, instead of the strongly favoring environments in which the Old World representatives of the type have lived, we substitute the poverty of that to which the Eskimos, in historic times at least, have been confined, we may have a partial answer. Yet we do not know that the Eskimos have always been held within the Arctic, and for the small groups of people of this type which reached the southern continent no claim of the repressive influence of environment can of course be made. The paradox therefore remains.

The same anomaly finds expression in the history of the Proto-Australoids and Proto-Negroids as summed up by Dixon:

> Not only have the Proto-Australoids and Proto-Negroids not been able to reach out and hold any considerable portions of the world outside the tropics, they have had to give ground within their own territory, and half of Africa, the whole of which they seem once to have held, very early passed out of their control. In the great struggle they have, almost from the begin-

ning, been losers. . . . In the New World, where their apparent presence is so surprising, with few exceptions the peoples showing the characteristics of these types were of low culture, generally marginal in distribution, and largely extinct when Europeans came to the continent. One exception, however, is a paradox, for the Iroquois and southern Algonkian tribes were among the first of those north of Mexico in ability and prowess, and neither in outward appearance nor in culture betrayed what seem to be their affiliations.

In attempting to explain the distribution of racial character in America let us begin with the fact that in America, if Dixon is right, the Caspian-Mediterranean or Nordic mixture, as exemplified among the Eskimos and the Alikaluf, stands very low in civilization. On the contrary among the Iroquois and Algonquins a mixture of Proto-Negroid and Proto-Australoid stocks stands relatively high. It may be that the ancestors of all the tribes here mentioned were subjected to a repressive evolution during their journey from Asia to America through Siberia and Alaska. That in itself may have retarded their mental development as compared with the corresponding people in Europe, but it does not explain why they differ among themselves. So far as the Eskimos are concerned, it is quite clear that they live in an environment which is still repressive. Not only does their country fail to provide the raw materials which are so convenient and so helpful in the advance of civilization, but their sufferings from hunger, from long cold winters, and from nervous exhaustion, resemble those already described among the Siberians. If we were right in a previous chapter as to the effect of pronounced and prolonged low temperature upon the people of Siberia, it seems reasonable to suppose that the more nervous and progressive types among the Eskimos have largely been weeded out by nervous diseases or by migration. Hence, no matter what race they may belong to, one would hardly expect the Eskimos to achieve much.

As to the Alikaluf, they are a primitive hunting people who use the bow and arrow, build large canoes, and have a reputation for treachery. They live in what most people would call a decidedly repressive climate. Their home is near the shores of

THE ANOMALIES OF ABORIGINAL AMERICA 91

the Pacific, among the mountainous islands of the western part of Tierra del Fuego. There they are exposed to the full fury of an interminable series of gales which make the weather of this region "probably the wildest on earth." (Kendrew.) About half of the gales come from the cool Pacific and are the southern part of the "Roaring Forties," the terrific westerly winds which the mariner dreads off Cape Horn. Although no records are available in the land of the Alikaluf, the reports of the nearest meteorological stations to the northwest and southeast indicate that the warmest month averages about 49° F., while the coldest averages 36° F. The temperature in itself is no worse than that of Iceland, whose high civilization we shall discuss in a later chapter. But the monotonous repetition of gales with scarcely any respite, and the tremendous rainfall make the environment very depressing. The annual precipitation is probably well over 100 inches, being 117 inches at Evangelists Island, where about 8 inches fall even in the driest month. Worse still is the fact that about two hundred and fifty days in the average year are wholly cloudy, and only fifty or less are clear. As to the vegetation I have been able to find no exact data, but Professor Roland Thaxter of Harvard gives the following information: "As that end of the Strait is a focus for all the winds of Heaven, I imagine that what vegetation there may be is similar to that of the other islands of the region which I saw, stunted and blown horizontal. . . . I am glad I am not an Alikaluf and have not got to live there. I never saw it blow so anywhere." Agriculture is impossible, game is scarce, and the natives have never had any domestic animals like the sheep. Add to this the extreme isolation of the region and one no longer wonders that the Alikaluf are backward. They are like the Yaghans of whom we read that they "live under conditions of extraordinary rigour. In order to obtain food, they venture naked in small canoes into the treacherous seas; their life is a constant battle with starvation and a rude climate, and their character has become rude and low in consequence."

The energy and ability of the Iroquois and southern Algonquins of the eastern United States seem to be in harmony with their environment quite as much as does the backwardness of the Eskimos and Alikaluf. It is true that the Proto-Negroid and Proto-Australoid blood of these finest tribes of the primeval United States may have given them a smaller innate endowment of mental capacity than that of many other North American tribes, although as to this we have no data. It is likewise true that the experiences of their ancestors in the north may have helped to develop the stoical, unsympathetic side of their natures and may be one reason for their lack of the nervous energy and initiative which make men inventors and reformers. Yet one would hardly expect them to be dull or inactive like the people of similar racial composition in central Africa, southern Asia, and the tropical islands. The same whirl of chance which made the Caspian-Mediterranean blend backward in America, though progressive in Europe, gave the Negroid-Australoid blend a wonderful opportunity in the New World. It drove them into a region having one of the most stimulating climates. They responded in the way that the geographer would expect. Perhaps they never had the innate capacity of the Nordics, but the fact that they stand out so strongly among the peoples of America, while the seeming "Nordics" of aboriginal America stand so low, suggests that environment may be as potent as race in determining people's capacities and achievements.

Let us now turn to the question of why people with so high a degree of activity as the Iroquois and Algonquins, and with no mean endowment of mentality, made so much less progress than the Pueblo Indians of Arizona, for example, and were almost immeasurably more backward than the primitive Mayas, or than the early Peruvians who originally developed the arts which were later practised among the Incas. A part of the answer seems to be that the Mayas, Aztecs, and Incas, as well as the pueblo people of New Mexico and Arizona, practised agriculture. There has been much discussion as to the condition of agriculture among the primitive Americans, especially in the northeast. Corn, beans, and squashes were an important ele-

THE ANOMALIES OF ABORIGINAL AMERICA 93

ment in the diet of the Indians of New England and New York, while farther south potatoes, sunflower seeds, and melons were also eaten. The New England tribes knew enough about agriculture to use fish and shells for fertilizer. They and their neighbors had wooden mattocks and hoes made from the shoulderblades of deer, from tortoise-shells, or from conch-shells set in handles. They also had stone hoes and spades, while the women used short pickers or parers about a foot long and five inches wide. Seated on the ground they employed these to break the upper part of the soil and to grub out weeds, grass, and old cornstalks. The regular method of preparing a field was to burn over an old patch each year and then replant it. Sometimes the seeds were merely dropped in holes, and sometimes the ground was dug up and loosened, each hill being treated separately. Clearings were made by girdling the trees with stone implements, that is, by cutting off the bark in a circle at the bottom and leaving the tree to die. The brush meanwhile was hacked or broken down; and both trees and brush were burned when dry enough.

All this gives the impression that agriculture was fairly well developed, but such a conclusion is scarcely warranted. Before the days when the white man had introduced iron, the primitive Indians of the northeastern United States derived only a small part of their living from agriculture. To cut a large tree with an axe of stone is extraordinarily difficult. To girdle the trees and cut down the brush of several acres, as each man would need to do in order to support a family, is by no means out of the question. But it takes a long time, for the American Indian not only despises steady manual work, but is unfitted for it by the very qualities which make him a good hunter. Moreover, in order to keep his family alive, the Indian who is trying to make a field must frequently interrupt his farm work to go hunting. Then when the trees are girdled and the brush cut, another long period of waiting is necessary until the vegetation is dry enough to burn, and the crop can be planted, grow, and ripen. Hence the regular method was for a man to find an opening in the forest where the trees happened to have died and where he could easily break down the brush. There he made his little field, planted it, and left it to itself. Almost never could he stay near by and let his wife tend the crop steadily, for only rarely was the game in any one place sufficient to support people several months.

In spite of all these difficulties the northeastern Indian might have made progress in agriculture, if only he had been able to retain the same fields permanently. In five or ten years, however, he was generally obliged to give up his fields. Three causes probably co-operated in this: first, the partial exhaustion of the soil because of repeated plantings of the same crop without plowing or spading; second, the accumulation of toxic bacterial material because the soil was not properly aerated by plowing; and third, the invasion of grasses and other perennial weeds. In the Southern States, the relative scarcity of humus causes the soil to become exhausted more rapidly than farther north, while bacterial infections seem to be more common. Yet the Indians cultivated the land about their villages for long periods. Tribes like the Creeks, Cherokees, and Natchez appear to have occupied the same sites for generations. But the grass in the South is more bunchy and less turfy than in the North, and can be more easily uprooted with primitive tools. When a clearing is first made in the northern forest, it is usually almost free from grass, but little by little the grass encroaches, especially when the fields are left largely to themselves. Any one who has cultivated a garden knows how rapidly the weeds grow. He also knows that almost no weed is so hard to exterminate as grass. When once it gets a foothold hoeing seems only to make it grow faster. Some authorities believe that the growth of tropical grasses in the fields of the Mayas was the chief cause of the destruction of their promising civilization. When grass has become well established or when the soil has become infected with harmful bacteria the easiest way to insure a crop is to plow the fields and start over again. This the Indian could not do, because he had no iron tools and no beasts of burden. Hence, even if there were no other adverse conditions, the Indian would presumably have obtained smaller and smaller crops from his fields until they became meadows. Then he perforce either gave up agriculture or cultivated new land. Hence we infer that in regions like New York and New England, and still more in the prairies, the grass joined with the bacterial infection of the soil as a main deterrent to agriculture.

But why did the Indians not clear other pieces of forest close to their ruined fields, and thus continue to live in the old place? One probable reason is that the labor of making a clearing with stone axes and by the slow process of girdling and burning the trees is so great that the Indians usually chose favored spots where by accident the growth was less dense than usual. When once a clearing became grassy or the soil became toxic, the regular procedure was to hunt for a new site, and move the village. This was apparently one reason why the Iroquois, although successful in other ways, failed to establish permanent towns. Hence their advancement not only in architecture but in many of the most important elements of civilization was greatly delayed.

We have seen that two of the great disadvantages of the American Indians were the lack of iron tools and beasts of burden. Were these lacks the fault of the Indians? Would more competent people have discovered how to smelt iron and tame wild animals? A positive answer is impossible, but it seems to me that even if the Indians had been far more intelligent than they were, they might still have had no iron tools and no animals wherewith to plow the sod. That the Indians were familiar with copper is well known, but they could not fashion it into hard tools with a cutting edge, nor did they learn from copper the art of smelting iron. The discovery of this art requires at least four important conditions. First, there must be bits of ore which accidentally, or with intent, are heated in a very hot fire until they melt or at least soften. When this happens there must be at hand a man of unusual genius whose attention happens to be directed to the results of the melting process. Thirdly, the man who saw the capacities of iron ore and of its molten product must have the zeal, the determination, and the leisure to bring to fruition the ideas engendered by his observation. Finally, a single mind was presumably not enough to consummate the great discovery. It was necessary that the generation of men who lived with the genius should have a sufficiently active mentality to assimilate and use the new process and pass it on to future generations. All these conditions might have occurred in the New World as well as the Old, but as a matter of fact they did not. It may be that the relatively duller mentality of the Indian had something to do with the matter, but it may have been mere accident.

Up to a certain point civilization can evolve without iron tools. In America the Maya civilization of Guatemala and the pre-Inca civilization of Peru made marvelous strides which suggest that the mental handicap due to migration through the cold north, if such there was, had been largely overcome. They reached a point almost as high as that of the Mesopotamians, Egyptians, and early peoples of northern India and China before iron was introduced. If iron had been available, and still more if there had been a good beast of burden such as the horse or ox, civilization might possibly have moved northward and southward from its early homes in Central America and Peru. As the climate of the glacial period gradually changed to that of to-day and as man's ability to conquer cold regions increased, there might have been a poleward movement of the centers of civilization in the New World as well as in the Old.

As to beasts of burden the case is clearer than as to iron. The earlier Indian tribes may perhaps have been less competent in taming animals than were the people of Asia. It is far more likely that they failed to domesticate any animals except the dog, turkey, alpaca, and llama, because there were none of the right kind in America. Horses existed in the New World down to the glacial period, but were then exterminated. Our present horses and asses are descendants of stock introduced from Europe. The camel likewise was extinct. In the ox tribe the only available animal was the bison, but he appears to have been too

THE ANOMALIES OF ABORIGINAL AMERICA 97

large, and possibly too stupid, to be domesticated. Nor was there any type of sheep that could compare in value with the one domesticated in Asia. Wild pigs or peccaries might indeed have been domesticated, but they would have been of no value for plowing. In South America the most available animals, the llama and alpaca, were actually domesticated, but the alpaca is of little use even for burdens, and the llama is useless for agriculture. Thus among the wild animals of America after the last glacial epoch, there was none which man could domesticate and use for the difficult work of plowing the sod.

Consider again how handicapped the Indians were by this twofold lack of iron tools and beasts of burden. In the southeastern United States, where the grass grows in bunches and can be pulled or dug up with comparative ease he could indeed carry on permanent cultivation. Still more was this the case in semiarid and tropical areas where the grass grows in isolated clumps. But in the parts of America which are climatically best for human progress there was no such possibility. In the central prairies cultivation was feasible only in the bottom-lands which are flooded by the rivers part of the year, and which support bushy growths rather than turf. Elsewhere the rich grasslands waited untouched until the white man brought beasts that could draw an iron-shod plow. In the eastern forests temporary cultivation was possible in chance clearings where the trees had recently died, or been burned in forest fires, for there the soil is at first free from turf. But as soon as the grass had an opportunity to grow, the fate of the fields was sealed until plow animals from Asia came to the rescue. Thus where man ought to have progressed most rapidly in America he was compelled to be a savage hunter roaming the woods, able to tend his fields only sporadically and forced to move his home at frequent intervals as his gardens became meadows.

The contrast between the effect of agriculture and of hunting upon character can scarcely be overestimated. Even the care of domestic animals requires a certain degree of steadiness and concentration which is not demanded of the hunter. The family that has many domestic animals must look after them at least to a slight extent each day. This is often done for the sake of the milk if for nothing else. Nor can such families wander so freely as hunters; hence they acquire a greater feeling of possession of the land. The careless type tends to be weeded out more than among the hunters.

When agriculture begins to be practised, a far more important selective principle appears. The qualities that make a man successful in hunting and in farming are quite different. For the hunter, craft, cunning, the power to glide silently through the forest, the ability to detect the recent passage of an animal by the slight breaking of the twigs, the power to find his way through the forest in storm and snow, the capacity to endure long periods of fasting alternating with feasts when he gorges himself with some large slain animal, the capacity to endure prolonged hunger, thirst, and great fatigue while pursuing a wounded animal, and the power to recover from tremendous exertion by long periods of rest—these are the qualities that make a hunter successful and which enable him to bring up his children in health. But for the farmer none of these things is necessary. What he needs is the ability to work steadily day after day, to come back to the same field time after time and clean out the weeds. He must have the capacity to plan storehouses for the winter, to foresee what difficulties the cold or the dry season will bring forth, how much food his family will require, and how a given supply can best be divided among several mouths during the days when no new food can be procured. The more completely he becomes a farmer the less he needs the qualities which enable him to find his way in the dark or in the forest. On the other hand, the greater becomes his need of the ability to plan for the future, to note the kind of food which grows best in a given place, the amount which is produced in one soil or on one slope compared with another. Ten thousand years of life on the farm could scarcely leave a people with the same character as ten thousand years of the life of a hunter. And it is the farmer's life which gives the

THE ANOMALIES OF ABORIGINAL AMERICA

opportunities for progress in civilization. It holds him to the soil; it gives him an interest in developing one special piece of land; it enables him to lay up capital; and it puts a premium on foresight for the future. This perhaps explains why in early America the agricultural people in relatively unstimulating climates made far greater progress than did the hunters who lived in the most stimulating environment and had the greatest energy.

An understanding of the relative part played by race, climate, inventions, natural resources, and culture or folk-ways in determining human character is so important that I shall add another illustration taken in part from *The Red Man's Continent*. On the west coast of America the state of civilization in pre-Columbian days was, on the whole, lower than on the east coast. On the west the most advanced people were the Haidas, living in the Queen Charlotte Islands of British Columbia. Most authorities appear to agree with the statement of the Encyclopædia Britannica that they "constituted with little doubt the finest race and that most advanced in the arts of the entire west coast of North America." According to the *Handbook of the American Indians*, they displayed, for example, great skill in canoe-building, wood-carving, and the working of stone and copper as well as in making baskets. Canoes of uncommon size were hollowed out of logs of cedar with the aid of fire. Houses which were sometimes forty by one hundred feet in size were built of huge cedar beams and planks. The planks were first shaped with stone and then put together at great feasts not unlike the "raising bees" at which the neighbors gathered to erect the frames of houses in early New England. The totem poles in front of the houses represent a fairly high development of primitive art, while the permanence and size of the villages, and the complexity and thorough development of the social usages indicate a high type of barbarism rather than the savagery which prevailed among the Iroquois, for example. Another notable feature of the Haida life was commercial activity. The Haidas were the great traders of the west coast, and sent their wares hundreds of miles by sea among the bays and islands of their indented coast.

Eastward from the Haidas the character of the Indians declines very rapidly. A few hundred miles away, among the Rocky Mountains and on their east side, there dwelt some of the lowest tribes of the whole continent. To-day, as always, these tribes seem to be so occupied with the severe struggle with the elements that they cannot advance out of savagery into barbarism. When the white man came they were homeless nomads, whose movements were largely determined by the food-supply. Southward from the Haidas along the Pacific coast a great deterioration of the Indians is likewise noticeable, especially in California. Only to the north among the Tlingits and related tribes do the standards of character and achievement show any similarity to those of the Haidas, but even these tribes fall much below the level of that tribe.

Several factors probably combine to explain the relatively civilized and progressive character of the Haidas. In the first place the Queen Charlotte Islands are located in a climate much like that of northwestern Scotland, a climate which is cool, moist, and invigorating although not especially good for agriculture. The average temperature and rainfall suggest, indeed, those of the depressing home of the Alikaluf in about the same latitude in Tierra del Fuego. But there is a great difference, as appears in the fact that the Queen Charlotte Islands are clothed with a splendid forest, whereas in the windy part of the Tierra del Fuegan region the trees are mere stunted shrubs. The difference in climate depends on three facts: first, the Queen Charlotte summers average perhaps 8° F. warmer than those of the home of the Alikaluf; second, the winds are far less violent, being in fact quite gentle most of the year; and third, although the northern islands have about as much rainfall and cloudiness as the southern during most of the year, they enjoy a relatively dry summer, with just rain enough to be almost ideal. Moreover, in winter there is relatively little snow and no severe cold such as still benumbs the Indians of the interior, and often led to starvation in the past. Although the Haidas dwelt where it was perhaps never quite warm enough for the best physical development, they lived in a climate where mental activity is stimulated and where life is not unpleasant.

Fully as important as the climate is the fact that the Haidas had a large and steady supply of food close at hand. Most of their sustenance was obtained from the sea and from the rivers, in which the summer runs of salmon furnished abundant provisions, which rarely failed. In Hecate Strait, between the Queen Charlotte Islands and the mainland, there were wonderfully productive halibut fisheries. These excellent fish were dried and packed away for the winter, so that there was almost always a store of provisions on hand. The forests in their turn furnished berries and seeds, as well as bears, mountain-goats, and other game. The Haidas likewise had the advantage of not being forced to move from place to place in order to obtain food. They lived on a drowned shore where bays, straits, and sounds are extraordinarily numerous. The great waves of the Pacific are shut out by the islands so that the waterways are almost always safe for canoes. Instead of moving their dwellings in order to follow the food-supply, as the hunting tribes were forced to do, the Haidas and their neighbors were able without difficulty to bring their food home. At practically all seasons the canoes made it easy to transport large supplies of fish from places a hundred miles away if necessary. Another advantage was that the Haidas lived where fine forests not only furnish splendid building material and abundant fuel, but grow close to waterways which make it easy to transport the wood to the villages. Because of their permanent villages, good shelter, and abundant and assured food-supply, the Haidas could accumulate property and acquire that feeling of ownership and stability which is one of the most important conditions for the development of civilization. Perhaps biological mutations, racial mixture, and natural selection had made the Haidas intellectually superior to many other tribes, but even without this their surroundings would probably have made them stand relatively high in the scale of civilization.

Another possible reason for the superiority of the Haidas is that their ancestors may have come to America by sea rather than by land. Among the islands and bays of the Asiatic coast from Chosen and Japan northeastward there have presumably long been fishing tribes with some knowledge of boats. Such people would naturally spread northward along the coast, and might reach the Aleutian Islands. The distance from Kamchatka to Bering Island is only one hundred miles, although the island is not in sight from the mainland. The next step to Copper Island is easy, but then comes a stretch of over two hundred miles. If people were constantly using boats, some of the more adventurous might be blown within sight of the islands on the American side. Later a band of savages, including women, may have deliberately sailed across the open water, but only the bravest and most adventurous would be likely to do so. This may have happened during an interglacial epoch when the land stood higher than now with respect to the sea, and the islands were more numerous and nearer together. Even if it occurred in post-glacial times when the climate had not yet moderated to its present condition, it would not be very difficult to cross from Asia to America in boats along the line of the Aleutian Islands.

That any such migration from Asia to America ever occurred by way of the seacoast and the islands is, of course, pure speculation. Dixon, indeed, thinks that the Haidas came to their present home from the east instead of the northwest. But if such a maritime migration did occur, it would enable people to come to the New World without passing through the depressing experience of the Siberian "pole of cold." Under present conditions of climate no part of their route would have a January temperature averaging much below 15° F., or about like that of Montreal in Canada and New Chwang in Manchuria. If the ancestors of the Haidas, Tlingits, and their relatives by any chance came to America this way, their freedom from the nervous strain of the long land migration may have something to do with the relatively active mentality of their descendants.

THE ANOMALIES OF ABORIGINAL AMERICA 103

So, too, may the natural selection which presumably kept the less adventurous people on the Asiatic coast while those with more bravery and initiative came to America.

Southward from the Haidas, around Puget Sound and in Washington and Oregon, we find a gradual decline in the civilization of the aboriginal Indians. The Chinook tribe of the lower Columbia had large communal houses occupied by three or four families of twenty or more individuals. Their villages were thus fairly permanent, but there was much moving about in summer since the food-supply consisted chiefly of salmon together with roots and berries. The Chinooks were extremely skilful in handling their large, well-made canoes, which were hollowed out of single logs. They were also noted as traders. In disposition they are described as treacherous and deceitful, especially when their cupidity was aroused. Slaves were common and were usually obtained by barter from surrounding tribes, though occasionally by successful raids. These Indians of Oregon by no means rivaled the Haidas, perhaps because their food-supply was less certain and they did not have the advantage of easy water communication, which did so much for the Haidas.

As to the tribes farther south, we are told by an eminent authority (Kroeber, in the *Handbook of American Indians North of Mexico*) that:

> In general rudeness of culture the California Indians are scarcely above the Eskimo, and whereas the lack of development of the Eskimo on many sides of their nature is reasonably attributable in part to their difficult and limiting environment, the Indians of California inhabit a country naturally as favorable, it would seem, as it might be. If the degree of civilization attained by a people depends in any large measure on their habitat, as does not seem likely, it might be concluded from the case of the California Indians that natural advantages were an impediment rather than an incentive to progress.

We are likewise told that some groups such as the Hupa had no tribal organization and no formalities of government. Formal councils were unknown, although the chief might and often did ask advice of his men in a collected body. In general the social structure of the California Indians was so simple and loose that it is hardly correct to speak of them as divided into tribes. Whatever solidarity or cohesion they possessed was due in part to family ties, and in part to the fact that a given group lived in the same village and spoke the same dialect. Between different groups the common bond was similarity of language, as well as frequency and cordiality of intercourse. In so primitive a condition of society there was neither necessity nor opportunity for differences of rank. The influence of chiefs was small and no distinct classes of slaves were known.

Extreme poverty was apparently an important cause of the low social and political organization of these Indians. The Maidus in the Sacramento valley were so poor that, in addition to consuming every possible vegetable product, they devoured all birds except the buzzard, and ate badgers, skunks, wildcats, and mountain lions. They even consumed the bones of salmon and the vertebræ of deer. They gathered grasshoppers and locusts by digging large shallow pits in a meadow or flat; then, setting fire to the grass on all sides, they drove the insects into the pit. Their wings being burned off by the flames, the grasshoppers were helpless and were collected by the bushel. Again, of the Moquelumne, one of the largest tribes in central California, it is said that their houses were simply frameworks of poles and brush which in winter were covered with earth. In summer they erected cone-shaped lodges of poles among the mountains. In favorable years they gathered large quantities of acorns, which formed their principal food, and stored them for winter use in granaries raised above the ground. Often, however, the crop was poor, and the Indians were left on the verge of starvation.

Finally in the far south, in the peninsula of Lower California, the tribes were "probably the lowest in culture of any Indians in North America, for their inhospitable environment which made them wanderers, was unfavorable to the foundation of . . . even . . . the rude and unstable kind [of government] found elsewhere." The Yuman tribes of the mountains east of San-

THE ANOMALIES OF ABORIGINAL AMERICA 105

tiago are described as "very dirty on account of the much mescal they eat." Others speak of them as "very filthy in their habits. To overcome vermin they coat their heads with mud with which they also paint their bodies. On a hot day it is by no means unusual to see them wallowing in the mud like pigs." They were "exceedingly poor, having no animals except foxes of which they had a few skins. The dress of the women in summer was a shirt and a bark skirt. The men appear to have been practically unclothed during this season. The practice of selling children seems to have been common." In other words, although the mode of life among these poor Indians was quite different from that of the Chinese, whom we shall study in a later chapter, the great elemental fact was the same. Dire poverty and recurrent famine must often have caused some of the Indians to wander away in search of a better land. Those who went away were presumably the ones with more initiative, originality, and energy. Here as in the case of the Haidas, we have no definite knowledge, but we seem to have a clear example of a common and widespread type of natural selection. The principle involved is this: a bad environment, especially one with recurrent periods of acute distress, seems to drive away the more capable inhabitants; a favorable environment, especially one where the means of subsistence are not only abundant but vary little from year to year, attracts people. Thus such regions become the object of strife among the more capable people from the less favored regions, and are likely to be inhabited by a race in whom natural selection has concentrated a higher ability than that of the people in the poorer districts.

In the preceding paragraphs two sharply contrasted views have been suggested. According to one view the natural environment of California is highly favorable to Indians as well as to white men. But inasmuch as the Indians have been in an extremely low stage of progress and of mentality ever since anything has been known about them, it is inferred that the environment has nothing to do with the status of the people. According to this view the "natural advantages" of central California seem almost to be "an impediment rather than an incentive to progress." The degree of progress, therefore, must depend either on the absence of certain social usages, inventions, and other cultural aids to progress, or on inherent defects of character.

According to the other view the environment of central and southern California and likewise of Lower California is highly unfavorable. Its direct repressive influences together with its selective effect in causing the more energetic types to migrate elsewhere perhaps explain why the people, especially in Lower California, were "the lowest in culture of North America."

These contrasted views sum up the sharply defined antithesis between what we may call the anthropological or ethnological hypothesis that the fundamental differences between different races are largely due to cultural conditions and historical accidents, and the geographical hypothesis that they are due primarily to physical environment and to the action of that environment in causing migrations and natural selection whereby races become inherently different. Of course there is much truth in both hypotheses. That cultural advantages are of enormous importance and that they lead to the selection of certain types of character for preservation I have already tried to show in discussing the importance of iron tools, beasts of burden, and agriculture. Each of these represents a great cultural advance, which carried with it an almost unlimited train of consequences. And these consequences include not only human habits, but also human inheritance. But in the case of California and its Indians it seems to me that we have an instance where physical environment comes first and cultural conditions second. The climate of California may be good for health, but the absence of rain in summer and its great variability in winter produce conditions of vegetation and animal life which kept the Indians in the direst poverty. Such conditions subjected the Indians to what we have called repressive evolution, and presumably led the more competent Indians gradually to migrate into regions more highly favored.

One reason for this view is found in the advanced Mohave tribe which had its home on the lower Colorado River.

No better example of the power of environment to better man's condition can be found than that shown as the lower Colorado is reached. Here are tribes of the same family [as those of Lower California] remarkable not only for their fine physical development, but living in settled villages with well-defined tribal lines, practising a rude but effective agriculture, and well advanced in many primitive Indian arts. The usual Indian staples were raised except tobacco, these tribes preferring a wild tobacco of their region to the cultivated. (Hodge, *Handbook of American Indians*.)

In order to appreciate the force of this quotation, it should be remembered that, although California is famous for its agriculture, its crops are European in origin. Even in the case of fruits, such as the grape, which have American counterparts, the varieties actually cultivated were brought from Europe. Wheat, barley, and the chief vegetables for which California and similar subtropical regions are noted, were unknown in the New World before the coming of the white man. So far as we are aware no crops of any kind were ever cultivated by the primitive Indians in California west of the Colorado valley. In pre-Columbian America corn was the only cultivated cereal. The other great staples were beans and pumpkins. All three are pre-eminently summer crops and need much water in July and August. In California there is no rain at that season. Wheat and barley, on the other hand, need rain only from October to April, for in regions as warm as California they can grow slowly during the winter. The comparatively dry weather of May and June is just what is needed to ripen these grains of the Old World, and the complete drought of July and August does them no harm, whereas it is fatal to crops which depend on rain in summer.

Crops can of course be grown during the summer in California provided they are irrigated. But irrigation is far harder for primitive people in a region of summer droughts than in one where the winter is dry and there are summer showers. If irrigation is to be effective in California, it cannot depend on the small streams which dry up almost as soon as the rains cease, but must depend on comparatively large and permanent streams. Most of these flow in well-defined channels; and during the summer at least, they do not afford much natural irrigation like that of Egypt and Mesopotamia. With our modern knowledge and machinery it is easy to make canals and ditches, and to prepare the level fields needed to utilize this water. A people with no knowledge of agriculture, however, and with no iron tools, cannot suddenly begin to practise a complex and highly developed system of irrigation. The lower Colorado fortunately floods broad areas every summer. While elsewhere in Arizona and New Mexico, that is in the old Pueblo area, summer rains cause similar floods along the rivers or on flat "fans" of sand and silt. Along the rivers, as on the Nile, the retiring floods leave the land so moist that crops can easily be raised. This natural irrigation helped the Mohave Indians to practise agriculture and to rise well above their kinsmen not only in Lower California, but throughout the whole region. Perhaps we may look upon them as a selected group of people who broke off from their kinsmen in other parts of California, and by reason of unusual mental ability and certain favorable geographical conditions were able to utilize a new cultural method, namely agriculture. Environment, inheritance, culture—how closely the three are intertwined.

In this review of civilization in America we have considered only a few of the primitive tribes, and have not discussed those most advanced in civilization. We have, however, seen enough of the continent to discover that the character of the aboriginal Americans is not explained by any one factor. That the original racial inheritance with which the different migrating bands started from Asia still persists and influences their descendants, seems probable. That certain qualities were intensified, eradicated, or otherwise changed while the migrants were on their way from Asia to the different parts of America, and while they passed through such environments as the Siberian pole of cold, the American deserts, the great plateaus, or the tropical forests, seems equally probable. But on the original inherited character-

THE ANOMALIES OF ABORIGINAL AMERICA 109

istics which have thus arisen, as described in a previous chapter, there have presumably been grafted still other characteristics arising from the mode of life. People like the Iroquois, whose ancestors have depended on a precarious hunting existence for tens of thousands of years, cannot be expected to have a character like that of the Haidas, who have long depended on a fairly constant and easily available supply of food which they can procure without migrating. Nor can either of these be like the Pueblo Indians, who for many generations have carried on agriculture by means of irrigation.

Consider once more what a contrast there must be between people who practise agriculture and those who cannot practise it either because the climate, the vegetation, the absence of iron tools and beasts of burden, or their own low stage of culture makes it impossible. When a race settles down to agriculture a great selective process takes place. It can be seen to-day in Asiatic Russia where some of the Khirghiz and Turkomans have taken to farming since the occupation of the country by Russia. Practically all nomads hate farming. But some hate it worse than others. The ones who hate it most are likely to be the most active, adventurous spirits, whereas those who take to it most readily are the ones who are predisposed to a sedentary life, and who are least averse to steady work, a thing which nomads and hunters usually find extremely irksome and often impossible. Thus if the pressure of population and the presence of a favoring environment cause part of a race to adopt a sedentary life, and part to remain as hunters or cattle-raisers, there is likely to be a clear-cut line of fission. As time goes on the line will presumably become still clearer, for among the children of those who have taken to farming, there is a strong tendency for some to go back to the nomadic life. Those who thus return are almost sure to consist of two groups: first, the more adventurous, and second, those who are least competent to carry on steady work. Moreover, many of those who take to agriculture are almost sure to cultivate their fields most carelessly. The weediness and unkempt character of the farms of the Khirghiz in Siberia are almost incredible. This condition also may indicate that a selective factor is at work. Those who are most careless will not only tend to go back to the old life, but their children will in many cases be poorly nourished, so that an unusual number will die. At the end of a hundred generations there must apparently be a strong hereditary difference between the people who have taken to farming and those whose main means of livelihood has been hunting.

Thus we would sum up the whole problem by saying that in America, as in the Old World, the beginnings of civilization took place in regions where the winters were warm and the summers hot, and where by natural irrigation or by means of summer rains it was easy to raise crops with a minimum amount of work. These beginnings took place so long ago that the climate was cooler and more stormy than now. Because of the difference in the position of the various planetary belts of winds, and especially in the southern margin of the belt of westerlies and of cyclonic storms the area of early civilization lay about ten degrees farther south in North America than in Asia and Africa. Then, as time went on, early man in America developed his civilization as far as was possible without iron tools and beasts of burden. Perhaps he developed more rapidly in the Old World than in the New because of a better racial inheritance. We have already seen that Europe and its glaciations, and Asia and its deserts presumably exercised a remarkable selective effect upon the people who in prehistoric times began the development of civilization in western Asia and northern Africa. The racial mixture which happened to occupy the best regions may also have been more favorable in the Old World than in the New. Perhaps for that reason, and perhaps for some other, the discovery of how to smelt iron was made in the Eastern hemisphere and not in the Western.

If iron tools and beasts of burden had not become a part of man's cultural inheritance, civilization would probably have stagnated in both hemispheres. It would have stagnated because there seems to be a certain level of achievement beyond which it is impossible to rise without iron tools, or at least beyond which

THE ANOMALIES OF ABORIGINAL AMERICA 111

no race has ever risen. It would also have stagnated or perhaps retrograded because as the glacial period passed away the climate of the places where civilization first developed became less favorable, while the favorable places became those where primeval forests are dense and the grass takes the form of sod. The steps that were needed were the discovery of how to smelt iron and make it into tools, and of how to tame and use animals for draft purposes. The first discovery was made in the Old World and not in the New. Perhaps this was mere accident, and perhaps it was the result of greater intelligence in one place than the other. Fortunate racial mixtures, the helpful selective effect of glaciation in Europe, and the harmful selective effect of the migration of the aboriginal Americans through Siberia and Alaska may have had something to do with it. As to beasts of burden the Americans were not to blame. They had no available beast, whereas the ox and horse were both at hand in Asia, as were also the donkey, water-buffalo, sheep, goat, and camel. These propitious circumstances helped the old civilization of Eurasia to spread northwestward, while that of America stagnated and died. The aborigines of the northeastern United States were highly energetic, but the difficulties of agriculture doomed them to remain savage hunters almost indefinitely. Yet who can say what might have happened had iron tools and beasts of burden found their way to America before the white man appeared on the scene?

CHAPTER VIII

THE ASIATICS WHO DWELL IN TENTS

In our attempt to understand why human character differs from race to race and place to place, we have thus far dealt largely with primitive people of the past. We have thereby obtained some idea of the part which climatic and topographic changes, migration, racial mixtures, human inventions, and natural selection have played in modifying racial character. But character is influenced not only by mutations, racial mixture, and natural selection whereby new types of inheritance are established, but by the direct effect of climate, food, occupations, and diseases upon people's energy, health, and activity. Hence we shall now consider the daily life and habits of certain chief types among the people of Asia, and shall attempt to analyze the effect of that life and of the other factors such as education, government, religion, and the like upon natural selection. In this we must limit ourselves to a few examples, for in this book we can merely illustrate certain great principles.

The herding of animals is one of the most important of human occupations. It is important as a stage in the development of civilization, and likewise as a mode of life which is still necessary if man would occupy certain enormous sections of the earth. Civilization, as we have seen, may indeed rise to a fairly high level without the aid of domestic animals, as among the Mayas of Central America. But its higher development and its spread into many of the most favorable regions, such as the prairies and temperate forests, seem to have been impossible until people somehow obtained domestic animals able to bear burdens and draw the plow as well as to furnish clothing, meat, and milk. It was the good fortune of Asia, Europe, and north Africa to have a long and important stage of pastoralism, whereas North

THE ASIATICS WHO DWELL IN TENTS 113

America and Australia never had such a stage, South America merely had a faint hint of it in its llama-raising people among the high Andes, and Africa south of the equator probably had few or no domestic animals until a relatively late period.

The regions where the pastoral mode of life still persists must not be thought of as in a stage of transition. Although pastoralism was probably a step in the evolution of the civilization which now dominates the earth, it is by no means an indispensable precursor of high civilization. The ancient Egyptians may have begun to practise agriculture without first being herders. Indeed there is much to be said for the idea that farmers could domesticate animals better than could wild hunters who had no permanent abiding-place. Pastoral nomadism, as it exists to-day, is generally a mode of life which persists because the environment and the stage of development of the people make any other mode difficult or impossible. Most nomads live in deserts or among mountains where there is an abundant growth of grass at certain seasons, but where agriculture is for the most part precarious. Because of drought, low temperature, or excessive moisture, the use of the grass by means of animals is the only sure means of making the land yield a living. Where the grass is scanty, or lasts only a few months, as in deserts, or where it is accessible only during a short season free from snow, as in high mountains, the people who wish to get a living from it must be nomads, travelling from place to place with their animals.

The great majority of pastoral nomads have many traits in common. This is likely to be true even when such people represent reversions from a high stage of civilization. For example, during the summer a good many sheep-herders frequent the high plateaus of the western United States. The only one of this kind with whom I ever stayed reminded me strongly of the nomads of central Asia. He was overjoyed to see myself and my companion because he had seen no one for several weeks, and we brought news of the outside world. His equipment consisted of one knife, one fork, one spoon, one cup, a stewpan, and a

Dutch oven. He set before us a meal of roast lamb, bread, stewed apricots, and coffee. The meat we ate with our fingers; we took turns in dipping the apricots from the stewpan with the spoon; and we passed the coffee-cup from hand to hand. He was an excellent host, and took great pains to see that our horses were properly hobbled in a good patch of grass. He likewise was most solicitous that we have comfortable places in which to spread our blankets on the ground. I can readily imagine that if he and his descendants continued to practise the nomadic life of American sheep-herders for a few generations they might become much like some of the people of Asia.

Let us consider a group of people who have been nomads for an indefinite period. One of the most noteworthy features of Arabia is the great uniformity of the habits and character of all its nomadic people. Because the country is so dry, nomadic pastoralism is the only mode of life possible in the major part of the country. The settled inhabitants, to be sure, greatly outnumber the nomads but they occupy a relatively small space. An account of some of my own experiences is perhaps as good a way as any of giving an idea of certain important characteristics which stand out strongly in the character of the nomads.

One day in April, 1909, I left my caravan and rode off among the gentle hills of the plateau of Moab to visit some ruins. The caravan leader failed to halt at the appointed rendezvous, and my guide and I lost the trail. A few miles to the west the plateau ended in the steep escarpment which drops 4,000 feet and more to the bitter blue brine of the Dead Sea; eastward the sunset light gave us a fine view of the gentle slope which descends by insensible degrees to the great Syrian desert, a waste of gravel and sand. The region, ten or fifteen miles wide between the escarpment and the desert, forms a narrow strip of grassland, whose ownership is fiercely debated between settled farmers and wandering nomads. In 1909 the nomads had the upper hand, for the rains of early spring had failed, the grass had withered, the springs had ceased to flow, and the Beduin had been forced to take refuge in any and every place where they could find a scanty bit of herbage for their flocks and camels, and a little water to assuage their thirst. An hour or two after sunset the barking of dogs guided us to a Beduin encampment, a group of about twenty long, low tents of black haircloth set in an ellipse a hundred yards long, and lighted by small fires at which the evening meal had been cooked. With the practised eye of the traveller my guide, an Arabic-speaking soldier in the Turkish service, picked out the tent of the chief or sheikh. One end of the tent was devoid of any front wall, and was readily recognizable as the men's quarters or guest-room. It was crowded with Arabs muffled in woolen cloaks and clustered around a fire of dry weeds whose intermittent flames lighted up a slab of limestone bearing the rudely scratched insignia of the tribe of Beni Sakr.

We dismounted silently, as men do in a land where no man knows whether the stranger is friend or mortal enemy. The Arabs, grouped cross-legged or asquat around the blaze, received us without a word, although the chief men rose and motioned to us to seat ourselves in places of honor. A quilt was brought to spread on the coarse woolen rugs where the rest were seated; since I was a foreigner and hence a guest of distinction, another was rolled up so that I might rest my left elbow upon it. Only when we were comfortable did conversation slowly begin. At first the subject was roads and villages, but soon we touched upon the Arab's perennial topic, his quarrels with the government. As the sheikh told how he and his people had fought with the Turkish soldiers not many years ago his eyes grew so fierce and his gestures so violent, that I should have feared we would be murdered had I not known that an Arab never does violence to his guests.

"Why do you wear that uniform?" he said to my guide. "You are a good man, but how I hate your clothes! Why does the government take taxes from poor Arabs who come from the desert in times of drought? Have not the Arabs the right to feed their flocks wherever there is grass? Some day soon the Turks will see what my people will do."

It almost seemed as if his eyes shot fire. He little thought that in his hatred of the government he was simply reflecting his environment. Almost all nomads are hard to rule and are often at enmity with their government. One reason is that when they have made a raid or otherwise incurred the displeasure of the officials, they can easily slip away into the desert and disappear. Moreover, they live in such small groups that it is impossible for the government to maintain officials among them. So they have their own patriarchal form of self-government, and bitterly resent any attempt to force anything else upon them.

While we conversed, green coffee-beans had been roasted on a long-handled iron spoon. Now they were vigorously and rhythmically pounded in a large wooden mortar. The grounds were then boiled in a copper coffee-pot, poured into another and boiled again, then back into the first, and so on. After about half a dozen boilings a strong black decoction was poured into a little china cup from which the host drank first, to show that it was good—unpoisoned. Then this cup and another circulated among the guests who now numbered about twenty. Next the evening meal was served, a tender lamb freshly killed and boiled for our benefit. We pulled it to pieces with our fingers, as the custom is in the desert. With it were thin sheets of unleavened bread, good not only as food but as scoops wherewith to ladle up sour milk and soft butter. This was no ordinary Arab meal such as is commonly eaten in times of drought, but a feast for guests; so we were offered still another dish, cracked wheat boiled with a little fat, and eaten with the hands. In all essential details this meal was remarkably like the one with the sheep-herder in Utah.

When the sheikh was not excited he was an admirable host. Thinking that I might be chilly that April night, with the temperature close to forty, he thoughtfully threw over my shoulders a thick Arab cape of wool. When we went to bed he took great pains to see that we were warmly covered, especially our heads. Not till he had seen the guide and myself, together with two Beduin guests, comfortably stretched on the ground around the

THE ASIATICS WHO DWELL IN TENTS

ashes of the coffee fire and apparently asleep with our heads resting on saddles, did he retire to his own part of the tent.

As the fire faded the beauty of the clear moonlit night was more and more apparent. The occasional faint bleat of a young lamb or the suppressed bark of a dog only emphasized the stillness. Then, suddenly, the sound of guns rent the air, dogs began to bark wildly, men shouted, children screamed, and the shrill cry of women arose. The camp jumped to its feet in a moment; the men flung their striped white and brown cloaks over their shoulders, slipped their feet into low shoes, if they had any, and gun in hand hastened away on foot or on the camels and horses which had been tethered near the tents. Before I could put on the clumsy boots evolved by our type of civilization, they had gone off toward the southeast, followed by a crowd of women. Then there was silence once more; almost before I understood what had occurred the excitement had subsided. In the bright moonlight the women, with queenly gait and haggish tatooed faces, streamed back down the hillside in their trailing garments of dark blue. Nothing unusual had happened; it was merely a "ghazzu," or raid. The Howeitat Arabs, enemies of the tribe of Beni Sakr, and likewise suffering from the drought, had come in from the desert to the region where water and grass were to be found. That night they raided their neighbors and drove off a herd of a hundred or more camels which had been rounded up at a little distance from the tents.

To me that raid was exciting, even though it was not my first experience of the kind. I felt that the raiders were guilty of a great crime. Not so the Arabs. Nothing about the raid impressed me so much as the attitude of the two gray-bearded Beduin who were my fellow guests that night. When awakened by the guns, they merely sat up, realized what had happened, and lay down once more to sleep. What was a raid to them? A mere part of the day's work. It was nothing to them if other people's camels were stolen. Perhaps their own tribesmen had done the deed. But anyhow, here was no matter of right or wrong, nothing demanding redress. If they should help recover the camels, they might themselves incur enmity and even become involved in a blood-feud with the raiders. Perhaps it would be their own turn to be raided to-morrow; then they would exert themselves. Since raids were in progress, they doubtless had their minds on some in which they themselves might take part. Neither their help nor ours was expected nor wanted by our hosts.

Next morning in the chill, damp air that precedes sunrise we four guests caught and saddled our hobbled horses and rode away without washing or eating, or speaking to any one. All the men of the camp were gone, and the women have naught to do with guests. Yet even in Arabia, the very heart of Mohammedanism, the nomad women are not secluded like their sisters of the oases. They cannot be, for who can wear a veil while milking sheep, managing a camel, or taking down a tent? They have their separate part of the tent, to be sure, but most of their work is done unveiled and out of doors. This helps to make them more self-reliant than the village women. Still more important is the fact that when their husbands go off on raids or in pursuit of raiders all the responsibility for children, tents, and animals falls on the women. Those who are not self-reliant are not wanted as wives by the bolder men. They fall to the men who shrink from raids, the stay-at-homes, who ultimately become villagers. Hence it is not strange that when a nomad girl is taken to a harem in a village or city, she and her children are generally dominant. Often such a woman dominates not only the other wives and their children but her husband. From such stock came Zenobia, the famous queen of Palmyra.

Let us return to the spring of 1909. Five times that season I came directly in contact with Arab raids on the border of the Syrian desert. Once in Jebel Druze I was walking on foot, unarmed, with two companions. Two men on horseback with rifles suddenly appeared from behind a great pile of stones and demanded our money. Seeing that I was the leader they concentrated on me. I argued the point and laughed, which seemed to puzzle them. For a couple of minutes one robber silently pointed his rifle at my head, while I looked up the barrel. Then they rode off. I judge that they could not understand our tactics, for they were used to dealing with villagers. Three raiders had recently held up sixteen villagers and stripped them of practically everything in that very region, so we were told. The villager is a coward compared with the son of the desert.

Another time we were returning to Homs from the ruins of Zenobia's home at Palmyra in the desert. We were driving in a carriage which easily traversed the smooth expanses of gravel even although there was no road. At one place we turned aside for some miles into the mouth of a mountain valley. After examining the ruins of a village, once well-watered, but now absolutely waterless, we were ready to proceed. It appeared that if we could get the carriage over a spur a mile or more ahead, a detour of six or eight miles could be avoided; and accordingly one of our two soldiers was sent forward on horseback to reconnoitre. Two hours and a half passed and he did not return, so finally the other soldier was sent to find him. At length the two horsemen returned after another hour or more. The first soldier looked much disturbed, and there were empty spaces in his cartridge-belt. At first he was uncommunicative, but finally he told his tale: "When I reached the ridge, I found that the carriage could cross, but there was another ridge a little farther on. I went to look at it, and found the path too rocky for wheels. When I turned back I saw ten or twelve Arabs down in the valley. I knew they were raiders. So I hid among the rocks. Then I was afraid that they had seen me and would climb up another way and catch me; so I got behind a safer rock and fired eleven shots at them. I did not hit anybody and they went off. I dared not leave my hiding-place till I saw my companion on the other ridge."

The Arabs were raiders, as the soldier thought, for later we heard of their depredations, but his action in shooting at them was idiotic. Luckily they did not know how many of him there were, and so did not attack him. If he had killed any one a blood feud would have arisen, which would probably have cost the soldier his life and might have cost ours. His conduct illustrates the bitter hatred which prevails between the wandering Beduin of the desert and the settled peasants of the same stock and language who live in the fringe of verdure.

I could tell of other raids that I have experienced among other kinds of nomads. Once, for example, I was eating lunch under the mulberry-trees at Pertag beside the upper Euphrates River. A party of mountain Kurds drove off the sheep of the village in broad daylight, and all the men went out to recover them. Times were hard, and the chief of the Kurds had died. So every man did what was right in his own eyes. I have heard the Turkomans tell of their raids before the Russians stopped them, and I have been warned of Afghan tent-dwellers, and have had them draw their guns on me. Such raids are an established part of the mode of life in many regions. They differ from the stage-coach raids of the frontier days in America or from banditry in China in that they are practised by the best of the community, men who are looked upon as the ideal type, whereas the modern highwayman and bandit are looked upon with contempt. In other words the camel-lifting raids of the desert nomads, like the warlike forays of the American Indians, are an established part of the "mores" or tribal customs of those who practise them, and hence do not go counter to the moral standards of the community as do the acts of stage-coach robbers, burglars, or bandits in settled communities.

Let us consider for a moment how such mores have grown up and how they in turn influence not only the culture and training of a people, but its inheritance. Picture to yourself the main events in the life of an Arab nomad during the course of a year. In the spring when the camels, goats, and sheep are giving abundance of milk he lays up a store of sour cheese and curds, dried hard as the toughest hardtack. Then when the summer comes he exchanges his surplus animals, chiefly the young males, for dates, wheat, and rice grown in the oases or in the borderlands around the desert where agriculture is possible. A few animals may be saved for future use as food, but only the most

THE ASIATICS WHO DWELL IN TENTS

wealthy can afford to eat meat often. In good years the ordinary Arab can lay by enough food to last himself and his family until the following spring. Yet even in good years many of the nomadic Arabs are unable to provide enough to last later than perhaps February, the time when the rains ought to come and when the young sheep and camels begin to be born, and milk is normally abundant. But if all goes well it is possible to live through the spring on nothing but milk. The Arabs do this frequently, although they do not enjoy it. Doughty, in his vivid account of *Arabia Deserta*, tells how the Arabs who have long been living on milk pine for something else. "Give us bread," is their cry. "For two months we have drunk this vile milk. Our stomachs are empty and we cannot fill them. Let us have something we can set our teeth into, something that has substance."

Suppose now that when the time comes for the young animals to be born and for abundant milk to supplement or replace dates and bread, no rain has fallen. No wonder the Arabs are anxious. No wonder the first question of the tent-dweller whom Doughty met was: "Where is the rain? Have showers fallen anywhere? Is there grass in the land? How far to the place of the shower? Did it extend wide over the country? Why does Allah withhold the rain so long?" Imagine the mental state of people who eagerly pack their tents and all their crude belongings upon camels and travel one or two hundred miles simply because they have heard that a little shower has fallen over an area no larger than that watered by a single summer thunder-storm. Yet this is a common occurrence in Arabia. Often, indeed, the showers are so scanty that the pools are not filled, and the grass springs up so sparsely that to the eye of any but the son of the desert it is almost invisible; and even to him it becomes invisible as soon as its scanty spears, two or three inches high, wither in the scorching sun. When such years come, as they often do, the mother animals can find no fresh pasturage; they may subsist themselves, but they have no milk for their young. Animal after animal dies. The Arabs see that they can lay by no curds and cheese for the winter; but that is far from the worst. They see also that the price of dates and wheat will be high because of the drought, and that they will have no surplus animals to exchange for food of any kind. And worst of all they see their children hungry and crying for food. This is no overdrawn picture. It is what happens when two or three dry years come in succession.

What is an Arab to do when his camels, his sheep, his wife, his children, and himself are all suffering the pangs of hunger? He cannot go off to some other land and get work. In the first place the better adapted he is to the nomadic life the more he hates the thought of steady work, and the less adapted he is to it temperamentally. In the second place there is rarely any work to be had, for when he is in distress the settled regions near him are also usually suffering from drought. And finally, unless he absolutely gives up his old life he cannot abandon or sell what animals he has left, but must care for them as the basis of any prosperity for which he may hope in the future. The only resource under such circumstances is plunder. The man who is starving has little thought of right or wrong. To have such thoughts would seem to him fatal. If considerations of humanity or any other moral ideas prevent him from engaging in raids upon the tribes around him, the doom of his family may be sealed, for his children may die of hunger. Thus through the thousands of years since nomads first lived in Arabia the hard conditions of climate have steadily forced the Arabs to frame a moral code which condones violence, and have at the same time weeded out those who withheld their hands from violence or were incompetent in robbery and raids.

The man who would succeed and who would keep his children in health must not only be ready to commit depredations and be utterly dishonest according to our standards, but must also be strong in the endurance of heat, thirst, and the weariness of long rides. Unfortunately he has little need of steady industry, or of strength to endure long physical labor. Laziness, according to our definition of the word, is no great disadvantage provided a man is able to summon up his powers in a crisis when the camels have strayed far away, when they have been

THE ASIATICS WHO DWELL IN TENTS

driven off by raiders, or when the man himself goes on a foray. Hence the Arab is lazy as well as disregardful of what we call common honesty. Just as he thinks of raids as a part of the ordinary routine of life, so he thinks of steady work as something scarcely to be demanded even of women and as fit only for slaves. "Shame enters the family with the plow." Thus do the Tuaregs, or Berber nomads of Algeria, sum up their idea of agriculture, as Gautier tells us in *The Geographical Review* (January, 1921). To them the towns are "nauseating." Nevertheless, a great many nomads are gradually drawn away from the life of the wanderer and become peasants. But "a nomad tribe ruined by losing its camels and consequently its mobility enters on sedentary life with rancor in its heart. It is the supreme humiliation, an irretrievable loss of caste." (Gautier.) The great historian Ibn Khaldun speaks of the degenerate descendants of ancient nomad tribes as "so abased that they pay the impost." The true nomad levies impost and pockets it.

Natural selection is very effective in this process of differentiating the nomads from the villagers. Some nomads may go to the well-watered lands as conquerors, but the ones who become peasants are generally those who either fail in the struggle as nomads or who have a particular aptitude for agriculture. It may be that they have not enough endurance and that the thought of having plenty to eat outweighs their dislike of steady work. Or it may be that by reason of physique or temperament the work itself is not so distasteful to them as to some of their comrades. So they leave the nomads and go to the oases or to the fringes of verdure surrounding the desert. Thus through the ages there goes on a steady process whereby one type of character becomes fixed in the desert and another is encouraged in the places where agriculture prevails.

The completeness with which the activities of the Arabs are controlled by the dry climate is illustrated by the fact that their raids, like the majority of human actions, are timed according to the earth's rotation and the inclination of its axis. As a rule the Beduin make raids upon one another rather than upon the sedentary population who are protected by their villages. On long raids the plunderers sometimes ride three or four hundred miles to the scene of operations. To be most successful they need camels to endure thirst and to travel hard on little food, and horses to use in the final dash when speed and docility are required in order to round up and drive off the camels or other animals selected as prey, or to ride down escaping victims. The mares, however, which are the only horses kept by the Arabs in any numbers, cannot endure long marches without drinking. Accordingly each pair of Beduin in a well-equipped party takes a milch camel and a mare which has been taught to drink camel's milk. They ride the camel on the long marches, and where water is scarce use its milk to supply themselves and their horse with both food and drink. Only when the camel foals are several months old can the mothers safely be taken from them. Since the young camels are mainly born in February and March, the months of May and June are the great season for raids. Earlier than this not only do the Arabs hesitate about taking away the mother camels, but the business of taking care of the young animals is too engrossing to permit of many raids. Later, during the hot, rainless summer, many springs and wells dry up, and this not only makes it hard to travel across the desert, but obliges the Arabs, both the plunderers and the plundered, to concentrate around the larger supplies of water. Raids thus become more difficult because many people are together and because the distance to be traversed from water to water is great. In dry years, however, the season for raids is greatly lengthened, and they occur almost everywhere.

Wherever we turn among the Arabs we find habits and traits which have either originated because of the desert environment or have been preserved by it through the process of natural selection. In addition to predatory habits, dishonesty, and laziness, other traits, such as democracy, as I have shown elsewhere, are dependent upon climatic environment. The utensils, ornaments, and dwellings used by the different kinds of nomads are very much alike. No one who packs up all his household goods

THE ASIATICS WHO DWELL IN TENTS

every two or three weeks and carries them on horses or camels can have large pieces of furniture. Bedsteads, bureaus, tables, chairs, and similar paraphernalia are out of the question. The way to sleep is on a thin mattress, quilt, or felt, spread on the ground. The way to pack one's belongings is in soft bags or small boxes that can be slung on the backs of animals. A piece of crockery is a continual nuisance for it has to be packed with great care at each migration, and ten to one it will be broken before long when some animal runs away. Iron pots, pans, and kettles are reduced to a minimum because of their weight.

Because life is so simple and because people must carry all their paraphernalia with them at each migration the outward difference between the rich and the poor is reduced to a minimum. The tent of the richest man may be two or three times as large as that of the poorest, but in general style the two are identical. It is utterly impossible to have any such contrast as that which exists between the palaces of our multimillionaires and the hovels or tenements of the poor. This tends to make the nomad democratic. So, too, does the fact that the encampments are small, rarely more than a dozen or twenty tents, so that every one is thrown into intimate contact with every one else. Moreover, the rich and the poor alike realize that at any time they are likely to lose everything. In a raid the poor man who has three camels and the rich man who has three hundred both may lose them all. This complete loss of property, this levelling of all grades of society to the plane of direst poverty is so common that there has grown up among the Arabs a peculiar kind of generosity. When a man has been robbed it is the common custom for all of his immediate clan to contribute camels according to their means and thus start him once more toward prosperity.

Among pastoral nomads there is likewise great uniformity in the dominant types of art. In the business of life the things that the nomad needs are wooden bowls, skins in which to store milk, bags, saddles, rugs, quilts, and similar small, soft, or unbreakable articles. Hence the art of all those people turns primarily to the making of clothing and of ornamental rugs and felts for the floor and walls of the tents, and ornamental leather-work for bags, saddles, and the sides of boxes. Practically everything in the way of art and handicraft employs wool and leather as the chief raw materials. Even wood takes a secondary place.

Another almost universal trait of nomads is hospitality. Even where the nomads and the people of the irrigated villages are of the same race, there is an almost incredible difference in their attitude toward strangers. The tent-dweller receives every one who comes, gives him of the best, and sends him on his way rejoicing. The villager looks upon the traveller with suspicion, tries to send him somewhere else, and at best gives him a grudging hospitality. All this is the inevitable result of the fact that the nomad has to wander and that his life often depends on the hospitality of other encampments. Suppose the animals stray away. The man who starts in pursuit of a stray animal rarely goes back to the tents for a supply of food, or for an extra garment, in spite of the bitterly cold weather which is common in many Asiatic deserts. It is his business to pursue the animal as quickly as possible, even if he has to go a score or more miles. At nightfall he makes for the nearest encampment, whether it be his own or some other. Every one knows that he may be in dire need of hospitality himself at any time, and those who fail in this respect are quickly made to suffer in their own hour of need. The habit of hospitality has now become almost instinctive, if such an expression is allowable.

Among the Arabs and among pastoral nomads in general few qualities are more important than the capacity either to lead or to be led. East of the Caspian Sea, for example, before the Russians put an end to such habits, the regular mode of starting a Turkoman raid was for some one to drive his spear upright into the ground and say: "I am going on a raid. Who will go with me?" Others came forward, thrust in their spears, and said that they too would go. When a raid is once under way three conditions are absolutely essential to success: Some one must lead; the rest must obey; and each man must have implicit confidence in his comrades. Failure in any one of these respects may spell not only disaster but death. To the hunter or farmer it makes little difference whether he can lead other men or not. Nor is it essential that he should obey on the instant. He usually works alone and depends almost entirely on himself. But when an encampment of the desert people is raided, the only chance of recovering the animals is for some one immediately to issue orders and all the rest to obey. The farmers of the oases and elsewhere are sometimes subject to raids, but not nearly so much as are the nomads. It is far harder and more dangerous to raid a village than to raid a camp. Then, too, animals—not grain—are generally what the raiders want.

Again, during the course of a raid it may happen time and again that the raiders are in sore straits. Often two men ride a camel. Or if each has his own camel one man may care for several animals while the rest are busy in other ways. If the one who has the camels flees at sight of danger his comrades may perish. If they escape, the man who failed them is fiercely hated and driven from the camp. So necessary is it that the comrades in a raid or in the chase after stray animals hang together that a nomad scarcely dares come home alone if some accident has befallen his companions.

We have spoken chiefly of the Arabs, but in practically all desert regions and among mountain nomads, such as many of the Khirghiz, almost the same habits prevail. Let us sum up the qualities which make a nomad successful and which thereby enable him to marry the finest girls in the community, and to bring up many healthy, sturdy children. First and foremost stands the ability to exert himself strenuously and without stint in an emergency. It matters little whether the animals have strayed, have been attacked by wild animals, or are threatened by a sudden storm or flood. Or it may be that the man himself is going on a raid or his encampment is being raided. In any case he must not spare himself. He must not think of food, drink, or his own sufferings. For a while he must do and dare everything, and exert every ounce of his strength. After the emergency is over he may be absolutely idle. In fact he generally is. His wife and children can do the milking and care for the animals so long as there is no emergency. Hence, as already stated, laziness in the ordinary sense of the word is no disadvantage. In fact the man who is good at steady industry is very likely to be weeded out of the community either because he does not succeed or because he finds the sedentary life better adapted to him. A second great quality of the nomad is the power of leading, and a third is the power of being led. These are perhaps as essential as the capacity for sudden and violent exertion. Without them a nomad camp would quickly disintegrate and its members would perforce take to agriculture or become mere hangers-on in other camps.

Another fine quality of nomads is self-reliance. The nomad meets sudden emergencies far oftener than does the farmer. Coupled with this is the necessity of self-reliance on the part of the nomadic women. Little by little the type which is not bold and self-reliant tends to disappear from among the tent-dwellers. Either their children meet with accidents, or they themselves are not wanted as wives by the bolder and more energetic men, and hence are likely to drift with the less adventurous men into the sedentary life. Still other qualities, such as hospitality and faithfulness to one's comrades, are at a greater premium among nomads than among those who live in settled villages, for failure in these respects is much more likely to spell disaster to the nomad than to the farmer. Thus through thousands of years, in north Africa, Arabia, central Asia, and to a less extent in certain other deserts, there has grown up a type of bold, forceful, active, dominating wanderers who nevertheless are lazy and have little skill in the handicrafts. They are inherently different from their patient, timid, and submissive cousins who have become inured to the quiet life of the villages, but who are also industrious and skilful. The interplay between these two types, and the repeated conquest of one by the other have been among the dominant factors of history.

CHAPTER IX

JEWS, ARMENIANS, AND TURKS

If the ideas advanced in this book are correct, racial character is plastic. Not only has it changed repeatedly in the past, but in many cases it is still changing and may alter greatly in the future. The character of the Jews, for example, seems to have passed through a gradual evolution. Old traits still remain, but new ones have been developed or at least have sprung into prominence. The first Jews of whom we have record were Israelite nomads not greatly unlike the Arabs discussed in the last chapter. Jacob's fear of a raid on the part of Esau was almost identical with the fear of the modern Arabs during periods of scarcity, such as 1909. In the book of Job not only was the hero a rich nomadic chief, but he was raided repeatedly under circumstances not essentially different from those of to-day. Although some, at least, of the Israelites stayed for a while in Egypt, they had not lost the nomadic character when they departed from that country. The forty years in the wilderness were apparently merely the normal wanderings of a tribe of Beduin. We are told that Joseph and his brethren settled in Egypt because famine drove them from a dry region into lands that were better watered.

Without going further into this question it seems fair to conclude that when the Hebrews first appear in history their character was much like that of the modern nomadic Arabs. For example, Caleb was one of the spies sent by Moses from Kadesh in southern Palestine to spy out the land of Canaan. He and his brother Jerahmeel are called sons of Hezron, which signifies "the nomadic life." In the same way we are told that Caleb's first wife was Azubah, which means "an abandoned desert region"; Jerioth or "tent curtains" is mentioned as a second wife, which is probably another way of expressing the desert origin of the Jews. During the period of the Judges several stories reflect the habits and characteristics of a desert people. When Sisera fled to the camp of Heber he probably hid in the women's part of the tent, for in no other part would he be screened from the view of the passers-by. There he was killed by Jael, perhaps because no strange men may with impunity enter the women's quarters.

How far there was a tendency for those Israelites who possessed the more strictly nomadic character to drift back into the desert during the days of the early Judges, when they were not yet well established as farmers, we cannot say. At an earlier date Ishmael, who "dwelt in the wilderness," almost certainly represents the type of nomad who clings to the desert. He was "as a wild ass among men; his hand was against every man and every man's hand against him; and he dwelt to the east of all his brethren." Lot, on the other hand, apparently stands for the type to whom settled life made the strongest appeal, for he chose the cities of the plain of Jordan, "well watered everywhere." Presumably this same tendency toward a differentiation in character between the adventurous, active, nomadic type and the more conservative, industrious, agricultural type continued during the long period when the Hebrews were conquering Palestine and driving out or mingling with the older Canaanite population.

Then there began another selective process which is one of the main themes of large sections of the Bible. In the books of Kings, Chronicles, and the prophets it is described as the lure of strange gods and strange women. To the geographer it is the lure of the great highways, the lure of fertility and prosperity, and the menace of accessibility, that is, the danger that arises when people are unduly exposed to the influence of the chance passer-by and of the great routes of travel. In most parts of Palestine the lure and the menace went so far that the Israelites disappeared. Only in the rocky, isolated little plateau of Judea was there an environment which enabled two tribes to resist the lure and face the menace, so that a remnant of "God's chosen people" stood fast. Hence, there and there alone, in a tiny plateau no more than a day's walk in width, and two days' walk in length, the Hebrews as a race persisted and gave rise to the people who later were called Jews. The rest of the Israelites are the "lost" ten tribes. It was the tribe of Judah and its little adjunct, Benjamin, that held their own in the relatively infertile plateau and thus gave the world many of its noblest ideas.

The disappearance of the ten tribes took place partly because they mingled with other races whom they did not completely exterminate, as is so vividly described in the Bible. It also occurred partly because they were in lands that were open to invasion. East of the Jordan in periods of stress the nomads of the desert swept in again and again. The record of this is preserved in vivid accounts of raids in the land of Moab. Elsewhere, as in Galilee and Samaria, the Israelites were in close and easy contact not only with the great road from Egypt to Syria and Assyria, but with the Phœnicians of the Syrian coast and the Philistines of the plain of Sharon.

The importance of caravan roads is rarely understood by Westerners. To us who travel on fast trains most of the villages through which we pass are wholly unknown. Thousands of people in scores of trains may pass through a village each day, and yet the only ones who affect its life may be the handful who get on or off the one or two local trains which stop there. But in a land where every one travels on horses, camels, or donkeys, or else on foot, a great road has a very different effect on the people. Every caravan must stop each night, and must frequently rest a day or two. The average distance traveled by a camel caravan is not more than fifteen miles per day, and even with horses a twenty-mile march in one day is the usual limit. So every village on or near a caravan road is frequently called upon to provide for the wants of great caravans. Often the traveling merchants turn aside a few miles to avoid the crowd at certain points, or in the hope of finding cheaper food or more abundant grazing. Every villager with aught to sell, if he cannot find ready sale in his own home market, bethinks him of the caravan stations. He loads his donkeys with panniers of grapes, rope crates of melons, coarse bags of barley, or sheaves of half-dried hay, and wends his way to the great highroad. A day's journey thither and a day's journey back is no great matter. No part of Lower Galilee or of Samaria north of Shechem lies farther from a great caravan road than a man might drive a slow donkey in a day. The Oriental, even more than his Occidental brother, must gossip and get the news when he waits for buyers or sits idle of an evening. Thus in the days of active caravan traffic the low portions of Palestine between the plateaus at either end were permeated through and through by outside influences. They could not remain secluded. Therefore the prophets wailed over "the wickedness of Samaria"; and the northern province was "Galilee of the Gentiles."

Judea, unlike Samaria and Galilee, lies off the main routes of travel. It is rugged, hard to traverse, infertile, and relatively undesirable. Hence the people of Judea were like children living in a retired house among the woods, near a great road, but far back out of sight and sound of it. Those of Samaria were like children whose home stands close upon the busy turnpike, whose friends are the teamsters and hucksters, and whose playground is the dust of the street or the vacant lot across the way. Sheltered among the rounded hills of their plateau, the Judeans were close to the other parts of their little country, and yet apart in safety. They felt the influence of the highly varied districts round about them but retained their individuality. The traffic of the world passed on the west and north, and even on the south, giving them some knowledge of the busy life of the great world, but not really entering into the daily routine which is the heart of human existence. They shuddered as hordes from the eastern desert pressed over Moab on the east, or the Negeb on the south, but their isolation saved them from the scourge. So for over a millennium they dwelt in seclusion and developed noble ideas of God and truth and justice, until the greatest of men, a Judean in ancestry and birth, came up from Galilee, and, taking the truths which had been fostered and preserved in the rocky plateau, transformed them into the peerless rules of conduct which form the basis of Christianity.

But why did the Judeans develop such lofty ideals, such a stern moral fiber, such great strength of purpose? It was partly because the ruggedness and aloofness of the plateau isolated them just enough to permit natural selection to work with great effectiveness. Because of the physical character of their homes the Judeans did not mix freely with the followers of strange gods who surrounded them. Yet always from the beginning of their settlement in the land there was a tendency for some to leave their rugged home and seek their fortunes in the richer regions round about. Samson is a good example. It was the Philistine women who tempted him to come down into the plain which was his undoing. The prophets are full of denunciations of the Israelites who married heathen women, followed strange gods, and took up heathen practices. It must be remembered that although Judea seems a great and potent center to us, it was not a center of any importance in the ancient world. It was a mere little backwater in the ebb and flow of the great civilized nations of Egypt, Syria, Assyria, the Hittites, and others. It was too poor to be greatly desired by its neighbors, too rugged to be easily conquered, and too difficult of passage and too far away from the main routes to be crossed by armies or caravans. Hence it was left to itself like the highlands of Scotland, or like the Alleghany plateau in Kentucky and Tennessee. The regions around it including Samaria and the Philistine plain would by ordinary standards have been counted much more advanced, progressive, and prosperous.

So the Judeans were constantly tempted to go out into the civilized world around them, a world which was within sight from hundreds of the roofs on which the villagers sit every day toward evening in the long, dry summer. The thing which drove them out was the desire for gain, ease, pleasure, luxury, and culture. The thing which kept them in Judea was love of home, a fondness for their mountains and hills, the lure of a shepherd life, and especially the strong promptings of religious and racial loyalty. If we may judge by the Bible, the religious element was of uncommon importance in holding some Jews in Judea while others drifted away. So it seems that here in this stony little plateau a great process of selection went on which picked out for preservation the people who were tenacious of their race, tenacious of their homes, and, above all, tenacious of their religion. Thus Judah and Benjamin gave rise to a chosen people, while the other ten tribes were lost in the great maelstrom of migrations, wars, and surrounding civilizations.

During the course of Jewish history the selective processes did not always work in one way. For example, when the kingdom of Judah was finally conquered by the Babylonians there were several deportations. In each case the tendency was to carry away the most influential and able parts of the community. This may have markedly weakened the moral and intellectual fiber of the people who remained in Judea, but some compensation was perhaps found in further selection that occurred in the lands of the captivity. While the Jews were in Mesopotamia a great many doubtless gave up their old religion and amalgamated with the Assyrians. The book of Daniel is a vivid account of an attempt by the Assyrian authorities to force four young men to eschew their old religion and accept that of the people among whom they dwelt. The book of Esther presents a similar picture of unyielding racial loyalty in the face of strong compulsion. To yield would have been the easy course. Such a course was presumably pursued by all except those who were most strong-minded and most tenacious of their racial and religious inheritance. Thus when a brighter day arrived and there was finally a chance to go back to Judea, the people who were still called Jews were a highly selected remnant. But the process of choosing a certain type as ancestors of the later Jews went still farther. When Nehemiah called for volunteers to return to the land of their fathers, it was the most religious, the most earnest, and those most proud of their race who flocked around him. Like the Puritans of a later time, those who fared back across the desert had been picked because of their religious as well as racial zeal. How far their return made up for the deterioration due to the earlier deportations it is impossible to say. This much, however, seems clear: the whole episode of the captivity tended to reinforce the previously strong selective process which made the Jews a peculiar people intensely devoted to their race and religion, disregardful of the sneers and buffets of the people around them, and ready to go their own way and stick to their own faith no matter what happened.

It would be interesting to trace the development of the Jews still farther. Throughout their whole history they have been subjected to indignities and persecutions largely because of the very traits which make them Jews, the traits which were first selected by the environment of the rocky plateau, and were later intensified by captivity. Simply because the Jew has been either forced or self-impelled to migrate so widely, the temptation to give up his old faith and his racial identity has been peculiarly strong. At the same time because the Jew has such pronounced characteristics and looks with such disdain on other religions the tendency to persecute him has been great.

Without taking time to trace Jewish history, let us consider what happens in persecutions such as the *pogroms* which have been so frequent in Russia even within our own memory. According to common report two types of people are likely to suffer. In the first place the authorities are likely to connive at or even cause the death or exile of any Jews who stand out as leaders in war or politics. To-day, of course, this tendency is passing away, but we are talking of what has happened in the past. In those times a Jew who achieved eminence in either military or political affairs was particularly an object of enmity. On the other hand, if he distinguished himself in medicine or science, or in his own rabbinical lore, he incurred no special hatred among the Christians and Moslems who lived around him. In fact, if he were a physician, for example, he might be highly revered. Oftentimes such men were so deeply respected that they were carefully shielded at times of persecution. Between the political and military leaders on the one hand and the scientific leaders on the other stood the merchants. They were hated because of their power and extortion, but were also feared for the same reasons. Moreover, they were peculiarly able to purchase protection. Thus the first kind of selection during Jewish persecutions has been to weed out the active, aggressive leaders who tried to give the Jews military power or political unity, while the capable merchants have suffered far less and the leaders who have benefited mankind by the arts of healing or by scientific knowledge have had a still better chance for preservation.

Consider now what persecution has done to the Jewish rank and file, as contrasted with the leaders. All over the world it seems to be the rule that the mobs who engage in massacres are largely a rabble of the baser sort intent on revenge, plunder, and excitement. Such people live in the poorer sections of the cities, which usually means that they are near neighbors of the poorer Jewish quarters. Those poorer Jewish quarters as a rule are the first to be attacked. For that reason and because of the relative poverty and incompetence of the inhabitants, the Jews in those parts of the cities suffer more than do those of the more prosperous quarters. Not only are they less able to escape death, but those who survive often lose home, property, and friends, and have nothing to fall back on. Inevitably many of the people who are weak in mind and body are weeded out, as are those of less tenacious spirit who under the stress of persecution give up their faith. Moreover, certain types of Jews have a great advantage under persecution. One such type consists of people who can skilfully dissemble; another of those who can wheedle their persecutors into leniency; and a third of those who know how to place attractive bargains before their oppressors. The farther we go in studying the way in which oppression and persecution have weeded out some types of Jews and preserved others, the clearer it becomes that in general the Jews who possess the supposedly typical traits of that race are the type best able to weather the storms of persecution. Doubtless the training of the Jews enters into the matter, but that does not alter the case appreciably. What we are here concerned with is the fact that

JEWS, ARMENIANS, AND TURKS

persecution, oppression, and the indignities to which the Jews have been subjected from time out of mind tend to place a handicap on certain inherent types of character and a premium on the kind of character which is typically Jewish.

Let us turn now to another race. The Armenians are often spoken of in the same breath as the Jews. For thousands of years their ancestors appear to have been settled villagers who cultivated the plateaus and basins among the mountains of Armenia. There they acquired something of the tenacity of spirit which is so often ingrained in highlanders like the upland Scotch and Swiss as well as the Jews. Among them the traits that were most valuable as a means of preserving the race were the ability to cultivate the fields, to work steadily, and to carry on the common handicrafts.

Let me tell some of my experiences among the Armenians. One day in company with an Armenian sheep-dealer I was riding on horseback across a fertile plain set among the great mountains of Armenia. My host that night was a Turk of the more prosperous village type. While the sheep-dealer was tending the horses my host and I fell to talking of the massacres in the years from 1895 to 1897 in which perhaps 200,000 Armenians perished. "What kind of people are these?" said the Turk. "Three years ago we killed half of them in this village, and took everything away from the rest. And now," as the Turkish idiom puts it, "they eat better than we do. What can we do about it?"

Unconsciously that Turk was asking why the inexorable process which Darwin called natural selection has given these people such remarkable unity and such clearly defined traits. The answer is suggested in such experiences as this: During the four years that I lived and taught in Turkey I several times visited the little plain of Bermaz south of Harput. The people of its dozen villages are fanatical Moslems who are sometimes spoken of as Kurds and sometimes as Turks. They have the curious custom of making the sign of the cross before their meals. Why? Because once Bermaz was inhabited by Christian Armenians. During a period of massacres and persecution many generations ago, the majority of the surviving villagers gave up the struggle and "turned Turk." Those who would not become Mohammedans fled or were killed. The renegade Armenians who remained in the villages tried to pattern their lives after the surrounding Mohammedans, but the habit of years persisted in the sign of the cross. Perhaps at first they clung to it in the secret hope that some day they might again become Christians.

I have known other Armenians who gave up Christianity. One was a professor in the American college where I taught. To save himself and his family from outrage and death he agreed to become a Moslem, but after the massacres were over he turned back to his old faith. He was a good man, a friend of mine, and lived thereafter a useful and honored life until he was foully murdered during the massacres of the World War. But he had less tenacity of purpose, less power of passive resistance than most of his fellow professors of the Armenian race. He was not lost to his race, for he had the courage to return to the faith of his fathers even while it was still dangerous to do so. But many others who "turned Turk" failed to re-enter the Armenian fold. Thus it has been for centuries. During massacre after massacre a certain portion of the Armenian race has separated itself from the remainder and gone over to the Turks. Not all kinds of people have done this, but only those who had less will-power than the average. A strenuous process of natural selection has gone on whereby those Armenians who have less tenacity of purpose, less devotion to their race and their religion, less ability to endure hardship and obloquy, and even to face death, have gone out from among the Armenians and become part of the Turks. Always the bargain has tended to raise the average level of determination, will-power, and perseverance among the Armenians who remained true to their race.

The qualities which have thus been concentrated in the Armenians by hundreds of years of persecution sometimes have a disagreeable as well as a good side. An Armenian boy once came to a friend of mine who teaches in an American school in Turkey.

The boy wanted not only to be admitted to the school but to get some help in defraying his expenses. He did not appear promising and my friend sent him away unsatisfied. A month or so later he came again, with the same result. A third time he appeared with his plea, and was so persistent that not only did he have to be shown the open door, but to be given a little push to get him out. A Turkish boy in the same situation would have hesitated about asking help, he would have been polite and agreeable, and after one refusal would never have been heard from again. But what finally happened? The Armenian boy came again; he got what he wanted because of his much speaking. And more than that, he made good. Crude as he was, there was sterling stuff in him. He was laughed at by the other boys, and his teachers were alternately vexed and amused by his ludicrous blunders, but he worked intently, and finally won the respect of his mates and teachers, and proved that it was worth while to educate him. He was the normal product of a process of natural selection which puts a tremendous premium on persistence.

Still another kind of natural selection has helped to give the Armenians their peculiar character. It is estimated that about a million Armenians were driven south into the Syrian desert during the World War. I have heard a high-bred Armenian girl tell the story of that deportation to a committee of Congress. She brought tears to the eyes of those seasoned politicians. With superb self-control she related how she was driven from home at the age of twelve by a band of rough Turkish soldiers; how her father had already been carried off and presumably killed by the Turks; how her brother and uncle were taken from her; and she and her mother were left unprotected. Some of the few remaining Armenian men were shot almost within sight of the women and children who heard the screams of the wounded. Then the helpless survivors were forced onward over rough mountain trails until their bodies ached with weariness, hunger, and thirst; their limbs sank under them, and they fainted from utter exhaustion. After a week of such travel the girl was sold for sixty cents (I think that was the sum) to a Kurdish chief, and was taken to his harem along with other girls. They tried to make her become a Moslem, but she would not renounce Christianity. They tortured and killed another young woman in her presence to make her yield, but she would not. Then, when she was unconscious from blows, they tattooed her face with great blue marks and said: "We have made a Moslem of her." At last there came a day when she could stand it no longer. Jumping from a window, she fled along a mountain path. But she miscalculated her time; some one saw her, and a band of horsemen gave chase. With shouts and jeers they tied her by a rope to the tail of a horse and dragged her back, bruised into unconsciousness. And still she clung to Christianity and to her race, until at length a British force arrived and freed her. If a girl between the ages of twelve and fifteen can have such tenacity and courage, is it any wonder that the Armenians are one of the most tenacious races on the face of the earth? She was only one of thousands who suffered similarly. It was the ones most strongly endued with her qualities who escaped from captivity and came back to their own people to be the mothers of the Armenians of the future. And this same thing has happened not only within the present decade, but time after time in the past.

We have spoken of the selection of moral qualities which is a feature of persecutions, massacres, and deportations, but there is likewise a physical and mental selection. Among the million Armenians, more or less, who were driven into the Syrian desert during the World War, about a quarter drifted back, chiefly through Aleppo. There they were cared for by the American Red Cross. Doctor Lambert, who was at Aleppo two years and who had charge of the medical work for a year, has written a significant report on post-war medical conditions among the refugees. The returning sufferers had been half starved and half frozen; they had suffered almost intolerably from the heat of summer and the unmitigated attacks of vermin. Many were not only weak but sick. Yet the physicians were continually impressed by the organic soundness of the stricken refugees. The weaklings had been relentlessly picked out for destruction. Although no exact figures are available, it seems almost certain that the mentally weak were destroyed or otherwise lost to the Armenian race even more effectively than the physically weak. When the Armenians were driven from their homes almost without warning, those who were mentally most alert went out best prepared. They knew how to wheedle their captors into giving them a chance to procure food and shelter, while the stupid merely angered their savage herdsmen and brought blows and violence upon themselves. The mentally dull were likewise the easiest to persuade into giving up their religion. Having once become Moslems they were less likely than their more quick-witted kinsmen to return to their old faith and race when better times at length arrived. Among the children and especially the girls who were carried to Moslem houses, it was the quick-witted, the resourceful, and the tenacious who ultimately escaped. By processes such as these, during one dire massacre after another, the dross has been consumed and the gold refined. Thus natural selection, combined with other circumstances, has produced in the Armenians a character closely similar to that of the Jews.

But mark another feature of this process in both Jews and Armenians. Almost on the very day when the Congressmen shed tears over the story of the bravery of an Armenian girl, I heard some young Americans berate a highly educated and highly competent young Armenian, an American citizen, because he alone of a group of a dozen associates had failed to enter the army during the World War. No ties of family or kindred held the Armenian back, he was physically fit, and the fate of his own people was at stake. He stayed at home, as all his comrades believed, because he was afraid. Is this consistent with the character of the race as illustrated by the girl? I think it is. Both the girl and the man had an unusual tenacity and capacity for passive resistance. But the man, and perhaps the girl, lacked the sort of aggressive boldness which is characteristic of the soldier. This is in harmony with all the later history of the Armenians and Jews. Both races seem to be deficient in military qualities, and especially in the power of military and political organization. As a leading Armenian professor once said to me: "We want our own government, but I am not sure how long we could run it unaided. We do not trust each other." What he really meant was that, although the Armenians are unusually competent in certain lines such as business and many scientific and literary pursuits, they have relatively little capacity for political as well as military leadership. The reason, in part at least, seems to be that for a thousand years the military and political leaders have been the first to suffer at times of persecution.

Now contrast the Turks with the Armenians. Almost every one who knows the two races intimately agrees that the Turk, so long as he is under control, is far easier to get along with than the Armenian. He is more hospitable, more courteous, more suave, less assertive, less likely to make a sharp bargain. Most of the time he is a mild, gentle sort of person, a delightful host, who seems never to have anything to do except look after the interests of his guests. But there is another and more important phase of the matter. If one wants a good carpenter, a good mason, a good accountant, or a good business man, only rarely does one employ a Turk if an Armenian is available. And if one makes a bargain with the Turkish Government, one may be almost sure that nothing but constant watchfulness and prodding will cause it to be carried out. Moreover, the Turkish peasant is by no means industrious in the fashion of the Armenian. Put the two side by side and soon there almost inevitably arises the condition already expressed in the Turkish idiom: "The Armenians eat better than we do."

Compare this with a case which, though perhaps extreme, is in many respects typical of the Turkish peasant. Before the World War a German company was installing an irrigation system whereby to bring the water of Lake Bey Shehr to the plain of Konia. The Turkish peasants looked askance at the work of the foreigners, but one day one of them was impressed by the arguments of a German engineer.

"When we turn the water onto your fields," said the engineer, "you won't have any more poor crops and hard times. You'll get four times as much grain as now."

The peasant thought deeply for a few minutes. Then he stood up, adjusted his fez, and gave voice to a brilliant idea:

"Do you know what I'll do when that day comes? I'll sell three-fourths of my land, and cultivate only a quarter. Then I'll only have to work a quarter as hard as I do now."

If the Armenians are really so much more competent than the Turks, why have they for hundreds of years been so completely at the mercy of these ruthless masters? The answer, I believe, lies largely in the selective power of the original modes of life of the two races, plus the further selection that has gone on through massacres and otherwise since they have been in contact. The Armenian has for untold centuries been a tiller of the soil and a villager. For him the qualities that have made it possible to bring up large families in health and comfort have been steady industry, skill in agriculture, good artisanship, and business acumen. Since there has been little restriction of families, success along these lines has meant that parents have brought many children to maturity, whereas lack of success has meant poverty and poor health so that great numbers of the children have died in childhood.

Among the Turks, on the contrary, there has been no such premium on the substantial qualities which tend to develop a strong, capable middle class. From time immemorial until they finally settled in Asia Minor the Turks were nomadic keepers of sheep, camels, and horses. They had no special need of steady industry, of skill as artisans, of ability in agriculture, or of capacity for business. Those qualities did not help in supporting wives and children. If a nomad had the capacity to lead a raid successfully, and to direct the pursuit when wolves stampeded the herds, or if he obeyed his leader implicitly at such times, he had a far greater chance of success than if he was steadily industrious, or had the qualities which make a good carpenter or a good trader. The nomad must be able to ford mountain streams, face the leopard, or endure hunger, thirst, and fatigue during a raid or a hunt for stray animals. The rest of the time he may be utterly idle and yet be successful and prosperous.

In Elliot's *Turkey in Europe*, one of the most penetrating books ever written on the races and problems of the Near East, a striking passage sets forth the idea that the Turk is still a nomad. He never seems to feel that he is permanently settled. At least he acts as if that were the case. If he wants a sheet-iron stove in his house, he does not take the trouble to build a chimney, or even to carry the pipe to a window. He simply knocks a hole in the wall, sticks the pipe through, stuffs the cracks with rags to keep out the wind, and lets it go at that. What is the use of bothering to do more? To-morrow it may be the will of Allah that we move on.

All through the Turkish mentality there can be traced the old nomadic life. For instance, the Turk, as I have said, is a mild, pleasant man at most times, but push him beyond a certain point and he becomes a fiend incarnate. When aroused to attack his Christian neighbors, or to fight, he does it with a complete abandon which makes him most terrible. This is perhaps merely a remnant of the old spirit whereby the nomad would sit idle for day after day, and then undergo almost incredible exertions and privations during the short period of a raid, or more often a search for stray animals.

Again, there is no middle class of Turks comparable in size or quality with that of the Armenians. One reason is that when the Turks came to Anatolia from central Asia they split into three strongly contrasted sections: a small ruling class; those who continued to be nomads; and those who became farmers, which happened largely because they were not sufficiently competent to maintain a position in either of the other classes. Another reason for the absence of a strong Turkish middle class is that in receiving people from other races, including the Armenians, the Turks have generally gotten either the worst or the best of the bargain, but have rarely exchanged on equal terms. It has been the weaker and more incompetent Christians who have succumbed to the urge of Mohammedanism, but it has been the finest and most attractive girls who have been taken permanently into the harems of the Turkish leaders.

At first thought it might be said that the strong middle class of the Armenians ought to have given them the upper hand. But here appears another factor. Not only have leaders of the political and military type been killed off unmercifully among the Armenians during massacres, but the farming mode of life does not tend greatly toward the production of such leaders. Thus the Armenians were relatively weak in this respect before the arrival of the Turks, and have grown weaker with the lapse of time. Among nomads, on the contrary, as we saw in an earlier chapter, success and even preservation depend largely upon the quality of leadership. And that is where the Turks are strong. They came to Turkey as bands of nomads, the leaders and the led. The leaders became the rulers, they took to themselves wives from among the best of the native races, they strengthened their racial stock. Their main road to success and thus to large harems and many children was through military and political leadership.

The political leadership often takes merely the form of playing off one party or one power against another, but nevertheless it is very real. Consider the events of recent history. The sultans have often been inefficient, but it is surprising to see how frequently they have had qualities like those of the crafty Abdul Hamid which somehow kept them on top when by all the ordinary laws of the game they ought to have sunk. Turkey has been called the Sick Man of Europe until we are tired of the phrase. But that sick man has kept the upper hand for a century. After the World War he defeated the Greeks, kept all Europe at bay, and for a time at least made himself the only one of all the warring nations which came out a marked gainer by reason of a decade of war. All through the period after the war, just as in the earlier years, the Turks have had leader after leader—not great men, perhaps, but competent enough to hold their people together, outwit the statesmen of Europe, and maintain armies in the field when any other nation would have been likely to give up in absolute bankruptcy. The Armenians, on the contrary, have had no real leader for many generations.

Here then is where the matter stands to-day. When fighters like the Turks who know how to lead and how to be led, and who are temporarily capable of enormous exertion, meet men like the Armenians, accustomed to steady industry, but unused to being led, not given to surges of sudden activity, and lacking military leadership, the result is certain. The side which is usually counted the more valuable from the standpoint of civilization goes down before the side which is peculiarly adapted to primitive war. Ever since the first encounter of the two races the disadvantage of the Armenians has increased. During ordinary times the Turk, because of his nomadic nature, neglects the details of government, business, and every other occupation. The Armenian gradually gains power. Then the Turk rouses himself and tries to regain by violence what he has lost by sloth.

The much-persecuted Armenians are peculiarly homogeneous, peculiarly distinct in racial character, and peculiarly strong in racial coherence. They are extraordinarily persistent, patient, and tenacious even to the point of being disagreeable. They have great capacity in business and in the handicrafts and arts. They have no mean standing in the more intellectual pursuits. They are conspicuously free from criminal inclination and the tendency to become a public charge. And with all this they are of a strong, tough, enduring physique. These high qualities have been bred in the race by centuries of agricultural life and of the most cruel persecution that can well be imagined. But unfortunately by reason of lack of one special type of genius this sturdy, self-reliant, middle-class people are a prey to the Turks with their capable military and political leaders and their relatively stupid, docile lower class who can easily be aroused to fight like their nomadic ancestors. So the Armenians, who under a wiser government might have been the solid backbone of Turkey, her essential middle class, are wanderers on the face of the earth, seeking a home in which they can rally under the leadership which is their sorest need. If the Turks and Armenians were of one religion and could form a single race, that race would apparently possess a peculiarly strong combination of qualities. It would have a docile peasantry, partly Armenian but containing a larger proportion of Turks, a peasantry not highly intelligent or extremely active, but good-tempered and easy to get along with so long as it was treated justly. There would likewise be a strong middle class, largely Armenian, very industrious and with good mental capacities. And there would be strong leadership furnished by Armenians in the lines of business, science, and most of the professions, and by Turks in military and political lines. To-day, by their expulsion of the Armenians, the Turks seem definitely to have put an end to the prospect of such a race, at least for a long time.

CHAPTER X

CYCLES OF CHINESE HISTORY

THE history of China consists of a repeated and dramatic cycle of invasion, migration, progress, decay, anarchy, and again invasion. The invaders practically always have come from the northwest or north, that is, from the deserts of Turkestan and Mongolia, or from Manchuria. The first invaders may have been the original Chinese, but as to that we have no certain knowledge. The later invaders have been Hiung-nu, Mongols, Tartars, and Manchus, and have been either nomadic keepers of animals, or the descendants of such nomads. "There is excellent ground," as Parker says in his vivacious history of China, "for believing that the Scythians, Huns, and Hiung-nu were practically reshuffles of one and the same assemblage of people—the Turks and Mongols of later date." The Manchus belong to this same group.

As Parthians, Mamluks, Mongols, Seljuk and Ottoman Turks, to leave the lesser breeds unnamed, the distant congeners of the Manchus have not only invaded but repeatedly controlled all the civilized nations of the continent. The history of China cannot be properly understood unless due notice is taken of the impact of her northern neighbors from the period of the great Ch'in [249 B. C.], to recent times, nor can we afford to neglect the fact that her own great dynasties and governing element have come from those northern provinces which are chiefly peopled by descendants of a Tartar-Chinese intermixture. (F. W. Williams.)

In spite of great walls and armies the settled Chinese have never been able to keep these competent barbarians of the desert out of China for more than a few hundred years. Time after time they have swooped down upon the agricultural Chinese and overwhelmed them. The greatest of such disasters (or were they blessings?) have occurred at times when unusually severe and prolonged periods of increasing aridity appear to have prevailed in central Asia; for example, in the third century before Christ, the fourth and fifth of the present era, and again in the twelfth and thirteenth. Thus the stress of economic want and the capacity for leadership which we have found to be characteristic of nomads have combined to cause the desert peoples to conquer China.

In China itself the periods of invasion appear to have been times of internal chaos. Droughts there as well as in the deserts apparently co-operated with floods and other natural disasters and with internal degeneration due to other causes to reduce the Chinese to a state of almost hopeless anarchy. At such times the Chinese of the north have been under the twofold compulsion of foreign invasion and economic distress. The natural result has been that many have migrated southward or southeastward. Sometimes, without doubt, such migrations have also been impelled by the desire to obtain new and more fertile lands or to acquire the wealth of more prosperous people. But on the whole the outstanding fact is that when famines and barbarian invaders have afflicted north China, the Chinese have streamed out toward the south. Wave after wave of migration has surged outward, first from the original Chinese centres far up the Hoang-ho near its great southwestward elbow in eastern Shensi, and later from all the northern provinces. Not till almost the time of Christ, perhaps two thousand years after the beginning of recorded Chinese history, did these waves finally reach the coast near Canton.

In the north the places left vacant by the migrants were filled by barbarian invaders, who promptly adopted the advanced culture of the Chinese and soon mingled with them to form a new and more virile race. Most of the Chinese dynasties have either been genuine forcigners or else of mingled Tartar-Chinese stock. For instance, "from 309 to 439 A. D., there was a bewildering succession of Hiung-nu, Tungusic, Tibetan, Tibeto-Tungusic, migrated Tungusic, and rebel Chinese 'Dynasties' ruling in various parts of the north." (Parker.) Only in 581

A. D., did a pure Chinese dynasty rule both the north and the south and it almost certainly had a large admixture of Tartar or Hiung-nu blood from earlier migrations. Again for three hundred years prior to the Ming dynasty (1368 A. D.) the Peking plain was in Tartar hands. The Mings were the first native Chinese dynasty to rule north China for four hundred and fifty years. Counting all these foreign dynasties together, it appears that the northern part of China has probably been ruled by avowed foreigners for longer periods than by real Chinese, while most of the so-called Chinese rulers have probably had a good deal of the blood of the northern deserts.

Coincident with the early stages of barbarian rule in north China, the Chinese migrants to the south and southeast have advanced into regions peopled by primitive aborigines. At first this happened in the Yangtse valley, and then farther and farther south until Canton, Yunnan, and the island of Hainan were reached. In later times this same process of migration has carried the southeastern Chinese to Formosa, Indo-China, Java, and the Malay peninsula, where their type of culture would probably have become dominant had it not met the equally active culture of Europe. In the southerly migrations the Chinese have intermarried relatively little with aborigines and often have exterminated them. Nor have the invaders adopted anything of importance in the culture of the aborigines upon whom they have forced themselves, thus reversing the conditions in regions where the northern barbarians forced themselves upon the Chinese. It is not strange, therefore, that time after time "the true Chinese were not . . . to be found in Old China, but in all those parts which, as immigrants, their ancestors from Old China had populated." (Parker.)

After a cycle of Chinese history has passed through the stages of invasion, anarchy, and migration, it generally enters a stage of progress. Again and again this has happened under the brilliant leadership of an invading dynasty as in the case of the Mongols and Manchus. One of the most significant facts in Chinese history is the extent to which the regions and periods where invaders have been dominant have been characterized by new ideas, inventions, and noteworthy steps in the progress of civilization. It seems not improbable that the selective action of migration has helped to bring into north China successive groups of unusually competent people who have been important agents in producing the upward swing in the successive cycles of history.

The Manchus furnished an excellent example of such invaders. Their "decadence—apparently an inevitable result of their contact with a higher culture—should not blind us to the extraordinary success of their great performance." (Williams.) They conquered not only all Manchuria and China proper, but Mongolia, Turkestan, Tibet, and Korea. They were accepted as suzerains in Nepaul south of the great mountains on the northern flank of India, and in Indo-China where Siam listened to the authority of Peking, while Burmah and Annam were among the tribute-bearers. They even crossed the sea and took possession of Formosa.

As a land power—the Manchus [were] even more solidly established than the Mongols;—for although the immediate successors of Jenghiz commanded the personal attendance before their desert throne of Russian, Armenian, and Persian princes, the most powerful Mongol Emperor, Kublai, really ruled in an effective sense over the Eighteen Provinces alone, and was at perpetual loggerheads with his vassal relatives of Persia, Mongolia, and Manchuria; moreover, the Mongols were not the intellectual or literary equals of the Manchus, and never had either the same prudence or the same financial grasp of the country's resources. (Parker.)

Even in their latest days the Manchus produced truly great personages like the famous Empress Dowager, in whose old age and weakness the dynasty finally collapsed. No thoughtful person can view the new summer place, built by this empress after the inexcusable destruction of the old summer place by the English and French in 1860, without feeling that people who can plan and execute so magnificent a piece of landscape-gardening are extremely capable. The workmanship, to be sure, is in many places shoddy, but parts are fine and the general conception is splendid. In the same way the old Chinese literary examinations may be a poor way in which to choose officials, but success in those examinations certainly required an unusual power of concentration, unusual determination, and many other high mental qualities. If the Manchus had not possessed those qualities in considerable measure we should scarcely find large numbers of them holding high Chinese literary degrees. Yet in 1910, at the end of the old regime, among 410 major officials who held such degrees, 50 are recorded as Manchus and 3 as members of the Imperial Clan.

The history of the Manchus suggests that they acquired their unusual ability at least in part through natural selection. Of course, it must be recognized here, as in every other case, that before there can be selection there must be differences among which to select—mutations or deviations from the normal, to speak biologically. And those mutations or deviations may trend in a specific direction so that evolution pursues a definite course, or they may be induced by environmental conditions as to which we are still ignorant. That, however, lies beyond the scope of our present inquiry. We simply accept the fact that the people of any given group may differ widely, and hence some may be selected by their physical or social environment or by their mode of life for preservation and others for destruction.

The first that we hear of the Manchus is in the beginning of the seventh century. Under the name Kitans, which is apparently the origin of Marco Polo's Cathay, they invaded north China. At the beginning of the tenth century these same Kitans, although driven out after their first inroad, established themselves as a ruling dynasty. Two centuries later they were overthrown by another allied dynasty, the Nuchin or Nuchens, the direct ancestors of the Manchus. A century or so later the Nuchens were expelled from China by the Mongols under Genghiz Khan. Then for three centuries they remained almost unknown in southeastern Manchuria, a wild barbarous people, but possessed of great vigor and ability. In the middle of the sixteenth century a chief named Nurhachu came into power and rapidly established a Manchurian kingdom which eventually conquered China.

For our present purpose the most significant fact about the Manchus seems to be the long period of struggle and migration during which they necessarily underwent a rigid natural selection. The Tartar ancestors of the Manchus appear to have been pastoral nomads for thousands of years. Their wandering life, with its sudden calls for exertion and co-operation, as explained in a previous chapter, apparently gave them full measure of that power of leadership which we see so strongly in Arabs, Mongols, and Turks, as well as Manchus, and which may have been evolved in the ancestors of the Nordics under similar conditions. When the ancestors of the Manchus invaded China as Kitans and Nuchens they again apparently suffered selection, especially when they were ejected under the Mongols. Just how they migrated back to their old home in Manchuria we do not know, but judging by modern examples it was quite surely the ablest and most energetic who returned. Many of the leaders were doubtless killed, but on the other hand it is almost universally true that in such cases the weaker, less efficient, and more submissive among a conquered people remain in their old homes. It is the people with initiative and individuality, as well as with physical and mental vigor, who refuse to endure humiliation, and are willing to suffer the loss of their property and the hardships of migration in order to escape from their conquerors. It was probably such people who retired to Manchuria after the Mongol conquest. And it was the most competent and active of their descendants who founded the great Manchu dynasty. Nurhachu, their great leader,

though he never entered China, stands as an exponent of the highest qualities of his race, a creative genius not only in strategy but in politics, the founder of a great tradition capably maintained for two centuries by his descendants, the establisher of a line of monarchs which have been surpassed by no other ruling house during an equal period in China. Yet they succeeded through sheer force of character, as the Ottomans have succeeded during a much longer period in western Asia, in dominating a people that were superior to them in every important quality except that of leadership. (F. W. Williams.)

In the case of the Manchus, as in that of other similar conquerors, we find an interesting interplay between ability on the part of the invaders and natural disasters acting upon them and upon the people whose lands they invade. We also find an interplay with more strictly human causes such as the degeneracy of the rulers by reason of luxury and dissipation, and the weakening of the people by internal dissensions and rivalry. Here is what F. W. Williams says of the period of the Manchu invasion:

The internal condition of the Chinese empire had become desperate under a long series of famines and rebellions which had utterly paralyzed its economic resources and brought about a general anarchy. It is impossible to decide whether under such loosely organized agencies as that of China the general prevalence of distress is a cause or a consequence of political disturbance. When thickly populous agricultural communities are reduced to starvation the people will inevitably break up into robber bands and prey upon each other to the confusion of all civil administration. No government can reduce the disorder unless provisions can be obtained to satisfy the needs of those made desperate by want; but a bad government may by its inefficiency aggravate the starving people and succumb to the forces of disruption thus let loose. It is notable that in the history of China no great upheaval has occurred without its concomitant of famine. In the third decade of the seventeenth century the northern provinces were visited by an unusually severe drought which was so badly met by venal officials that multitudes took to the mountains and attacked the roads and villages. In addition to these natural causes weakening authority in an imperfectly articulated domain, increased taxation and recurring levies of troops to meet the Manchus began in 1621 to arouse angry opposition in the western provinces. Revolts broke out which were painfully and only partly subdued. By 1631 the robber bands throughout all the inland provinces had swelled to great armies under redoubtable captains, whose successes encouraged the able-bodied to enlist under their banners and live upon the spoil of captured cities. At the end of another decade Li Tsu-cheng, a Shansi leader, after many vicissitudes, had become the greatest of them all, and with an army composed of nearly a million needy adventurers he was swarming, in 1641, over the famine-stricken province of Honan toward Peking. Despite the impotence of the imperial government in this score of years of carnage it is remarkable that the various rebel armies met with obstinate resistance in many cities. There was no systematic opposition, yet owing to the indomitable spirit in defending their own which characterizes the Chinese people, as well as the lack of organization among the rebels, the agony was long continued. The contrast between the Chinese rebel Li and the Manchu Nurhachu is suggestive as typical of the differing genius of two races. It has often been said that the Chinese were conquered because they were unwarlike. They showed, on the contrary, a persistent fighting eagerness both before and after the Manchu irruption that ranks them among the martial people of the world. They failed both in rebellion and in defense because they could produce no leader capable of consolidating and fixing an orderly system of control. The Manchus succeeded, though they had to borrow and adapt the system of their enemy, because they know how to make themselves obeyed.

In the normal cycle of Chinese history the stages of anarchy, invasion, and migration, which we have just been discussing, are followed by a stage of progress. In the case of the Manchus we shall say no more about this latter stage because we have already dwelt on it for the purpose of emphasizing the great capacity of that northern dynasty. There remains for discussion therefore only the stage of decay. The facts as to the recent decay of China are too well known to need repetition. When the Manchus first ascended the Dragon Throne in the middle of the seventeenth century, China felt a wave of energy; this seemed to culminate toward the end of the eighteenth century when Chinese sway had reached its greatest modern expansion. China was then "The Middle Kingdom," the center of the Chinese world. But even at that time degeneration was already under way. I shall not dwell on it, for every one knows that it has now gone so far that China is politically impotent. Worse than this, she is, or at least has seemed to be, almost asleep. To-day the constant questions among the well-wishers of the Flowery Kingdom are: When will China wake up? Does the old ability still exist? It certainly seems to be still existent among the Chinese leaders, for intellectually they compare favorably with those of other races. Moreover, many of those leaders are actively and wisely trying to awaken China. In some lines, such as education, they are certainly succeeding, for in few countries is the desire for modern learning keener than in China. Yet, somehow, it still seems to the sympathetic foreigner that there is stagnation.

The China of to-day, to be sure, is by no means the China of fifty years ago. But the Japan of to-day, the Europe of to-day, and the America of to-day are perhaps still less the same as those of fifty years ago. Has the change in China been as great as in the other countries? How much real change in fundamental character, regardless of external appearances, has there been in any of these regions? Are the other great nations progressing so rapidly that in spite of some progress on her part the difference between them and China is increasing rather than diminishing?

I cannot answer these questions. The answer depends partly upon whether China is still in the stage of degeneration which has often formed so sad a part of her historic cycle. It is a common habit to attribute China's retrogression, and that of almost every other country, to a decline in the caliber of her rulers and leaders by reason of wealth, luxury, bad training, and vice. That these factors played an important part among the Manchus can scarcely be doubted, as is made clear by Backhouse and Bland in their *Annals and Memoirs of the Court of Peking*.

It seems to me, however, that certain great biological facts are of even deeper import. The significance of the intrusion of able people from the north has been realized by many thoughtful students of China; but few have appreciated the fact that after a process of selection has given a country an unusually able dynasty, the absence of further selection and the prevalence of intermarriage with stocks less highly selected causes a tendency toward reversion to the normal type. Still fewer realize that under certain conditions which frequently prevail in China there may be a rapid elimination of the more able people, especially where overpopulation and famine prevail. One of the greatest factors in the decline of China seems to have been the gradual deterioration of the people of north China as a whole and the increasing premium placed by famine and overpopulation upon passive qualities of economy, industry, patience, and endurance rather than upon active leadership. When a cycle of Chinese history culminates in especially severe famines, internal anarchy, and barbarian invasions an enormous extermination of the less competent parts of the population takes place. The remnant who survive or who migrate back into the depopulated areas, and whose descendants repeople the country, are apparently of distinctly greater ability than the mass of people who swarmed in every corner during the stages when degeneration was gradually taking place. Moreover, because the population is relatively scanty, these capable people have unusual opportunities—the opportunities that are characteristic of a new land. They likewise tend to be relatively well-fed and healthy, for overpopulation usually means poor food and poor health. When at last a cycle turns full swing and the land settles down under new leaders selected for their ability, the relative competence of the masses of the people and the opportunities due to freedom from the curse of overpopulation make it possible for the genius of the leaders to inaugurate a period of progress.

Thus the cycles of Chinese history run their course. Again and again they are interrupted by minor cycles and by all sorts of accidental events such as the birth of men of genius, the introduction of foreign ideas, or contact with new people. Nevertheless, the cycle seems to be a fundamental fact in Chinese history, and likewise in that of many other countries, although of course in greatly modified form. It seems to be so important, and also to be so little understood that I shall explain it in detail as it applies to China. The part that especially requires explanation is the decline of ability which begins even while progress is still being made, and which culminates in anarchy and wholesale depopulation.

CHAPTER XI

NORTH VERSUS SOUTH IN CHINA

AT the present stage in the last cycle of Chinese history one of the most notable features is the anomalous contrast between the north and the south. This contrast seems to show the results of degeneration in north China under the influence of adverse natural selection so clearly that I shall attempt to explain it fully. In the course of the explanation we shall find facts that bear on the whole course of the historic cycles. One of the strangest facts about China is that the northern and southern parts of that country invert the usual rôles. In most parts of the world a region in low latitudes is less progressive than a corresponding region in higher latitudes, provided the high latitudes are not so cold that life becomes difficult. This is true in Europe, western Asia, India, and North America. It is equally true in the southern continents, where Argentina and Chile lead South America, and South Africa leads Africa. But in China the opposite is true: the south is progressive and the north backward. Except where Europeans have recently settled, no region within twenty-five degrees of the equator probably shows so much real progress as south China, a fact which goes far to show that the Chinese are one of the world's most able races. But between thirty-five degrees and forty degrees from the equator there is perhaps no region save central Asia so backward as north China.

The Chinese themselves have long recognized the difference between the south and the north. The people of the south, they say, are fond of travel, quick to grasp new ideas, eager to learn, easy to persuade, ready to change their habits, fiery in action, and prone to adopt radical ideas in politics. Those of the north are fond of home, slow to accept new ideas, steadfast of purpose, hard to convince against their better judgment, tenacious of old habits, slow to act but very sure, and prone toward conservatism in politics. Those who are not so friendly to the Chinese say that the southerners are clever, but unstable, unreliable, and likely to lose their heads, while the northerners are unutterably slow and stupid, but relatively honest.

No one can study China with care without seeing that there is much truth in these statements. It was the south which led the revolution in 1911, and forced the retirement of the emperor. In its radical zeal the south imposed a republican form of government upon a country which is supremely unfit for such a form by reason of its almost complete absence of any feeling of public responsibility. Yet so far as there is any such feeling of responsibility, it is largely found in the south, or at least among people of southern origin. On the other hand, it was the north which supported the old regime. The northerners tried to put the emperor back on the throne, and failing that they converted the republic into the most conservative type of oligarchy. Of course the south, also, is highly conservative according to Western standards, but here we are speaking in relative terms in order to bring out the contrast between the two sections. Speaking in those same terms the south is rich, extravagant, pleasure-loving, generous, and immoral; while the north is poor, frugal, serious-minded, miserly, and moral. But these words must not be interpreted in terms of our civilization but of that of China.

This contrast between the two sections shows itself in many ways. For example, the southerners, especially the people of Canton and the surrounding coastal districts as far as Fuchow and even Chekiang are prone to travel. In the aggregate millions of them have gone to foreign countries and are found in Singapore, the Malay States, Java, Formosa, America, and many other parts of the world. The northerners migrate somewhat, to be sure, but only to Manchuria and Mongolia, where they are still practically in their own country and can return home at frequent intervals. Moreover, the southerners engage in large business affairs both at home and abroad, which is rarely true of the northerners. In the Malay Peninsula, Java, Formosa, and other parts of the East Indies many of the leading business men are Chinese who own plantations of sugar, rubber, tea, coffee, and the like, and who often have large interests in tin mines, sugar-mills, and other profitable ventures.

A conversation which I had at Shanghai with a Chinese business man illustrates one phase of the matter. He was born near Canton, lived there till the age of fourteen, and then went to Australia with one English pound in his pocket for spending money. For twenty-five years he did business in Australia, but did not like it because no matter how successful he might be, he was still a Chinese and was not treated with respect or on terms of equality by other business men. So he came back to China and established a large store which has a high reputation among foreigners. It sells all sorts of goods, both foreign and Chinese, at fair prices and of standard qualities. When this merchant started his store his more important assistants, such as heads of departments, were picked indiscriminately regardless of where they came from, and consisted about equally of men from the Shanghai and Canton districts. But little by little a change has taken place, and the great majority are now Cantonese. The reason? Simply because the Cantonese, so the merchant says, are more competent and at least as reliable as the others, although honesty is not a strong point in either group. So far as that particular quality is concerned, the people of Shantung are more honest than either the Canton or Shanghai people. But why are none employed here? They are too slow and stupid. And then this merchant, who is successfully employing European standards of fair dealing and uniform quality of goods, went on to deplore the fact that none of the Chinese seem to care for anything except what they personally can get. They do not even think far enough into the future to realize that strict honesty and devotion to public as well as private good would pay them personally in the long run.

Another evidence of the greater activity of the southerners than of the northerners is seen in the Chinese leaders whom I was fortunate enough to meet. I tried to meet as many as possible, but it did not at first occur to me to make an analysis of where they came from. Among seven Chinese leaders with whom I dined or otherwise spent considerable time in Peking, one was from that city, but was a Manchu, five were from the coastal regions south of the Yangtse River, and one I am not sure about.

A much better method of testing the matter was suggested to me by Mr. Price of the American Embassy, who lent me an official list of the higher metropolitan and provincial officials in October, 1910, just before the revolution. From this list I have taken all the men who held the Chinese degrees which are translated as "licentiate," "provincial graduate," and "metropolitan graduate." This of course excludes most of the military officers who were appointed by favor and not by means of the rigid old-fashioned Chinese examinations.

In addition to Manchus (50), Mongols (15), and Chinese Bannermen (14), who were surprisingly numerous among the degree-holders, there were 324 from the eighteen provinces of China proper. The distribution of these 324 is extraordinarily uneven in proportion to the population of the various districts whence they originally came. In Kansu, for example, there was only one high officer for every 10,000,000 inhabitants, whereas in Chekiang, the coastal province south of Shanghai, there were 39.7. A glance at the accompanying map (Figure 5) shows that among the six northern provinces only Chihli, in which Peking is located, gave rise to more than 7.1 of these highly educated officials per 10,000,000 people. Among the six southern provinces, on the contrary, none had less than 9.6, while among the central provinces the three on the coast had very high percentages of leading officials, while those in the interior had low percentages. Chihli is probably somewhat abnormal because many able men from other provinces have come to Peking from time to time during the course of the centuries, and thus have given that city a group of competent people such as do not exist in the outlying parts of the province. Barring the exceptional case of Peking, Figure 5 seems to suggest that in general the kind of ability which enables men to pass the severe Chinese examinations of the old style, and then to obtain high official positions in the face of keen competition, is least common in the northwest and increases toward the southeast. Of course a much larger body of facts is needed before this can be regarded as proved, but inasmuch as the place of origin of the officials agrees with a vast amount of evidence of other kinds, we may feel fairly sure of our ground.

The difference between the south and north of China is obvious in many other ways which are detected much more easily than is the distribution of men of ability. For example, in the streets of Canton one sees many women of all ages. At certain hours the streets are crowded with school girls as well as boys. Very pretty they look, with their loose white waists, black divided skirts coming a little below the knee, and white stockings and low shoes. Their fair round faces, only faintly yellow under the smoothly combed black hair which hangs down in a long braid behind, are often pretty, and show quite charming dimples when they smile. As they walk along and talk together one notices that they smile frequently. Among the older women, also, the majority are quite light in complexion, unless tanned by the sun, and have on the whole the appearance of leading lives that are by no means wholly unhappy. Many, though not all, of the older ones toddle along on the pitiful stumps of what would have been feet if they had not been cruelly bound in childhood. Among women perhaps thirty to forty years of age many have remarkably small feet, but not the little stumps of the older women. They were fortunate enough to have their feet unbound while still girls. But among the younger women not one has unnaturally small feet.

Look now at the Cantonese men. How many have queues? You will have to hunt some time to find one, although a few still survive. How about rags? Practically none. Beggars? Only a few, and those not especially persistent or disagreeable. The great majority of the people look relatively well fed, comfortable, and happy—at least according to Chinese standards. They walk

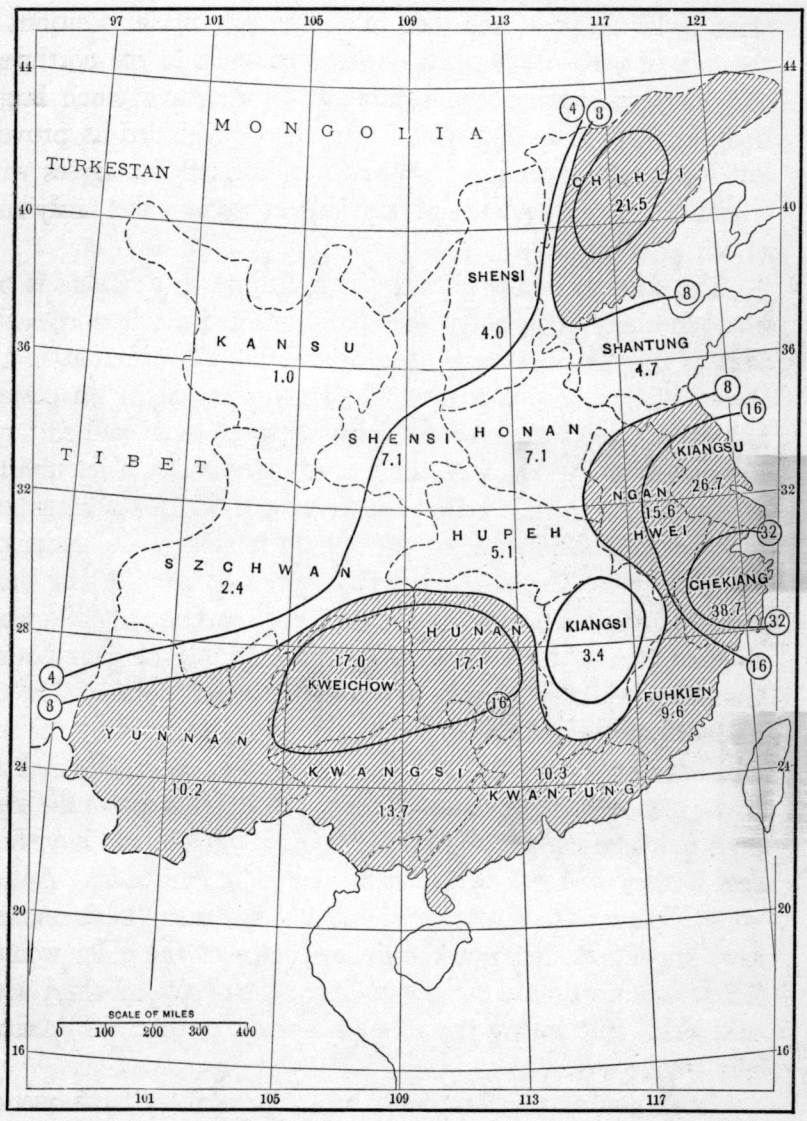

FIG. 5. BIRTHPLACES OF HIGH OFFICIALS IN CHINA.
Number of holders of high literary degrees among officials in 1910 per 10,000,000 inhabitants.

alertly compared with most of the people who live in equally low latitudes, for Canton is within the tropics. And, although they work slowly and spend much time in sitting down and waiting, and still more in apparently deliberating how they shall work, they are a remarkably industrious people in view of their climate. The women, as in so many tropical countries, seem to work harder than the men, and one sees them manning the innumerable boats, rowing hard against the tide, or walking under heavy loads balanced at the ends of poles across their shoulders.

Now come to Tsinan, the capital of Shantung. Scarcely one woman is seen where ten were visible in Canton. No such groups of pretty school-girls go laughingly along the narrow streets. The women of Shantung are kept secluded: the home is still considered the place for them by the conservative men of Shantung. Those whom one sees on the streets, whether rich or poor, are generally of darker complexion than are the Cantonese, as is true of the men also. And the women and girls are not so pretty or bright-looking as their Cantonese sisters. The American ladies in the missionary schools at Peking say that on the whole the fairest, prettiest, and most alert Chinese girls, although not necessarily those with the best minds, come from the south, or at least from families which have migrated to the capital from the south. Moreover, the women whom one sees in Tsinan practically all have tiny feet, and many of the girls, even the little ones, are still being tortured in this old-fashioned way. As for queues, as soon as I thought of the matter I began to count. Among the first hundred men whom I met exactly half had braids down their backs or coiled under their hats, while the other half had short hair. If I had gone into the villages, so I was told, I should have found 90 per cent of the men wearing queues. They are certainly conservative, these people of Shantung. And yet, curiously enough, they are physically far finer people than those of the south. They are big men and women with fine sturdy frames and with the appearance of strength and vigor. No, not vigor, for only a part appear vigorous. One sees in the northern streets many more signs of disease and suffering than in the south.

Rags, too, become a noticeable feature, and beggars a perfect pest. Go to a place like the famous Sacred Mountain of Tai Shan, where is the temple of Confucius, a few score miles south of Tsinan. The beggars swarm like flies: they follow one for miles up the great rocky road that leads to the mountain-top; they crowd upon one, displaying their sores, pressing their stomachs in a ghastly fashion to indicate their hunger.

It certainly seems anomalous to find these conservative people in a province that once was the very heart of Chinese progress. Previous to perhaps 300 B. C., Shantung was the home of Confucius and Mencius, the center from which culture radiated in all directions. So far as climate, diseases, and man's physical health and vigor are concerned, Shantung, with its cold bracing winters, seems to have a distinct advantage over Canton with its many months of damp heat during the summer. The foreigners who live in the northern parts of China sometimes have hard things to say of the dust-storms in winter and the damp heat in summer, but on the whole they rarely complain much of the climate, while those who live from Shanghai southward are always talking about its debilitating effect.

Another curious fact about China is that although the southerners are more progressive than the northerners, the northern cities have become more Europeanized than the southern. Tientsin and Peking not only have large areas where the streets are quite European in aspect, but enjoy the advantages of electric tramlines. Most of the other northern cities have a considerable network of modern roads where jinrikishas and automobiles can ply freely. But in the south some ports as important as Fuchow and Swatow have only a few miles of streets fit for motor-cars or even jinrikishas. In Amoy, up to the time of my visit in 1923, there was not a single wheeled vehicle except a few garbage-carts. One reason for this difference between north and south is said to be that southern officials have come to the north and have been open to foreign influence. It is certainly true that southern officials have held most of the important positions in the north aside from those held by the Manchus and Mongols, but that has also been the case in the south. I am inclined to think that the north owes its more rapid Europeanization in part to the fact that China is to-day undergoing an invasion which in some respects is like the old barbarian invasions. The invaders possess much more than the average ability of their respective races, and they enjoy better health and hence have more energy and initiative in the north than in the south. In other words, among these chosen invaders the conditions are just as one would expect on the basis of climate; the north progresses more rapidly than the south. But among the main mass of the people other factors cause the south to go ahead of the north.

The chief of these other factors is commonly said to be the longer contact of the south than of the north with Europe. Almost every one whom I asked in China expressed this view, and one can find it in books by the score. I doubt its truth, however, for several strong reasons. First, if the duration of contact with Europe were the cause of the relatively progressive character of the southern Chinese, why did those same Chinese spread into other lands and show their present spirit of enterprise long before they ever came in contact with Europe, as we learn from many old Chinese sources? Again, the people of Ceylon have been in contact with Europe far longer than have the southern Chinese, but they are not so progressive and competent and do not travel so widely and actively. On the other hand, the Japanese have not been in contact with Europe nearly so long as the southern Chinese, but they are far more progressive. Again, in Manchuria, especially the northern part, the Chinese are said to be the most progressive and active in any part of China. Even at Mukden I felt an atmosphere of life and activity such as I did not feel in any other Chinese city. Harbin is said to show the same thing to a still higher degree, and the regions north of that most of all. But these regions have had little contact with Europe aside from Russia, and Russian influence amounted to little until the Manchurian section of the Trans-Siberian railway was completed in 1901. If we compare the progressiveness of Ceylon, the Malay peninsula, south China, north China, Japan, and Manchuria with the length of time they have been in contact with Europe, we find no consistent relation between the two conditions. European influence can scarcely account for the difference between the north and the south of China.

When I persuaded people in China that this last statement is true, they generally tried to explain the peculiarities of the two parts of China by racial mixture, the southerners being supposed to have Malay blood and the northerners more Mongol and Manchu blood. There is certainly some truth in this so far as the facts of racial mixture are concerned. In the south one sees many Chinese with broad, flat noses and an almost negroid appearance. Many also are quite dark as if with Malay blood or perhaps an admixture of Dixon's Proto-Negroid or Proto-Australoid races. But nobody has produced any evidence to show that such a mixture causes mental alertness. The Chinese in whom these departures from the normal type are strongest are the peasants and coolies. The upper classes, those who give character to the country, are quite light, lighter than the people of the north. As I looked at them, especially the clerks and the women and girls who had not been tanned by the sun, I repeatedly said to myself: "These people are not really colored. They are scarcely even yellow. Look at that pale face and those cheeks with pink in them." Apparently the part of the population which gives southern China its distinctive character is the most purely Chinese.

On the other hand, it is undoubtedly true that the northern Chinese contain more Mongol and Manchu blood than do their southern cousins, and the amount of such blood may be considerable. But even if there is much Mongol and Manchu blood there is no assurance that this would make the northerners dull and stupid. On the contrary, the invading Mongols and Manchus, as we have already seen, are relatively capable and vigorous people, as appears, for example, from the number who have received the literary degrees among the officials in the list discussed above. Even at the supposedly Chinese game of studying and remembering ancient literature the Mongols and Manchus are by no means to be despised. Hence, while I do not question the importance of racial mixture in altering human character, I do not see any clear proof that this is the reason why the south of China is relatively progressive, the north backward, and the far north in Manchuria again progressive.

Another strong argument against the idea that southern China owes its peculiar character either to European influence or to racial mixture is found in the Hakkas. The Hakkas are a peculiar people, purely Chinese as that term is now used. They live to the number of 10,000,000 or more in the rugged mountains one or two hundred miles northeast of Canton and west of Fuchow. Their name means "guest people" or "strangers." They are also sometimes called by a term which means "wild old women," the word wild being used in the sense of barbarous. If you ask a coolie or even a merchant in Canton or Swatow about the Hakkas he will probably say that they are a barbarous, degraded set of people, little better than bandits, who take no care of their women and who are a danger to every one else. But inquire further, and you hear that in Singapore, Siam, Formosa, and the Dutch East Indies the Hakka coolies are considered especially competent. In Canton the native coolies fear them because they come down from their upland home and supplant the less competent Cantonese. The Hakka merchants come down likewise. In Swatow there are said to be about 15,000 Hakkas, and this number includes many of the most prominent business men. A similar condition prevails in Canton. Across the seas the same is true, for they are great wanderers. The leader of the great Taiping rebellion was a Hakka, as the general who seems to be squeezing Sun Yat Sen out of Canton while these words are being written is another Hakka. The fact is that the Hakkas are an uncommonly able people who little by little are moving out of their mountain home and displacing the coastal people to the south and east of them. That is why they are feared and reviled.

Talk to the Europeans who have lived among the Hakkas. It was my privilege to listen to several missionaries who have been among them—and missionaries are almost the only foreigners who really know the Chinese in their own homes. All these missionaries were essentially agreed. One of them was Mr. Spiker, who has lived among the Hakkas and also in Swatow and Shantung, and has travelled widely. "The Hakkas," said he, "are the cream of the Chinese people." He and the others back up this opinion with good reasons. For instance, the Hakkas are almost the only real Chinese who take daily baths and have never practised foot-binding. Their women not only are unusually pretty, but are held in unusual respect. The freedom which they are allowed is presumably interpreted by the Cantonese as lack of care for them. This does not mean that their standards of morality are like ours, for that is true nowhere in China. Moreover, in south China there is much greater laxity than in north China, which seems to accord with the conditions of climate. Among the Hakkas, the point to be emphasized is merely that the women are comparatively free and influential, and are not ground down by the terrible laws of custom so strenuously as in other parts of the country.

Another notable fact among the Hakkas is the prevalence of education. In their central district it is said that about 80 per cent of the men can read and write, a greater percentage than in any other part of China. At the present time in their main town of Kia-Ying, which is only a little place with 10,000 or 20,000 people, there are, if I remember rightly, some 3,000 pupils in what are known as middle schools, that is, above the primary grades. At any rate they gather there from the little villages for many miles around to such an extent that almost nowhere else in China do so many pupils come from so small a population. Again the Hakkas,

more fearless and self-reliant than the town dwellers, have all the love of liberty which characterize mountaineers the world over. They were the last to surrender to the Manchus and twice strove to throw off their yoke, first under the Taiping chief and again [in the present century].—It would not be easy to find an inland district where the people are as well housed as they are in Meichau (Kia-Ying). The artisans of Hing-Ning are as skilful as any in China. (Campbell.)

The Hakkas, as we have seen, are a purely Chinese race, practically unmixed with other elements, unless it be with an early Mongolic element long ago in north China. They live remote from the sea and far from European influence. Yet among them, more than among almost any other group of Chinese, one finds the highest development of those qualities which cause south China to be more progressive than north China. Since neither racial mixture nor contact with foreigners can be called upon to explain the high development of these qualities, it seems illogical to call upon those conditions to explain the lesser development of the same qualities in Canton and other coastal regions a few hundred miles away. Moreover, in northern Manchuria among people who originally came from backward Shantung and Chihli we again find a remarkable development of these same progressive qualities. There, too, neither racial mixture nor foreign contact seems to offer an explanation. In the next chapter we shall consider the one great factor, a beneficent natural selection, which appears to explain the qualities held in common not only by these two sets of people, but by the other progressive people of south China, and perhaps by the Chinese of Shantung in the golden age of Confucius, six or seven centuries before Christ.

CHAPTER XII

THE SCOURGE OF FAMINE

THE part played by famines and economic distress in Chinese cycles of progress and decay has been suggested in a previous chapter. These same conditions appear to play a dominant part in the contrast between the progressive south and Manchuria on the one hand, and the backward north on the other. The famines and distress have apparently played their part by selecting certain types of people for destruction, preservation, or migration, thereby leaving an incompetent residue in many parts of the north and sending competent people to the cities where they die off, and to Manchuria, south China, and other outlying but relatively progressive provinces. In this chapter, let us see how the famines exert their selective influence, leaving their effect upon character to the next.

Terrible famines are common in north China, but rare in Manchuria and almost unknown in south China. In north China the rainfall is concentrated largely in a few summer months; a delay in its arrival means that crops can begin growth only in fields that are irrigated. If growth is delayed till July, as often happens, and if July is dry, practically nothing can be reaped from the unirrigated fields. This practically never happens in the south, where the rainfall is less strictly concentrated in the summer months, and only rarely in Manchuria, where the temperature is lower so that evaporation is less, while the chance of showers in the spring and autumn is greater than in north China.

In north China great floods are another common cause of famine. They are likely to occur in the same regions and at the same general periods as droughts, for a dry climate commonly has torrential rains when the long droughts finally break. The high barren character of the mountains, their denudation of forests, the torrential nature of the rains, and the extraordinary flatness of the vast plains all combine to increase the ravages of floods. Nowhere, during many years of travel, do I recall a more vivid impression of the absolute levelness of the plains. For hundreds of miles scarcely a billow breaks the dead monotony. When the rain falls heavily day after day, as often happens in midsummer, the water may form a vast sheet a foot or so deep even without the help of the rivers. But the rivers are likely to break loose and add to the disaster. They are heavily loaded with mud because of the steepness and barrenness of the mountains, and many of them flow on a level with the plain or even above it, being kept in place by dikes. Only a moderate flood is needed to cause them to spill over and carry devastation and famine for scores of miles on either side.

Sometimes human selfishness makes the floods still more disastrous. North of Tsinan in Shantung, for example, some roughly parallel canals have been built at intervals of perhaps ten miles to carry surplus water from the Grand Canal to the sea during floods. In normal times the canals are not needed and their heads are closed. Such strips of good soil are very valuable, for farm land is worth perhaps two hundred gold dollars an acre, which is as much as one or two thousand dollars would be with us. Naturally the officials do not overlook such an opportunity for filling their pockets. They cultivate the canal bottoms, and make a very good thing of it. When a flood comes, they are not at all disposed to give up so great a source of profit, especially as other sources are sure to diminish. So for centuries they have done what they did in 1921, namely, keep the heads of the canals closed. The inevitable result is that the Grand Canal overflows its banks in many places and floods the surrounding plain. Thus within the canals the official fields are protected by dikes, while in the broad surrounding areas, a hundred times as extensive, vast shallow lakes are formed which kill the crops of hundreds of thousands of wretched villagers. Such an event is thoroughly Chinese, for one of the greatest weaknesses of China is selfishness: yet curiously enough that same selfishness, as we shall see, seems to be an inevitable means of self-defense against the famines.

Where the rivers break loose, the floods not only spoil the crops but ruin the mud villages. One would think that in areas liable to flood all the houses would be built on mounds high enough to be safe no matter how much the water might rise. But that is not the case even in our own country, and even less in China. Consider the circumstances. Floods are frequent. When the people come back after a flood they are desperately poor, the land must be restored as we shall shortly see, the crops must be planted, and new houses must be built. Every moment is precious. There is no time to build mounds for houses; that would take much labor, especially as the famine-stricken people rarely have any beasts of burden to help them. Doubtless they tell themselves that when a happier day arrives they will build new houses safely perched on mounds, but that day never comes; and after a year or two without another flood a false sense of security begins to prevail.

Another little item deserves mention in this connection. The Chinese plains are so densely populated and the people are so near the brink of starvation even in ordinary years that few of the peasants can afford to lose a single square yard of their fields. To build mounds for houses would involve stripping the fields of soil or else digging holes so deep that they would be filled with water much of the time. In either case good land would be wasted, and that would mean hungry mouths. We who live in a land where there is always a large surplus, have not the faintest idea of the miseries and the constant anxiety that result from overpopulation and a low standard of living.

Let us picture what happens when the rains descend, the plains become water-logged, and the rivers burst their banks. Little by little the water rises around the villages. By the time it is a foot or two deep all hope for the crops is gone. Then the houses begin to be in danger. Made, as they generally are, of mud bricks that have merely been dried in the sun, the walls soon become soft. Then it is only a matter of days before the

THE SCOURGE OF FAMINE 173

heavy, thatched roofs press the mushy walls out of shape and the whole house—the whole village—collapses. Such utter destruction of villages is almost the rule in most of the areas subject to floods by the rivers. Dikes around the villages sometimes help for a time, but the inertia of the Chinese allows such dikes to fall into disrepair. Moreover, that selfishness which is so strongly developed among them makes it difficult to secure cooperation in the attempt to keep the dikes in repair when there is no immediate threat of flood. Thus in flood after flood, for hundreds of years, millions of people have lost their crops, while hundreds of thousands have also lost their homes and all their worldly goods. Practically all of them ultimately lose also their animals, for they have to sell them to buy food.

Many of the millions who are thus in danger of starvation become "wanderers." That word looms large in China to-day, as it has for two thousand years. Those who have lost houses as well as crops become wanderers at once. Those whose houses are safe may have some store of food remaining and may also be able to salvage a little from the drowned fields. Hence they stay at home at first, although winter sees many of them also on the road. In many cases the wandering is prolonged. After the fields have been flooded they do not dry quickly. Many remain wet until the dry season is well under way in the autumn, and then of course it is too late to plant any crop, even cabbages, before cold weather sets in. In this fact lies one of the reasons why famines are so much worse in the north than in the south. From the Yangtse valley southward some kind of crops can be raised even in winter.

Even when spring arrives the sufferers from flood in north China are often unable to raise crops. In many cases the standing water dissolves the large percentage of alkali which the soil contains. When the water finally dries up, this is left as a whitish deposit on the surface. People with higher skill might devise a way to wash it out, but to the Chinese it is an insuperable obstacle. The only remedy is to wait until nature redistributes the alkali and once more makes the land fit for cultivation.

Hence the people whose land becomes alkaline may be forced to remain as wanderers for years. Still another group shares the same fate. These are the people whose lands happen to lie in the main path pursued by a river as it breaks its way through the dikes and flows off in a new course. After the flood has abated, such a course is often marked by a broad band of newly deposited fine sand which may have a width of several miles in the case of the Hoang Ho. Sometimes the sand is only inches deep, but elsewhere there may be feet. In no case is it fit for the great majority of crops. If the sand is only a few inches deep it may be turned under and mixed with the old soil. That is a very laborious process, especially for people without domestic animals. Moreover, when it is completed the soil is not so rich and productive as formerly, and hence can feed fewer mouths. In other cases a greater depth of sand forces the returning peasants either to dig it off bodily from part of the fields and deposit it upon other parts, or else to plant fruit-trees or some other relatively unproductive crop which will grow in the sandy silt, but which yields relatively little food. Hence some mouths are unprovided with food, and people must still wander unless they have already died. Such injury to the fields by floods is by no means confined to China, although elsewhere it rarely produces such dire results. For example, Mr. R. V. P. Coleman tells me that in the potato-growing Kan valley, in which flows the Kansas River west of Kansas City, the spring floods sometimes deposit a layer of sand so deep as to spoil the potato-fields for two or three seasons. The sand may cover the bottom-lands to a width of five or ten miles.

The great scourges of overpopulation and famine seem to have prevailed in China intermittently for over two thousand years. A little tale told by Doctor Wilder of Peking may give some inkling of how the pressure of population weighs China down to-day. Not far from Peking a considerable number of villages do not expect to raise food enough to support themselves more than perhaps nine or ten months even in good years. Their land is not sufficient, or, to put it in another way, they

THE SCOURGE OF FAMINE

have allowed their numbers to increase too rapidly. After the crop is harvested in the autumn these wretched people complete all the preparations for the agriculture of the next year and then prepare to leave home. In each house is placed enough grain for seed, and enough to support life while the spring work is being done and before the earliest crops can be relied upon for further food. When all the rest of the food is consumed, the houses are sealed up with mud bricks filling the doors and windows. Then the houses are simply left unguarded. The villagers fare forth in the bitter winter weather to other more prosperous towns and villages. They would work if they could, but there are twenty men in China during the winter for every job, no matter how trivial or disagreeable. They have no alternative except to beg, steal, rob, and even commit raids, or else to migrate. But for this last alternative they seem to be too conservative. Hence they eke out a miserable existence until spring, and then go back to their villages, sure that they will find the houses unopened. That is the most curious feature of the story. So common is this degree of poverty, and so well recognized is the need of wandering, that it is a point of honor in all that region not to break into a sealed house. You may steal on the road, you may enter a house that is occupied, you may dig through walls in other places, and you may commit banditry, but you must not enter a house that is sealed. I tell the story as it was told me by Doctor Wilder, a man who knows China as only a missionary of many years' standing can know it. He adds that an investigation during a recent famine made it seem probable that the people of these villages are as a whole subnormal mentally. They are little more than morons, apparently. This is not surprising. Even without the adverse selection which I shall shortly mention, the constant underfeeding from earliest infancy must stunt them in mind and body.

Such conditions are nothing new. Here is a quotation from one of the many authors, cited in her *Economic History of China* by Miss Mabel Ping-Hua Lee, who, by the way, is one of the few Chinese women who have taken the degree of doctor of philosophy in America. The following passages are given substantially as they occur in her excellent doctor's dissertation. The first one dates from the year 1530 A. D., as nearly as I can ascertain, and was written by an official in Shensi who made a journey to Anhwei a hundred or more miles west of Shanghai, and then to Yunnan in the far southwest. From the seventh month onward, he says, that is, from the time when the early crop is usually harvested,

> I saw that the rice [millet?] crop had been almost entirely eaten by worms, especially in such places as Shensi and Honan. When I passed through Tung Kwon (in Shensi) I noticed that there was no late rice left for harvest. The wandering people filled all the roads. Once in a great while I saw people who were harvesting something. Quite delighted, I made inquiry, but found that they were simply harvesting a kind of grass. There are two kinds of grass, one called Mein Poon (soft grass) and the other Tzu Poon (thread grass). Both can be used to make noodles, and the hungry people have depended upon these grasses as food for five years already. I tasted them myself, and found that the grass stung my mouth, upset my stomach, and caused discomfort for days. How can we possibly describe all the bitter and exhausting conditions which the poor (literally, the small people) have to endure!

The famine here referred to was presumably due to drought rather than flood. That is the commonest cause of famine in Shensi. Moreover, as a rule, it is only the dry famines which last year after year. The famines due to floods are much more complete and severe while they last, but the same district is rarely afflicted two years in succession. Moreover, during such years the fields in the neighboring areas too high to be flooded are likely to produce good or even unusual crops. On the other hand, the famines due to drought are rarely so complete as the flood famines and do not destroy the houses or injure the land. But they often come year after year and affect vast areas, as in the case cited above, where the famine had already lasted five years. The most unfortunate places are those where famines due to drought and flood alternate, and that unfortunately has been the fate of considerable portions of north China. That is

THE SCOURGE OF FAMINE 177

why China has probably suffered more from famine than any other country in the world, not even excepting India.

In order to appreciate the effect of famines upon racial character it is necessary to realize how numerous and how devastating these terrible scourges have been. One of the worst series of early famines occurred about two hundred years before Christ. At the beginning of the Han dynasty, according to Miss Lee's translation:

> One load of rice was sold for 5000 (cash?). Human flesh was eaten and more than half the population perished. Then Kau-tso (the first Han Emperor) allowed the people to sell their sons in order to go and get food at Tso Hou (Szechuan).

A hundred years later, after various other famines, the arable land had largely passed into the hands of rich landowners, just as happened in Italy at the same time for the same reasons. Each period of famine caused many of the poor to mortgage their land, and ultimately the rich "ate it up."

> As a result the rich possess land from field to field, while the poor have not even enough to accommodate the point of an awl.—At this time Shantung suffered famine from the Yellow River trouble, the harvest being poor for several years. Sometimes the people even ate one another. This was prevalent over an area of two or three thousand li (700 to 1000 miles). The Emperor pitied them and permitted the hungry people to travel to get food from places along the Kiang and Wai Rivers. And they might stay in these places if they so desired. There was a great flood, and the people who died from starvation in Kwang Tung (Shansi, Shensi, Chihli) amounted to thousands. So an edict was issued ordering the transport of grain (millet) from Pah Sauh (Szechuan) to Kiang Ling (Hopeh), and the hungry people were allowed to go to get provisions in the region of Kiang Wei (Yangtse and Wei Rivers—Kiangsu, Anhwei, Kiangsi, Honan, Hupeh, etc.).

Half a century later, that is, about 48 B. C.:

> On account of a poor harvest, the suffering people were exempted from taxes. Also ponds, lakes, gardens, reservoirs were lent to the poor. [This suggests that the famine was due to drought.] On account of a great flood in Kwan Tung (Shansi, Honan, Shantung, Chihli) grain and money were transported from other countries to save the people in Kwan Tung.

The accounts of what happened about 160 A. D. throw an interesting light on the conservatism which by that time had begun to be prominent among the Chinese peasants.

> There was a famine in the state of Tsi (Shansi and Chihli) and the people ate human flesh. So the different states and counties were ordered to relieve the poor and weak.

A government adviser at that time said:

> At present in the states of Tsing, Hsu, Yen, and Chi the population is dense and the fields are narrow, so that the people cannot get sufficient provisions. But [in certain other regions] the population is sparse and many fields are left vacant and uncultivated. As a rule the natural tendency of the small (poor) man is to be satisfied with his native land, and to weigh carefully the advisability of moving. He prefers to suffer hunger where he is, rather than go to some more happy land. So the government ought to remove to the unoccupied country those poor people who cannot make a living by themselves.

This little comment on the unwillingness of the Chinese peasants to change their homes should be remembered. It helps to explain their present poverty and backwardness.

Time and again in the old records one finds accounts of other hideous practices as well as cannibalism. About 297 A. D.:

> There was a famine in Kwan Chung (Shensi) and the price of rice was as high as 10,000 cash per load. So an edict was issued that those wishing to sell their relatives would not be prohibited by the government.

The pitiful thing about it all is that the same troubles keep recurring. In one of the hundreds of accounts of famine we read that about 575 A. D.:

> Many people, carrying the young and supporting the old, wander around with grass shoes. Having already lost their own occupation [presumably because they have sold their land under the stress of famine], they have fallen into the class of people without work. And with the coming of famine and plague, they cannot but wander and leave their native places.

Shortly afterward, in 616 A. D., a brief period of prosperity and of increase in population came to an end, and

people began to leave their work and flock to the cities, and so were unable to provide for themselves. At first people took the bark of trees for food. Then gradually they used leaves. When both of these became exhausted, they cooked soil and bran into flour. Finally they ate human flesh.

Elsewhere we read that when bran, chaff, bark, and leaves were exhausted, the people stayed their stomachs with a kind of clay. For a few days the clay gave a feeling of comfort, but soon those who ate it fell dead. At the end of this period, about 620 A. D., two-thirds of the population had perished (or perhaps migrated) in the areas infested by famine. This, too, like the second century before Christ, was a time when the countries around the Mediterranean also suffered terribly from famine.

Throughout the long and painful record of Chinese famines there is constant evidence of an attempt on the part of the government to combat these dire disasters. Often, but not always, the attempt was prompted largely by the desire to improve the curtailed revenues. For instance, about 700 A. D.:

The wandering people left their native places for a long time; so public income and savings are lacking. The land is left barren and uncultivated, so we must immediately establish relief work to help the farmers to restore their farms. In case there is any unoccupied property which was formerly taxable, or any deserters and fugitives from military stations, all are to be excused and not investigated, and all are to be relieved from taxes. Those who ought to go home but are so poor that they cannot afford to do so will be given provisions for the trip so that they can reach their native places.

No matter what the government might do, however, the rich constantly "ate up" the poor, for every famine gave them a new opportunity. About 990 A. D.:

The shoots of the millet grew several feet long this year, but for a great many previous years many crops did not give good harvests. Rich people speculated, while the poor had to pay double the amount of their debts. Even though there be occasionally a rather good harvest [this seems to have been a comparatively prosperous period], the rich collect their loans urgently.

Thus the earnings of the poor are gone before they can pay their rent and taxes. The government therefore prohibited the rich from charging more than 100 per cent interest on loans of wheat, rice, or money.

If 100 per cent becomes the legal maximum for interest during a relatively prosperous period, imagine the conditions during a time of frequent famine.

Within a century or two after this comfortable period there came another bad period, the worst apparently in Chinese history. About 1050 or 1060 A. D.:

> South of Kiang Wai there was great drought in the spring. In some places even the wells and springs were dried up; oxen and other animals died from thirst, and no chickens or dogs were left. As a result all nine kinds of farms (that is, all kinds) lost their occupation and the people are all complaining of hardship.

Again:

> Even parents, sons, and brothers cannot protect one another. And all the widows, widowers, orphans, and single persons cannot support themselves. All the strong have wandered and moved away; and the weak have died. The reason why conditions are so bad is because of floods and drought; and the reason why there are so many floods and droughts is that the Yin and Yang are not in harmonious relations.

The Yin and Yang are the two great principles of nature, which, according to the Chinese, control all natural events. When they are in harmony the world is at rest; when they are out of harmony all sorts of disasters occur.

From this time onward the conditions in China grew worse and worse for nearly two centuries. The worst is summed up in what seems to be rather a harmless statement dating from about 1185 A. D.:

> The strength of the people in the southeast has broken down and deteriorated, and even in the families of the middle class there are no savings for more than a few months ahead. . . . The states of Su, Hu, Shen, and Show (Kiangsu and Chekiang) used to have floods only once in a while, but now they frequently suffer disasters.

The significance of all this lies in the fact that hitherto the records of famine have all come from northern regions, chiefly in the Hoang Ho valley; now the famine area has expanded, and both drought and flood afflict the provinces around the mouth of the Yangtse and even as far south as Fuchow.

Because of these conditions an imperial officer in 1206 reports that:

The escaping and deserting families have become more and more numerous, and the return of the land tax has become shorter and shorter.

And then follows one of those strangely short-sighted comments with which the pages of history fairly bristle:

Both the government officials and the common people all feel the gravity of the situation more and more each day, and no one seems to know what has actually been the source of the trouble. Some district officers, who are well aware of the general alarming conditions, have the idea of correcting the boundaries of the fields. This of course would make the small families feel somewhat relieved, but the huge (rich) families would feel very agitated about the process.

The government actually tried to restore the fields to their rightful owners, "correcting the boundaries," as they put it. But seven years later we read that:

In case the harvests are very good we might be fortunate enough not to have much trouble. But if the year should be bad, several dozens of the tenants would group themselves together to rob the merchants and farmers in the neighborhood. Sometimes they even murder people, and some such cases are already on record. So I am afraid that these are the great sources of trouble in the years of famine. If we (the government) can clear up these places so as to give them no room to gather together, it will be one of the best policies to stop the evil of robbery and theft in a quiet way.

In a certain sense this official was right. What China needed then and now was *room*. She has far too many people. But in a country which is subject to recurrent natural disasters, it is the common practice to attribute the difficulties to the system of land-tenure. That has been done in ancient Italy, in Ireland, in Mexico, and in many other countries. But so long as hard times and famines recur, and so long as there are many years when the actual products of the farms are insufficient to maintain the accustomed standards of living, no form of land-tenure will ever bring prosperity. Under such conditions the poor inevitably mortgage their lands, their services, or their own selves to the rich, and no amount of lawmaking will prevent it. The only remedies are to overcome the vagaries of nature by means of irrigation, drainage, roads, and ultimately the prediction of the seasons far in advance, and especially to limit the population, so that in good years it may lay by a large surplus to tide it over the bad years that are sure to come. Where a trouble is due to the irregularities of nature, mere changes in land-tenure and taxation are only temporary pain-killers, not fundamental remedies.

How bad the conditions became in China, in spite of all the laws, is obvious from this account of the period from 1208 to 1234 A. D.:

> Although there were natural disasters (floods, etc.) and serious drought in the previous periods, the conditions in those days were never so bad as to-day. At present the granaries and treasuries are empty and the provisions are not enough for a month's supply. One sen is sold at 1,000 (cash?) and the price is still increasing without limit. Even the rich have been destroyed; nine out of ten families have died from starvation; and some have gathered to drown themselves in the Yangtse. In the villages and lanes people gather to criticise the government officials, and the complaints among the soldiers are too severe to be heard. The condition is serious. And it appears right in the heart of the city of great crowds (the capital) [which was then at Nanking, or Hangchow, both being royal residences]. The territory of Chehsi (Chehkiang) is the region where rice is supposed to be collected [it is one of the most productive regions in all China], but now the red land (that is, dry, untilled fields) covers 1,000 li. The people in the Wei region (Anhwei, along the Wei River) are wandering away and deserting their homes continuously, carrying their babies and wandering on the roads. They wish to find homes, but there is no such place for them; so they simply wander until, tired out and exhausted, they wait for death.

I have dwelt on these famines because they are one of the greatest of all facts in Chinese history. But I have not mentioned a tithe of those that are recorded. At no time does China ever seem to have been free from them for more than a few

THE SCOURGE OF FAMINE

score years, unless it be in the very early periods, the golden age. The worst periods seem to have been (1) the third and second centuries before Christ, (2) the fourth and fifth centuries after Christ, (3) the twelfth and thirteenth centuries, and (4) the sixteenth century. Other periods, however, have been almost equally bad, and it is not yet possible to speak positively.

Each great period of famines, as was stated in a preceding chapter, seems to have been accompanied by another great affliction, due probably to the same climatic cause. This other affliction took the form of barbarian invasions from the north. In the first of the periods mentioned above, the invasions were so serious that the Great Wall was built, bit by bit, and finally joined into a single magnificent structure about 214 B. C. In the fourth and fifth centuries the Tartars invaded China again, and swarmed over all the northern portion. During the twelfth and thirteenth centuries the same thing happened once more, ending in the Mongol domination. Finally in the sixteenth century there began a movement of the northern barbarians which culminated in the conquest of China by the Manchus. Apparently each of the great movements of nomads from the north was stimulated if not actually caused by repeated periods of deficient rainfall and consequent distress in those northern regions. Needless to say the southward raids of the fierce desert tribes added enormously to the sufferings which China was already undergoing by reason of its own floods and droughts. Small wonder that in China there have almost always been hundreds of thousands of wanderers, while at frequent intervals millions have suffered for lack of food.

CHAPTER XIII

THE SELECTION OF THE CHINESE

IN every Chinese famine thousands, or often millions of people perish. They may not die of hunger, but they die of something else. During the famine due to the flood of 1921, Mr. Torrey of Tsinan was distributing relief.

"Have any people died of hunger?" he asked in a village where there were at the time perhaps 300 people.

"No," was the answer, "we have food here."

As a matter of fact they had for months been subsisting largely on chaff mixed with just enough grain to support life, but they had "food" according to their definition. Then the missionary put his question in another way.

"How many people died during those ten cold days that we had a week or two ago?" That had been a very bitter period, with high winds and the thermometer near zero.

"Sixty," was the answer. Those people had not actually died of hunger, but they had died of other diseases and of exposure because prolonged undernourishment had sapped their vitality. That is the way it is in every famine. Our problem is to determine what kind of people die under such circumstances. Do any special qualities of character tend to cause death, or to prevent people from dying?

Two other similar questions must be answered. Innumerable people, as we have seen, wander from their homes in every famine. Some die, some settle elsewhere, and some finally come back to the old homes either the year after the famine if they are fortunate and the famine is mild, or many years later if the famine is severe or if their lands have been injured. What kind of people belong to each of these groups? What effect does natural selection have in differentiating the character of those who ultimately remain in the old home and those who permanently find homes elsewhere?

It needs no demonstration to show that the first effect of a Chinese famine is to kill the people who are physically weak. Any organic disease is almost sure to be intensified and ultimately to prove fatal under the stress of insufficient food, exposure, and mental strain. Hence we should expect to find that the Chinese are a peculiarly sound and sturdy race. Such seems to be the case by general consent, especially in the north where, as we have seen, the people of Shantung and the other famine districts are physically much superior to the quicker and more alert people of the south, where famines are rare or unknown. Nevertheless in the north one sees more sores, more deformity, and more minor defects than in the south. At least that is how it appeared to me, and physicians whom I consulted seem to think likewise. This, however, may merely be the result of the malnutrition which is so common, and which must have a most serious effect upon many of the young children.

Another selective effect of the famines would obviously seem to pertain to the qualities of thrift and economy. The family which consistently denies itself in order to lay by a little surplus is far more likely to survive than is the one which uses up all its resources as fast as they become available. During two thousand years of famine it seems safe to say that the ancestors of practically every Chinese now living must, time and again, have passed through crises when habits of thrift, economy, and even parsimony gave them a decided advantage over their neighbors, in whom such habits were less strongly developed. The wasteful and extravagant must have been killed off so relentlessly that to-day wastefulness is almost eliminated in China. Even in the south this is true, for the Chinese ancestors of the southerners all came from the north, and apparently had been subjected to many famines before they finally reached a land where famine no longer troubles them. Yet in them the elimination of extravagance would scarcely be as complete as in the north. Moreover, since there has been no such severe elimination of the extravagant in later times, it is scarcely to be expected that the spirit of economy and frugality should be as highly developed in the south as in the north. And that is just what we find. Yet according to our standards even the southerners carry their economies to a senseless limit. Chinese business men who have been trained abroad criticise their own selves because they are so anxious to save that they often fail to invest their earnings as profitably as they might, and thereby retard their own success.

This same quality of providing for a rainy day accounts in part for the constant "squeeze" which is so disagreeable a part of life in China. So anxious are the Chinese to lay hold of something to ward off the evil day of scarcity that they cannot wait for the legitimate fruits of their labor. Therefore they extract their percentage from everything that passes through their hands, regardless of the fact that if they did not do this their trustworthiness would tend to bring them much higher and more profitable positions. In the same way the officials begin to feather their nests the moment they assume office. In them, as in the vast majority of Chinese, the tendency to save for a rainy day appears to be an inborn trait which they cannot withstand except with painful struggles. Moreover, throughout the long centuries of famine, there has been a premium on the capacity to grab what can be gotten at once, regardless of consequences. Never can one be sure when a famine will come, and only among a few of the extremely rich is there any assurance that they, too, may not be plunged into the ranks of the hungry and empty.

Economy and thrift are virtues, but they easily pass into parsimony and selfishness. Now selfishness and self-centeredness, which perhaps are the same, are among the most prominent and most regrettable qualities of the Chinese. And they too, like a strong physique and the spirit of rigid economy, seem to have become inherent in the race, ingrained qualities innate in the germ plasm. They are innate because in the past those who showed the spirit of altruism and self-sacrifice have often paid the penalty with their lives. Let me explain this more fully.

THE SELECTION OF THE CHINESE 187

Here is a family, or rather a household, of fifteen souls. At its head is the grandfather, and it includes three sons and their wives and children. That family has food enough to last for a hundred days, but it will be a hundred and fifty days before any large amount of food can again be hoped for from nature. Suppose that several other households are in the same plight. If any one of those families gives anything to the others it merely diminishes its own chances of living. If it gave largely it would doom itself to prompt destruction. It is not as though one could give at one time and one at another. That happens in a community where the disasters are limited to individuals or to small groups, and not where whole provinces suffer at the same time.

This is not all that can be done by sheer selfishness and sheer indifference to the sufferings of others. Another important resource lies in selling the children, especially the girls, and also the young wives. So the elders in families such as we have described look over the family and decide that if the prettiest young wife and the pretty eleven-year-old daughter are sold they will yield enough to tide the rest of the family through the famine. But one family, being kind of heart, yields to the entreaties of the girls, and they are not sold. A few weeks or months later that family has reached the limit of its endurance. The beams from the house have been sold, for they are the only part that will bring anything. The bedding has been sold even from under the children, and the air is already showing the nip of winter. A few more clothes can be sold, a little more food can be obtained by begging, but that is all. Only three possibilities remain, provided the girls are not sold. One is for the whole family to take to the road as beggars or bandits; the others are death and cannibalism. If such a family is gifted with a little more than the usual degree of sensitiveness it may decide to follow the example of a family whose tale I heard from a missionary. When the last scanty meal had been eaten, the father tied all the family together with a rope, including himself and even the dog; then all threw themselves into the river and perished. In another case, the head of the house sold the last rags of bedding, bought a little food and some poison. "One last good meal," said he, "and then we will wander forth." But he put the poison in the food and they all died together. These are not imaginary instances: they are things that actually happened in a recent famine. That famine was by no means so severe as hundreds that have preceded it, and it was greatly mitigated by the fact that now there is a railroad in the district, while organized relief on a large scale was administered by the Red Cross. Yet two such instances and many more cases of suicide came to the knowledge of a single missionary. How many other such cases there must have been that he did not hear of, and how vastly many more in the hundreds of past famines!

But the family which was callous to the entreaties of its daughters and young wives survived. The ruthless sacrifice of the flower of the family saved the rest. Thus it has ever been. For more than two thousand years the sale of children, sometimes boys, but chiefly girls, has been one of the well-recognized means of survival in famine. Often it has been declared legal by imperial enactment. And in every such famine complete selfishness appears to have been one of the qualities that have given people the greatest prospect of survival. In the last analysis the man who fends for himself alone has been most likely to survive. But here another factor enters into the matter. Selfishness may go too far. By utter renunciation of all claims upon him a man may insure his own survival, but if he abandons wife and children he also goes far toward insuring that his kind shall perish from the face of the earth. Of course he may marry again, but unless he is unusually competent his poverty is likely to prevent it for some time. Moreover, even if he marries again, his family as a whole and hence his type in future generations is not likely to be so numerous as is that of the family where the whole household sticks together and the majority are saved by the ruthless sacrifice of a few. It is even possible that ruthlessness may go so far that one member of the family is not sold but is actually used as food by the others. Certain it is that cannibalism has occurred again and again. The outstanding fact is that utter selfishness helps to cause a family to survive in times of famine, provided a man makes his family and not himself the unit.

As I write this I wonder whether it can be possible that the selective power of famines can be so potent and can have given an hereditary impulse toward such a complex group of characteristics. But this much is quite clear, the characteristics which we actually find in the Chinese, especially the northern Chinese, seem to be precisely the ones that ensure survival in times of famine. The selection of such people for survival during two thousand long years in generation after generation can scarcely fail to have had a strong influence upon the hereditary complex of the people. A certain type of temperament must have tended to become fixed among the Chinese, even though social customs may have determined the details of how that temperament would act in a given crisis.

Let us return a moment to the children and young women who are sold. It may seem to the reader that I have exaggerated the importance of this. But recall how often it is mentioned in the records. Note also what happened in the last famine. When the foreign relief workers finally checked up the population which survived after the famine they found that the number of young girls as compared with the rest of the population indicated a shortage of something like 40,000 or 50,000. In other words, at least that number appear to have been disposed of so that they had been removed from the famine area. During the famine they could be bought anywhere. A missionary who bought some in order to save them, paid from $2 to $5 per head in Mexican money. A homely girl could be bought for a dollar in American money, and ordinarily attractive ones for $2.50; while the prices of really pretty girls ran up as high as a hundred dollars in gold. Inasmuch as this happened in a famine where relief was supplied on a large scale, what must have happened in far worse famines in the past? It is obvious that during the hundreds of famines the number of girls who have been sold must have been millions.

What kind of girls are these, and what happens to them? In the first place they are the prettiest, brightest, and most attractive, which means that they are the ones who combine the best physique with the best minds. Of course I do not mean that they are absolutely the best in these respects, for those who most excel belong to the most successful and well-to-do families which do not have to sell their children. But among the ordinary peasants, which often means all save an exceptional few, it is these prettier, brighter girls who are sold. How do we know that this is so? Partly because mere observation shows it, and partly because girls are so cheap that there is no market for the ugly, stupid ones, and partly because the prices of pretty girls run so far above those of the common ones that the self-centered Chinese heads of families cannot resist selling those that will bring the highest prices.

The effect of the sale of girls on the famine-stricken villages is obvious. Among the prospective mothers of the next generation those who are best in mind and body are removed. Is it any wonder that the north Chinese have gradually become more and more dull, and that pretty women and girls are far rarer among them than in the south? And is it any wonder that women are despised in north China to an almost unbelievable degree? Read Doctor Smith's *Chinese Characteristics* and his book on *Village Life in China*. His tale of the way in which Chinese brides are subjected to every possible indignity and treated as if they were the scum of the earth is most pathetic. The lot of women, especially in north China, is so miserable that every year great numbers commit suicide. And it is presumably miserable in part because in the famine areas many of the girls who might have had the force of character to make themselves respected and to uphold a certain degree of dignity have been mercilessly sacrificed. Even when there are no famines the sale of girls and wives is by no means uncommon. Perhaps this may be one of the great reasons why the people in some of the villages where famine is chronic appear to be little more than morons. Take away the finest girls from any community for half

THE SELECTION OF THE CHINESE 191

a hundred generations, and that community is sure to become degraded and stupid and to hold women in low esteem. These qualities become not merely a social but a physical inheritance.

But are these girls wholly lost to the Chinese race? Not wholly, but very largely. A part, and in some districts a considerable part, are bought as wives by men in regions where there is no famine, but the purchasers are likely to be the poorest and least competent. The families which have these cheap purchased wives are the kind who are most likely to be wiped out when famine attacks the region of their new homes. The more attractive girls may meet either of two fates. The most attractive are likely to be bought as secondary wives or concubines by the richer men in the larger towns and cities. They henceforth become part of the city population and never return to the country districts. Hence so far as rural China is concerned, they are utterly lost; and rural China is probably 85 per cent of all China. What becomes of them as city people, I shall discuss in a moment. As for the other girls, they are bought by the owners of houses of ill fame. Many of the prettiest and brightest as well as some who are less attractive meet this fate. So far as the future of the race is concerned they are as good as dead. Thus the sale of girls and young wives tends at all times to weed out the better types from the Chinese villages, and in times of famine this tendency rises to large proportions.

Thus far we have taken no account of the wanderers who form so large a proportion of the sufferers from Chinese famines. Even in a mild famine like that of 1921, great numbers of villagers take to the road. At first it is only the poorer people who do this, unless a village happens to have been wholly destroyed, as is not infrequent. But if a famine continues long, practically all except a few of the most wealthy may ultimately be on the road, and often the whole countryside is deserted by rich and poor, as we read again and again in the old annals. In the last famine the missionaries say that they saw bands of from twenty to thirty people, and sometimes fifty or more, wandering far from their homes, sometimes a hundred, two hundred, or even three hundred miles in the case of some who went to Manchuria. Picture the hardships of such a forced migration. The migrants rarely start until their food is exhausted. Then they travel from town to town, moving out from their own region into parts of the country where the famine has been less severe. If the famine is due to flood it is easy to find such places, but if drought is the cause, they may have to wander hundreds of miles.

At first it may not be extremely difficult to move from one village to another each day or at frequent intervals. And at first the villagers who are not famine-stricken may give fairly freely, although always in extremely small amounts. A missionary who observed many such wandering bands said that when one looks into the bowls of the beggars one usually finds nothing but tiny scraps of bread, for each person who gives anything throws into the bowl no more than a mouthful from the meal which he is eating when the beggar passes by. As time goes on it becomes harder and harder to get a living through beggary, for the wanderers grow weak, the other villagers become less and less inclined to give, disease breaks out, some of the beggars become footsore or otherwise incapacitated, and cold weather comes on with all its terrors. Then death runs riot, and there is a tremendous selection whereby not only the weak in body but those who are weak in spirit are pushed to the wall. Suicide is common, and the women have been known to throw themselves into wells until the wells were choked with dead bodies. But still the little band moves on. It is losing in numbers and losing in physical strength, and yet it is gaining in strength as a source of new generations.

As time goes on the wanderers divide into two groups. One set consists of those who keep trying to go back to the old homes. They are the ones for whom the old home has the strongest call, partly because they have lands there, and partly because their temperament is such that they feel the influence of their ancestral cult and love the old ways even though they be miserable. In a word, the more conservative type tends to go back home, just as it is the last to leave home. Hence the economical, thrifty disposition, the utter selfishness, the callousness to suffering, the conservative temperament, and the physical endurance of those who have stayed at home all through the famine find their counterpart in the similar characteristics of those who return. Like marries like as the years go on, and the famine centers become the focus of these traits.

The other group of wanderers consists of those who have more energy and initiative. They set themselves to do some definite thing in order to escape from the toils of hunger. Some take up brigandage, but they are a relatively small percentage. A larger number undertake a distinct migration either to the city or to some part of the country where they can find land and settle down. Doubtless most of them plan originally to return home in course of time, but that purpose often fades as time goes on. It is most likely to fade among those who are most successful. And those who are most successful are likely to be those who have some special qualification. The man who is quick and alert is more likely to make a successful coolie in the city than is the man who is dull. But all coolies have a hard time and are likely to return home if they have a bit of land which may by now be fit for cultivation. The man who is clever enough to be a good carpenter, mason, baker, maker of images, coppersmith, or other artisan is much more likely to find profitable work in the city, and the call of the country village appeals to him less strongly. So, too, with the man who has a clever brain and can live by one of the more intellectual pursuits. Thus whenever a famine drives great numbers of people to the cities, the more brainy, the ones with more skilful hands, and the ones best competent to take care of themselves are, in general, those who depart. The country districts are thereby drained of their best men as well as of their brightest, prettiest girls. In a certain way this same process is going on in other countries, but the Chinese famines greatly accelerate it.

This loss of the best would not be so bad if it were certain that the city people would continue to multiply as rapidly as those in the country. But this is almost never the case. In all parts of the world the birth-rate appears to be lower in the city than in the country, while in the past and in backward countries the death-rate appears also to be systematically higher in the cities than in the country. Indeed, it is often stated that in Oriental cities the death-rate is so high and the birth-rate so low that the cities would gradually die out if not replenished from the country. No exact statistics are available, but a study of the number of children per family by Doctor Lennox at the Peking Union Medical College Dispensary, and by Doctor Gray at the British Charitable Hospital of Peking indicates that the birth-rate among all the Chinese of the Peking district is much smaller than has generally been supposed. About a quarter of the city families appear to have no children, apparently due in considerable measure to venereal diseases. Counting these as well as the others, the average number of children appears to be only about 2.6 per family. Inasmuch as the death-rate is very high, it seems probable that in Peking the deaths appreciably exceed the births. If this is true, it means that when people migrate to the city they doom themselves to gradual extinction. In that case the constant draining of the more active and energetic parts of the population from the rural districts to the cities is slowly but surely lowering the general caliber, not only of the country districts but of the cities.

The most hopeful side of this dark picture seems to be the wanderers who migrate into the rural portions of new regions. The competent Hakkas of south China, previously referred to, are migrants of this sort. Their history affords a good example of the way in which people from the famine areas of north China have moved into other parts of the country. During the process they have apparently suffered natural selection in such a way that the weaker or more conservative elements have been left behind, while only the most able and energetic have finally settled in the new home. There has been much misapprehension as to the history as well as the character of the Hakkas. The most reliable account that I have found was prepared by Mr.

THE SELECTION OF THE CHINESE

George Campbell, a missionary who lived among the Hakkas many years. It was printed as a ten-page pamphlet for private distribution, after having been read at a conference of the English Presbyterian and American Baptist Missions at Swatow in 1912. It illustrates the way in which many missionaries have done most scholarly and valuable scientific work, much of which has never been made generally available because the time of the missionaries has been so occupied in their daily work.

One of the first facts to be noted about the Hakkas is that they speak a Mandarin form of the Chinese language, unlike the tonal languages of their neighbors, but much like that of parts of Honan, thus indicating their northern origin. According to Campbell's account, derived directly from old Chinese sources, the present Hakkas are the result of three migrations. First, in the fourth century after Christ north China suffered a period of terrible distress, due partly to famine and partly to invasions by the Tungus tribes, and especially the Huns. According to an old chronicle quoted by Miss Lee: "The population of to-day [in an unspecified area about 300 A. D.] is only one-tenth of that of the Han Dynasty.... At present the people who are not engaged on farms are innumerable." And again, "Good fields have been growing weeds and the people have been staying in swamps [presumably because those were the places where there is water or else as a refuge from invaders]. Both the dry and wet fields have lost their fitness for agriculture, pasturage has put an end to the raising of grain [which probbably means that because grain could not be raised, the former grain-fields had been used for pasturage], and the trees and woods have all become dry at once."

At about the same time the invading Huns captured two successive Chinese emperors, the second of whom was compelled to act as servant to the leader of the Huns, until that rough bandit became tired of seeing him around, and put him out of the way. "These insults and humiliations [together, we infer, with the famines] seem to have broken the spirit of the people. When the founder of the Eastern Tsin made Nanking his capital, many left their homes and took their families across 'the Great River' [the Yangtse]. This was to them a very serious step." (Campbell.)

Beyond the Yangtse, which at a much later date was regarded as the main barrier against the Mongol invaders from the north, the migrating Hakkas of the fourth century drifted southward, some trending southeasterly toward the coastal provinces of Chekiang and Fukien, and others south toward the province of Kiangsi. Their further history has not yet been traced, but the ones who went to the seacoast probably became the so-called Hoklos of the maritime sections, while the others may be represented by the Hakkas of Kiangsi, who apparently drove out the primitive Lau or Laos.

The next Hakka migration took place in the ninth century. At that time the condition of China was most distressing. According to Boulger as quoted by Campbell, the country was "desolate, the towns ruined, the capital reduced to ashes. Not a province that had not been visited by the horrors of civil war, not a fortified place which had not undergone a siege, and which might not be estimated fortunate if it had escaped a sack. With confusion in the administration, and the absence of all public spirit, it was not surprising that each governor should strive to make himself independent and fight for his own hand."

At this same time we have the usual accounts of famines of great severity. Whether they were worse than usual is not quite clear, although they seem to have been, for we hear that famine occurred several times in the province of Chekiang, where as a rule it is rare.

One effect of all these disasters was that a band of 5,000 men migrated from Honan. After much fighting and wandering they finally settled in the province of Fukien among the mountains back of Fuchow, about 885 A. D. That their families went with them is proved by the presence of the old mother of two of the leaders on a special expedition, when orders had been issued that no weak or infirm persons should be taken along, as the way was dangerous and provisions scarce. This is the foundation of one of those tales of filial piety so dear to the heart of the Chinese. The sons are called to account by their chief in these words:

"All soldiers have rules, there are no soldiers without rules. You have disobeyed my command. If I do not punish you discipline is broken."

The elder son replied: "All sons have mothers, there are no men without mothers?"

When the general ordered the mother to be beheaded, the son made answer: "We brothers serve our mother as we serve our general. Having slain their mother, how can you use the sons? Please let us die first."

Of course the faithful sons are saved by the intervention of the soldiers, and ultimately the elder of them replaces the general in the supreme command.

In the early part of the reign of Wang Chau, the faithful son, "there was undoubtedly frequent communication between his followers and their relatives in Honan. Doubtless his armies were largely recruited in this way. The sufferings of the people in Honan would dispose them to emigrate to the new kingdom, dominated by those of like blood with themselves." (Campbell.) Thus the Hakka settlements grew as did those of early America.

For nearly four hundred years after this migration the Hakkas kept themselves distinct from the surrounding people, whereas the Hoklos on the coast mingled with the earlier inhabitants. Campbell considers that the two cases are comparable to those of the Puritans in New England and the Spaniards in Mexico. The Hakkas, like the Puritans, found a few savages whom they drove out or exterminated, but with whom they did not mingle. The coastal Hoklos, like the Spaniards in Mexico, found a more numerous and more civilized people with whom they intermarried, and thereby probably lowered their innate ability.

At the end of four long quiet centuries a great disaster overwhelmed the whole Hakka region in the form not only of famines but of the Mongol invasions, about 1276 A. D. A large part of the population was exterminated. It is recorded that a Hakka named Tsok raised a regiment of nearly a thousand men among his own clan, and only one of them survived. The region in which the Hakkas now display their highest development became a wilderness. A native writer of that period speaks of the deserted houses and fields and asks if the people have all turned into foxes and birds. The chosen remnant whose skill, bravery, and endurance enabled them to survive this terrible period gave rise to the last migration. This involved merely a change of residence of a hundred miles or so from the Ning-hua district, in Fukien, which had not been completely devastated, to the Mei-chau district of Kwantung, where the devastation was complete. Some settlers also appear to have come from the survivors of the fourth-century migration to Kiangsi. In Mei-chau, the central area of the most progressive Hakkas, the new migration was probably very small. Even as late as 1390 the country, if so we may call it, is reported to have contained only 1,686 families, and 6,989 persons. In 1848, in spite of large emigration to other provinces, the descendants of these original settlers are reckoned as 268,193 persons. Thus the present Hakkas, especially those in the central area, appear to be the descendants of a small number of people from Honan, who in the fourth and especially the ninth century migrated southward. The later migrants settled in the mountains of Fukien and kept themselves unmixed with any other race for four centuries. Then they were again subjected to a terribly drastic process of selection by famine, war, and migration, and finally a small number settled in a new district which had been practically depopulated. There they have since maintained themselves almost unmixed. To-day, as Campbell well says, "the Hakkas are certainly a very distinct and virile strain of the Chinese race. The circumstances of their origin and migrations go far to account for their pride of race and martial spirit. It is safe to predict that the Hakkas will play an increasingly important part in the progress and elevation of the Chinese race."

In the Hakka district some movement is still going on, for

not all of the mountainous country is yet occupied. Ofttimes, so the missionaries say, it is the most progressive and most radical who move. As population increases, the fields are too small. Part of the men go to Canton, Swatow, and other cities. Others start new villages. Those who thus move to new villages are often, according to Mr. Spiker, the ones who want to adopt new methods, and hence are looked upon askance by their fellows.

The history of the Hakkas deserves careful study. Recall the fact that in the opinion of many good judges they are to-day "the cream of the Chinese." Their energy and cleanliness, their respect for women, and their high degree of education are almost unique. They differ markedly from the Chinese of earlier migrations who surround them, and the difference is the same kind as that which differentiates those same surrounding Chinese of the south from the less progressive and active Chinese of the north. The qualities of the Hakkas are in many respects like those of the energetic barbarian invaders of Tartar, Mongol, and Manchu stock from the dry northern regions to whom north China owes so much of its historic dominance. In the case of the Hakkas we have written evidence that they were impelled to leave their northern homes under the stress of famine and invasion. We get glimpses of the way in which hardship and war inexorably cut down their numbers and left only a chosen remnant of unusual capacity. We also find that this process of selection took place three successive times. Finally, when the Hakkas, especially those in the central and most typical area, were free from the difficulties and hardships which induced migration and natural selection, they kept themselves aloof from their neighbors and thus preserved their inheritance. The competent, wide-awake, progressive Hakkas, on the one hand, and the incompetent, dull, conservative people of the villages near Peking where the houses are left sealed while the villagers beg for bread, on the other hand, seem to represent the two extremes due to natural selection and migration in China. In the one case we have migrants in whom a high degree of ability has been concentrated; in the other we have the stay-at-homes from among whom most of the more able elements have gradually been eliminated.

Let us look at another and still more recent example of the selective progress here outlined. In Mukden, as I have said, there is more activity and life than in any other Chinese city that I have visited. Activity and progressiveness are said to be still more evident in Harbin, and most of all in the far north, where the town of Aigun on the Amur River, opposite Blagoveschensk, is reported to be inhabited by Chinese who seem quite unlike their countrymen in their modern spirit of progress and in their bustling activity. It is sometimes affirmed that this is because the Chinese have become Russianized by contact with the relatively large Russian population. In a certain way this is true. But why has not contact with the British at Hongkong done still more to give a British quality to the Chinese there? Nothing of the kind has happened, although Hongkong had a considerable British population for a generation or two before the Russians had much contact with the Chinese in Manchuria. The answer seems to be a recent and drastic selection in Manchuria, and only a mild selection in Hongkong.

Manchuria is inhabited almost entirely by Chinese who have recently come from the conservative provinces of Shantung and Chihli. In general the merchant classes and city people are from Chihli, and the farmers from Shantung. Here is what happens. Owing to the constant economic pressure, people from those two provinces migrate more or less at all times, but especially when there are famines. They go to Manchuria not only because that province is near and is under Chinese rule, but because until the nineteenth century it was only sparsely populated. The regular proceeding is for the Chinese men to go first without their families. A man newly come from Chihli works for a while for some one in Manchuria. Then if he is successful he starts a little business for himself. Once in three years, as a rule, he goes back home, usually staying five or six months. From time to time he brings with him other men from his village. For a while they live together as a single big family. But at last the merchant who succeeds, decides to have his family with him rather than several hundred miles away. He can afford to set up a house, and he does so. Because he is competent and successful —because he is the best man out of twenty, or fifty, or a hundred—his family comes to Manchuria and his children become permanent parts of the population. Of course he still calls Chihli his home, but the Englishman who settles in Australia calls England home. In due time the merchant's sons grow up. The more adventurous among them go farther north, just as their father did before them. When they succeed they likewise bring their families to the north, and still another stage in the selection of competent types is accomplished.

The same thing happens among the Shantung farmers. Each year toward the end of winter they come by the hundred thousand to Manchuria, some by rail, but many tramping hundreds of miles on foot. Spreading out into the country they are ready to work for the farmers as soon as spring breaks. In the autumn they go back to their families, only to swarm north once more at the end of the winter. But some are not content to be merely hired laborers. The more ambitious and energetic get hold of small pieces of ground. At first they cultivate these and at the same time work for others. But in a year or two they get enough land to support a family. Then a shack is built. Next year at the time of the northward migration there is a wheelbarrow on the road. On it sits the grandmother surrounded by a promiscuous heap of bedding, boxes, bags of rice, cooking-pots, and all the simple paraphernalia of a Chinese household. The proud owner of the Manchurian shack sways between the shafts of the wheelbarrow, his oldest son bends low in front, tugging at a rope over his shoulder to help his father with the heavy barrow. The wife walks behind bearing on her shoulder a bamboo pole with a basket on each end and a baby in each basket. And with her trudge one or two other children. Thus they toil along the snowy path to that new home, unconscious that they have been selected by their innate ability to people a new land. If the father is competent and ambitious his work soon makes his land increase in value. Then he sells out, moves north once more, this time in a cart drawn perhaps by two horses and a mule. Once more he succeeds, and then sometimes moves on a third time. Thus northern Manchuria is being peopled by the most competent of the inhabitants of Chihli and Shantung. The old, old process which apparently gave to the Hakkas their relatively good abilities is being repeated before our eyes. It is one of the most hopeful phenomena in all China, for it shows that even in provinces like Chihli and Shantung, where the ravages of famine have been especially bad, they have not weeded out all ability. Proper selection, free opportunity, and freedom from the depressing effect of overpopulation seem to be all that is needed to build up even in those regions a Chinese race of high ability.

In summing up this discussion of China one of the dominant facts in the history of the country seems to be the way in which each cycle has been characterized by the incursion of selected types from the north, especially from the deserts, the outward migration of the more able of the previous inhabitants, and the degeneration of those who remain in the areas where famines are most numerous. Thus in the early days we see the Hiung-nu breaking into northern China during the third century before Christ and bringing in a strong virile element. At about the same time other Chinese who had been pushed out from northern China were penetrating south of the Yangtse and ultimately reached the coast at Fuchow, Canton, and elsewhere. Although no records are as yet available it seems almost certain that these migrants went through a process of selection like that which later concentrated so much ability in the Hakkas. Thus it appears probable that the relative activity and progressiveness of south China had its origin in selective migrations of these first Chinese inhabitants of the lands south of the Yangtse.

In the fourth century after Christ another period of unusually severe famines coincided with new inroads of Tartar peoples on the north and new movements of the Chinese toward the south. These latter movements brought the first of the Hakkas and likewise seem to have carried other similar people called the Hoklos onward to the coast to reinvigorate the people already there and give them a new spirit of progress. Passing over the intervening periods we find in the twelfth and thirteenth centuries another example of the same kind. The able Mongols came in on the north, the Hakkas again suffered selection and moved southward. Even now they are still pushing coastward and thus maintaining in the southern cities that spirit of energy and progressiveness which distinguishes the southern Chinese from the northern. Meanwhile the Manchus have come and made their contribution to progress in the north, and are now in decline. At the same time there has occurred in the north a process of adverse selection whereby the abler people under the compulsion of poverty and famine move cityward and are eventually exterminated, or else go to other parts of China such as the relatively new provinces of Manchuria and likewise far Szechuan on the upper Yangtse.

Even among the abler people of the north there is presumably a tendency toward innate conservatism because many of the village girls, bringing with them their village inheritance, are taken into wealthy homes as concubines. Thus in the north we see a great antithesis: at one stage of the historic cycle a sudden inroad of able barbarians who in due time amalgamate more or less completely with the former inhabitants. Then comes a period when the impulse thus gained is lost through the natural selection due to overpopulation and selective migration to the cities or to other provinces. These contrasted events in the north are the central fact of Chinese history. Elsewhere, especially in the south, there is the same alternation between the invigorating inroads of new people, and the deterioration which almost inevitably follows. But because nature is more uniform and there is no such severe economic pressure as in the northern provinces where famine is most common, this deterioration in the south is far less rapid than in the north. To-day we seem to be at the stage in an historic cycle when the north has deteriorated to the point where it is more backward than the south. But already it is beginning to forge ahead under the influence of a new kind of selective migration whereby able foreigners—picked men in their own countries—as well as able Chinese from the south are peacefully assuming the direction of future progress.

CHAPTER XIV

THE THREE GREAT RACES OF EUROPE

HAVING considered the nomads of the dry parts of Asia, and likewise the Jews, Turks, and Armenians of the West, and the Chinese of the East, it would be natural at this point to consider also the Japanese and the people of Indo-China and India. I shall omit these, however, for lack of space and because it seems advisable to pass on to our own ancestors in Europe. Moreover, I wish to retain space for the especially unique and clear-cut case of Iceland where the principles of natural selection are illustrated with a sharpness found almost nowhere else. Japan, to be sure, illustrates the same principles, but not so clearly. India resembles China in many ways, but is vastly more complex, so complex that I hesitate to discuss it without again visiting the country.

Turning now to Europe, we have already seen that during each glacial epoch mankind must have been driven from a large part of Europe. When the ice retreated, however, vast areas which had previously been uninhabitable became highly attractive and hospitable. At the same time the dry regions of north Africa, and especially Asia, which had been comparatively well-watered and habitable while Europe was shrouded in ice, tended to assume their present condition of deserts. Thus the history of Europe for the last twenty or thirty thousand years, more or less, has been profoundly influenced by three great conditions:

First, we infer that previous alternations of glacial and inter-glacial epochs had rigidly weeded out the weaker human elements and mixed one selected race with another to a remarkable degree. Hence the tribes who dwelt in the Mediterranean lands, and in western Asia from Palestine and Asia Minor around through Mesopotamia to Persia and the Caspian region may be presumed to have been peculiarly competent, probably the most competent in the world.

Second, the change in climate from the rigors of the ice-age to the mildness of the present must have exerted a push and a pull upon these competent people. The push from behind was doubtless due not only to increasing drought and growing density of population, but to raids and invasions by less fortunate people nearer the centers of the great deserts. The pull from in front was due to the attraction of the new and unoccupied lands which were, little by little, passing out of the grip of the glacial climate in Europe. Whether the push or the pull was stronger we cannot tell, but probably the push often led to violent and rapid migrations, whereas the pull may have been especially effective in causing slow, quiet movements which, in the long run, may have been highly effective.

In the third place, the passing of the glacial period presumably caused migration into Europe to follow different paths at different times. This is well illustrated in Taylor's map of head-form and migrations. (Figure 3, page 76.) The earliest postglacial line of approach was probably northward across the Mediterranean, for the Sahara desert, lying relatively near the equator, probably felt the pinch of aridity sooner than its Asiatic counterparts. Moreover the southern peninsulas of Europe were then the most habitable parts of that continent, and presumably became more and more habitable during the early stages of the amelioration of the glacial climate. Another path of migration to Europe traverses Asia Minor and crosses the Hellespont, the Bosphorus, and the Ægean Islands. Movements in this direction presumably began later than those from the south not only because the highlands of Armenia, Asia Minor, and the Balkans even now are relatively cold, but because their fairly northern latitude presumably delayed the amelioration of their climate, while the deserts east of them probably did not become extremely dry so quickly as did those farther south. The third line of invasion from the Mediterranean-Asiatic belt of competent people to Europe runs north of the Black and Caspian Seas. Since the

Caucasus mountains form an effective barrier, this route was probably little used until the glacial climate had so far passed away that Russia in latitudes forty-five degrees to fifty degrees was moderately attractive to nomads, while the deserts to the east and south were dry enough to give an appreciable push.

This discussion of physical features and climate suggests that we should expect three main waves of post-glacial migrations into Europe. The first and most southern wave would be expected to start in Africa or east of the Mediterranean and spread into the Mediterranean peninsulas. It might likewise spread along the seacoasts to parts of western Europe, like Brittany and Ireland, for people who could cross the Mediterranean must have had some skill in boats, and the west coast of Europe, even as far north as Ireland, has a relatively mild climate. Our expectation seems to be fulfilled in the so-called Mediterranean race. People of that race were apparently among the first to practise agriculture in Europe.

In similar fashion, but somewhat later, we should look for an invasion of plateau people, whose route would take them from the Asiatic plateaus to those of Europe. The first waves of the great Alpine race may perhaps satisfy this expectation, although many people of related stocks have come in later times by more northern routes. In the rugged plateaus where the Alpines first dwelt the keeping of sheep has long been a main industry. Farther north, and perhaps somewhat later as befits the climate, the third great migration would be that of the people of the plains, keepers of horses and cattle. And the first stage of this seems to correspond to the fair Nordics, or to Dixon's Caspian folk. Thus the division of the people of early Europe into three great types of which the Mediterranean is the oldest is fully in accord with what the geography and climate would lead us to expect.

Of course the history of the great migrations into Europe is by no means so simple as the preceding paragraphs would seem to indicate. There have been all sorts of marchings and counter-marchings, and people of different races have moved over the same track, while race has mingled with race again and again. Moreover, the migrations out of Asia have continued till our own day, when Armenians and Greeks have been the last to move, pushed out by the broad-headed Turks during the World War; while those from Africa persisted as late as the days of the Arab invasion of Sicily and Spain. But this, too, is wholly in accord with the physical environment. It is almost universally agreed among geologists that since the height of the glacial period there has been a complex series of glacial stages. Sometimes the climate has rapidly become milder, again it has grown more severe. These changes, as I have shown fully in other places, appear to have continued to our own day. During historic times the general tendency of the climate seems to have been toward aridity in the deserts of Asia and in the lands of the eastern Mediterranean, but the tendency has been interrupted again and again by pulsations which have carried it back toward the glacial type, or toward a degree of aridity even greater than that of to-day. These stages and pulsations in themselves would be enough to cause innumerable movements of races both backward and forward. The complexities thus introduced are still further complicated by the varying character of the mountains, valleys, plains, and forests in different parts of Europe. Moreover, purely human causes like the racial fervor of the Turks are extremely important. Again, one racial movement interferes with another, and the effect of a movement in one place may be to displace tribe after tribe, so that the commotion extends thousands of miles, and endures for generations. The result is bound to be an extraordinarily complex series of migrations which give almost unlimited opportunity for natural selection and racial mixture, and thus for the creation of racial stocks with highly diverse characteristics.

It would be profitable at this point to undertake a general analysis of the post-glacial racial movements of Europe, but that would take us beyond the limits laid out for the present book. Accordingly we shall merely attempt to gain a general impression of the qualities of the three great races of Europe, and shall then take up certain specific problems pertaining to

THE THREE GREAT RACES OF EUROPE

individual stocks. In studying the character of the Mediterranean, Alpine, and Nordic races I shall set forth the opinions of two of the most recent investigators, Dixon and McDougall, and shall then express my own opinion. Dixon approaches the matter from the purely anthropological standpoint; McDougall is a psychologist. With neither do I agree entirely, but both present views that are well worthy of thought. Dixon, it will be remembered, has a new classification of races. Here is what he says:

Whereas the Proto-Australoid and Proto-Negroid types seem to have had their origin in the tropics, and the Mongoloid and Palæ-Alpine upon the great central Asiatic plateaus, the Caspian and Mediterranean types seem to be traceable to the Eur-Asiatic steppes surrounding the Caspian Sea and the regions adjacent to it, north of the plateaus. If the Proto-Australoid and Proto-Negroid types have been, except in the very earliest period, the most stay-at-home of types, the Caspian and Mediterranean, especially the former, have been of all the most adventurous. In late Palæolithic times spreading westward into Europe, and almost as early moving northeastward into America, in Neolithic times they forced their way across the eastern plateaus into the borderlands in China and Japan, and thence southward into Indonesia and far into the Pacific. Southward, also, across the eastern plateaus they made their way early into Arabia and northeastern Africa, driving from the whole northern part of the continent its older Negroid population, and infusing themselves along the East African plateau far to the south. Later yet they moved southward across the Iranian plateau to India. The Palæ-Alpines were on the whole content to be led; the Caspian and Mediterranean people were, on the other hand, leaders, the former perhaps, if one may venture so far in attempting an analysis, more in the affairs of the body, the latter in those of the mind. The Caspian was more a conqueror, the Mediterranean a thinker and artist. Each type had in it great latent possibilities, and when the two were blended, a people of great capability was the result. It was thus, among a Mediterranean folk in whom was a minority of Caspian, that the striking Minoan civilization of Crete arose, out of which grew more or less directly and among a largely kindred people the "glory that was Greece"; that, in a population where the two elements were perhaps more equally blended, there were evolved the great systems of Indian philosophy, whose influence has been so profound upon all the Orient; that among another people, mainly compounded of these same factors, that most militant of religions, Islam, arose, whose adherents have carried it with fire and sword into Europe, throughout northern Africa, almost the whole of Asia, and far out into the Pacific; and that in a related group of similar origin in Palestine we have the source of that faith which missionaries have carried to every land. Blended also of these two types, but with considerable elements of the older Proto-Australoid and Proto-Negroid, were the Baltic peoples, that "Nordic" race which wrecked the power of Rome, as their Caspian-Mediterranean kindred, the Hyksos, had conquered Egypt, or the Kassites had plundered Babylon, or the Persians had overthrown Assyria. Lastly, in modern times, it was largely the adventurous daring, the genius and the hardihood of these breeds which were responsible for the discovery, conquest, and colonization of America by Europe, an event which, in the development of the human race as a whole, was destined to be of great significance. . . .

If, in the history of the race as a whole, the Mediterranean and Caspian peoples have played a great part, that of the Alpines seems hardly less impressive; and there is not a little reason to believe that only where these types have met and mingled have the highest achievements been attained. Perhaps the idea is fanciful, certainly many, many other factors are likewise concerned, yet one may point to various cases in history which seem to bear it out. Thus Babylonian civilization grew out of the blending of the supposedly Alpine Sumerian with the Mediterranean-Caspian Semitic peoples who seem long to have been in occupation of the Mesopotamian plains; in Greece, before the florescence of Hellenic culture, the earlier Mediterranean population was reinforced by the immigration of the probably Alpine Dorians [elsewhere said to have been probably under Caspian leadership]; Rome rose to greatness only after the older Mediterranean-Caspian people of Latium had been half dominated by Alpines coming southward from the valley of the Po and the region where the older Etruscan culture had its centre. [But elsewhere we read that the Patrician aristocracy of Rome was Caspian in type.] In the East Chinese civilization had its rise in an area where strong Caspian elements were absorbed by the incoming Alpine folk; lastly, the marvellous development of modern European civilization has occurred in that region in which Alpine, Mediterranean, and Caspian have been more completely and evenly fused than elsewhere in the world. Is it perhaps more than mere coincidence that the reawakening of culture in Europe after the Dark Ages began at a time when, after a period of centuries during which wide shiftings of peoples had occurred, the new fusion of the elements had been begun? Is it mere chance that it was in the north of Italy, in Tuscany and the valley of the Po, where the influence of the Caspian-Mediterranean immigrants was strongest, that the Renaissance began; that in Germany it was in the south where the Baltic peoples had in large numbers blended with the older Alpine and Palæ-Alpines, rather than in the north where such amalgamation was less clear, that the revival of culture had its start; that many of the forerunners and leaders of the Reformation, such as Huss, Luther, Zwingli, Calvin, all came from regions where the fusion of types must have been vigorously going on? The complexity of the causes underlying all such great movements is, it need hardly be said, very great, yet I cannot but feel that, among the many potent factors which have determined or directed the rise of modern European civilization, this one of the fusion of Alpine with Mediterranean-Caspian elements has an important place. That the contact of two different peoples often produced a stimulating effect upon culture has of course often been noted; the point which I would make here is that this stimulation seems to be at its maximum when the peoples belong to the Alpine and to the Caspian or Mediterranean types. In the years before the war, Teutonic scholars were proving, to their own satisfaction, that most of the great names in the history of the European and Mediterranean world were those of men of Nordic race, and even Christ himself was claimed by some of the more daring as of "Germanic" blood. To no one race or type, however, can the palm be thus arrogantly assigned, rather to the product of the blending of those types which seem of all the most gifted —the Mediterranean-Caspian and the Alpine.

In the history of mankind there have been, from earliest times, many places, many occasions when amalgamations between two or more of the great fundamental types have occurred: and from these blendings, I am tempted to believe, have arisen again and again the cultures or civilizations which mark the progress of the race. From the fusions between types less dowered have come the feebler cultures; from those of types with larger, more richly endowed brains have come greater achievements; from those of the Alpine and Mediterranean types, whose brains in size surpass all the rest, have grown the greatest of them all.

This long quotation is inserted here not only because of its intrinsic value, but because it is well to remind ourselves that the mixture of races is one of the most important ways of producing new racial types. I do not agree with all that Dixon says, for in his enthusiasm over racial mixture as a cause of racial ability he almost disregards natural selection and environment. Nevertheless, there can be little question that some racial mixtures far excel others in providing a new, varied, and competent assortment of human types from among which natural selection has been able to choose, and which the environment has then helped or hindered.

Now let us turn to a wholly different line of thought. In his little book called *Is America Safe for Democracy?* McDougall gives a most interesting analysis of the mental characteristics of Nordics and Mediterraneans, together with some hints as to the Alpines. His discussion centers around five qualities: (1) curiosity, (2) individualism, (3) introversion, (4) self-assertion, and (5) acquisitiveness. In all of these he believes that the Nordics tend to go to the positive extreme, while the Mediterranean people are more negative, and the Alpines generally intermediate. Of course he deals only with the typical specimens of each race, and recognizes that the majority of Europeans are of mixed origin even though they may in their general character tend toward one type or the other.

Curiosity [says McDougall], with the emotion of wonder which enters as an essential element into all such emotions as awe, admiration, and reverence, may without exaggeration be called the mother of philosophy and of science. Now modern science is very largely a product of northern Europe, of those countries where the Nordic blood predominates; not exclusively so by any means. But note this fact: the Greeks who founded philosophy and science were probably, in their great age, compounded of the Nordic and Mediterranean races. The Romans were almost purely Mediterranean.* They produced great men, great lawyers, soldiers, administrators, and poets; but no philosophy and no science. For four hundred years they ruled absolutely the fairest part of the world, in a high state of civilization, but they invented nothing, discovered nothing, made no progress in science. Otto Seeck, the historian of the classical world, has drawn a vivid picture of this scientific stagnation. He points out how, even in the art of war, on success in which their whole empire was founded and maintained, the Romans made no progress, invented no new weapons, but fought with the same old weapons throughout the centuries of predominance. Note another evidence of the weakness of their curiosity. In spite of their supremacy, their navy and mercantile marine, they remained a Mediterranean power: their sailors penetrated hardly, if at all, beyond the pillars of Hercules; while the barbarous Vikings in their smaller ships sailed to Iceland, Greenland, and America, and perhaps landed on the banks of the Charles River. Here, then, is further evidence that in the Mediterranean race the instinct of curiosity is relatively weak.

Individualism, the second of the peculiar Nordic qualities according to McDougal, is diametrically opposed to sociability or the herd instinct.

The Southern Europeans are more sociable than the Northern. They delight in conversation, in coming together in large masses, in expressing

* Dixon does not agree with this. He holds that in the early ages of Rome, although the Mediterranean type was most numerous, the Caspian type formed the dominant aristocracy. By the time of Christ the Caspian ingredient had greatly diminished and the Alpine had become prominent. In this case, as in many others, Dixon's conclusions are based on a very small number of skulls, and hence cannot yet be considered conclusive, but they cannot be disregarded.

THE THREE GREAT RACES OF EUROPE 213

their emotions collectively, in great collective outbursts of applause, of admiration, or of execration. In all ages their civilization has been essentially urbane; the city has always been their natural habitat. Men of Nordic race, on the other hand, are taciturn; they take part in social gatherings only with difficulty and hesitation; they are content to live alone in the seclusion of the family circle, emerging from it only in response to the call of duty or ambition or war. The isolated home is their invention, their dearest possession; and the individualized home is one of their peculiar contributions to the culture of the world. The facts are all summed up in the phrase—"An Englishman's home is his castle."

The next contrast between Mediterraneans and Nordics is expressed as follows:

The Mediterranean peoples are vivacious, quick, impetuous, impulsive; their emotions blaze out vividly and instantaneously into violent expression and violent action. The Northern peoples are slow, reserved, unexpressive; their emotions seem to escape in bodily expression and action with difficulty. . . . [These two types correspond with what the psychologist calls extroverts and introverts.] The extroverts are the vivid, vivacious persons who charm us by their ease and freedom of expression, their frankness, their quick sympathetic responses. They are little given to brooding; they remain relatively ignorant of themselves; for they are relatively objective, they are interested directly and primarily in the outer world about them.—If they break down under strain, their trouble takes on the hysteric type—in spite of which they may remain cheerful, active, and interested in the world.—The introvert, on the other hand, is slow and reserved in the expression of his emotions. He has difficulty in adequately expressing himself. His nervous and mental energies, instead of flowing out freely to meet and play upon the outer world, seem apt to turn inward determining him to brooding, reflection, deliberation before action. And when he is subject to strain, his energies are absorbed in internal conflicts; he becomes dead to the outer world, languid, absorbed, self-centered, and full of vague distress.

Physicians who have studied nervous diseases in Europe say that in general the southerners suffer from the hysteric type and northerners from the neurasthenic type. McDougall shows that this psychological contrast is what would be expected from the extrovert and introvert types and may have much to do with the fact that while homicide is common in the Mediterranean parts, suicide and divorce are common in the Nordic parts of Europe. The southerner becomes quickly angry and vents his rage in violent physical action directed against the person who has wronged or insulted him. The northerner broods over his wrongs, and kills himself rather than face what he believes to be intolerable, or obtains a divorce when he has thoroughly thought over the matter. And the northerner likewise tends to become a Protestant because his curiosity leads him to investigate and his introvert temperament leads him to ruminate over what he learns.

The quality of self-assertion manifests itself in the Nordics in a highly developed capacity for self-assertion or leadership. An Englishman likes to work things out for himself, and is glad when an emergency throws him on his own resources. The Mediterranean and Alpine people, on the contrary, are much more docile, more willing to be led even when it is ultimately to their own disadvantage. It is not at all likely that Napoleon could have led the Norse as he led the French, or that the British would have let any one impose upon them the absolute obedience which the Nordic junkers imposed upon the Alpine remainder of the Germans.

McDougall believes that docility and self-assertion are rooted in two distinct and opposed instinctive tendencies, which he calls the instincts of submission and self-assertion. He thinks that the instinct of submission is

the root of all docility and suggestibility, that is, it is the principal factor in all those social phenomena which some authors have erroneously ascribed to the herd instinct. [On the other hand] the instinct of self-assertion is the most essential, the all-important factor, in what we call character, that complex organization from which spring all manifestations of willpower, all volition, resolution, hard choice, initiative, enterprise, and determination.

This instinct, he holds, is strongest in the Nordics, whereas the instinct of submission is strong in the Alpines.

It is this greater dose of self-assertiveness in the Briton which leads other peoples to complain that he goes about the world as though it belonged to him; it is this which, in spite of his lack of method and organization, has enabled him to muddle through the Napoleonic Wars, the Crimean

THE THREE GREAT RACES OF EUROPE

War, the Indian Mutiny, the South African War, and, lastly, the Great War. It is this which, in spite of his lacks, enabled him to subdue and govern the 300,000,000 of India. And it is this, in combination with his other qualities, that has rendered him the successful colonist par excellence.

Let us note in passing that the addition of this quality to the picture of the Nordic race completes, or makes more adequate, our explanation of the distribution of the Protestant religion in the world; for it shows us that the men of this race are by nature Protestants, essentially protesters and resisters against every form of domination and organization, whether by despot, church, or state.

The final quality ascribed by McDougall to the Nordics is acquisitiveness:

> The strength or weakness of this tendency is, I suggest, the main factor in determining that a man or a race shall be provident or improvident. And it is very easy to see how natural selection may have developed this quality in peoples inhabiting cold or arid regions. It seems, in fact, to be present in the principal races in proportion to the demand for it made by their habitat. It seems to be strong in the Alpine and the Nordic race and in the Chinese; less strong in most branches of the Mediterranean; but strong in the Semites, in the Jews and Arabs and the Phœnicians, who long inhabited the dry, desert regions. Its strength seems to be a quality essential to any people that is to build up a civilization based on the accumulation of wealth, on commerce and industry, as every higher civilization has been. Owing to this necessity, every communistic or socialistic scheme which would abolish private property is an empty dream, an unrealizable ideal, a Utopia. The strength of this impulse seems to vary widely even in nearly related peoples, and also from one family to another. It would certainly seem to be stronger in the lowland Scotch than in the Irish; and it is, I think, not improbable that its variations are a principal ground of social stratification which tends to arise in all acquisitive societies, that is to say, in all civilized peoples.

We have found reason to believe that, though the Nordic race has no monopoly of genius, though it does not excel, and perhaps does not equal, other races in many forms of excellence (as so extravagantly claimed by the race-dogmatists), it yet has certain qualities which have played a great part in determining the history, the institutions, the customs and traditions, and the geographical distribution of the peoples in whom its blood is strongly represented.

This last sentence brings us face to face with the central problem of this book. Do the innate qualities of a race determine its geographical distribution, or do the geographical, economic, and social conditions pick out certain types of character for preservation, and thus determine their distribution? In a certain way both are true. Unquestionably, other things being equal, a race with an innate acquisitive instinct, for example, will be better able to survive in a cold country or a desert than will a race without that instinct. But if we go back far enough in human or pre-human history we presumably come to a time when races had not yet been differentiated, and there were only individual differences. If a group of individuals wandered into a cold or dry climate the quality of acquisitiveness at once gave some of them an advantage. The children who inherited acquisitiveness would tend to survive, and ultimately that quality might become characteristic of the race.

Such reasoning leads one to believe that the qualities discussed by McDougall are not so much characteristic of races as of regions. Take the Norse and the Sicilians as extreme types of the Nordics and Mediterraneans. What differences in their history are likely to have given rise to differences of character? One of the most obvious facts is that the ancestors of the Norse were presumably hunters, cattle-raisers, and migrants for thousands of years after the ancestors of the Sicilians had settled down to agriculture. The Norse apparently could not practise agriculture in most parts of their present home much more than two or three thousand years ago. They occupy one of the latest parts of the world to be freed from a cap of ice. Even during the Christian era their home has been so stormy that times of climatic stress and repeated failures of crops have caused many of them to wander away, or else to depend on cattle-herding and seafaring. This seems to be one reason why the Norse have spread so far and so widely, for they settled Iceland, came to England, made a home in Normandy, ruled Sicily, and intruded themselves in many other places. Moreover, in order to get a living under the harsh conditions that prevail in Norway it has been necessary to cultivate little isolated tracts of land, herd cattle in small and almost inaccessible mountain valleys, and venture forth over some of the stormiest seas in the world.

THE THREE GREAT RACES OF EUROPE 217

Nevertheless, the environment is decidedly healthful and has no such terribly repressive conditions as seem to be found in the extreme cold of Siberia, the extreme dryness of the desert, and the malaria of Sicily.

In the pastoral migrations preceding their arrival in Norway, and in the stress of establishing themselves in such a land, it seems clear that the Norse must have experienced an especially strong dose of the kinds of selection which we have found characteristic of nomads and of migrants. That is probably one reason why the Nordics make such good colonists. Curiosity as well as boldness, initiative, individuality, and the power of leadership all seem to be among the prominent traits that cause men to migrate to new regions, and make them successful when they get there.

In later times the sea has acted as a strong selective factor, for the death-rate among the young men is very high when they first go to sea. This puts a premium on the quick minds and strong bodies which are needed to cope with sudden storms, high winds and waves, and long exposure. The scarcity of tillable land in Norway and its division into small and isolated parcels, and the fact that water can be procured practically everywhere, give an advantage to the people who have the temperament that can live alone; the introvert is at a premium. Self-reliance and the capacity to do things alone and on one's own initiative are especially valuable in a land where a large percentage of the families must live by themselves and must depend entirely on their own ability and initiative during the long periods when darkness and snow keep them isolated. Self-assertion is equally valuable, for a successful trip in a fishing-boat during a storm, or a journey across a snowy pass at a season when darkness may overtake one in the middle of the journey take much more of that quality than is needed by the man who lives in a Sicilian village and never goes to sea. The need of acquisitiveness is equally obvious, for the Norse winters are very long; and food, fuel, shelter, and clothing must be prepared in advance, or some of the family will perish. Most of these qualities are as necessary for the women as for the men, especially in view of the frequent and long absences and all-too-common deaths which the seafaring life and the care of the cattle in the mountains impose on the men. In fact the Norse families where the women lack these qualities are likely to suffer seriously. Hence no matter what the original character of the race which migrated into Norway, it seems as though the process of natural selection would tend to preserve to a high degree the qualities of curiosity, individualism, introversion, self-assertion, and acquisitiveness. It is quite possible that some races, such as the Negroes, would be exterminated under such circumstances because they have already acquired opposite qualities under another environment. Or an incompetent race might find the environment repressive where the Norse find it stimulating. Hence we infer that if some non-Nordic race had migrated to Norway its character would, to-day, be markedly different from that of the modern Norse. Yet in comparison with its near relatives who migrated to other regions I am convinced that such a race would show a marked tendency toward many Norse characteristics.

The Sicilians, on the contrary, have been permanent farmers since an indefinite time far in the past. Greeks, Arabs, Normans, and others have indeed settled in Sicily at various times, but judging by the modern inhabitants, these stocks, or at least the northern elements of them, have died out, as we shall see more fully in Greece. The great bulk of the present Sicilians seem to be descended from ancestors who for thousands of years have cultivated their little farms and tended their few sheep and goats almost as their descendants do to-day.

Even if the original Mediterranean people who first settled in Sicily possessed a considerable degree of the kind of curiosity, initiative, and self-assertion which led men to move on into new lands, the long agricultural life has presumably allowed these qualities to grow weak. The introvert capacity and fondness for living alone have presumably been a handicap in many cases, for Sicily and most Mediterranean lands have so long and dry a summer that the vast majority of the inhabitants are forced

THE THREE GREAT RACES OF EUROPE

to live near the main supplies of water. But in a compact village the sociable extrovert family gets along much better than the self-centered introvert. Likewise the man who is mild and easily led has an advantage, for he gets along without friction. This is especially true if a region is governed by strong intruders from outside, as has been the case during long periods of Sicilian history. Again, in a country so warm and mild as Sicily the acquisitive instinct is not nearly so necessary as in regions where the winters are long and severe, but where agriculture is, nevertheless, the main mode of life.

In addition to this it must not be forgotten that for a long period Sicily, far more than Norway, has been densely populated, and hence has suffered from poverty and even famine. Thus there, as in China, there has been a slow but perhaps important outward movement which in almost every generation may have taken away a considerable number of the Sicilians who had most of the pioneer qualities which are so strong in the Norse. And finally the climate of Sicily with its mild winters and long, hot, monotonous summers does not stimulate activity as does that of Norway, with its coolness, its storms, its variability, and yet its freedom from really severe extremes of temperature. It is not improbable, although we cannot speak positively, that the climate itself and diseases like malaria have a certain direct selective influence so that the less strenuous types have an advantage in southern Europe and the more strenuous active types in the northwest. Whether this is true or not, the facts here given suggest that natural selection has been one of the main factors in differentiating the Nordic and Mediterranean races and in making the Alpines more or less intermediate between these two. That racial mixture and other factors have also played a part, I do not doubt. The emphasis on selection in this book does not in the least mean that other causes of racial differences are ignored or minimized.

CHAPTER XV

THE CHARACTER OF MODERN EUROPE

WE have looked at the character of Europe from the point of view of the three great races. Let us next attempt to get a composite idea of the character of the various countries in quite a different fashion. Let us get away from the subjective method where we deal with our own impressions and opinions, and let us look at the matter objectively in such a manner that our own opinions have no weight. One way to do this is by means of a consensus of opinion such as is found in a book like the Encyclopædia Britannica. The eleventh edition of that famous publication contains biographies of about 8,600 Europeans who were born since 1600 A. D. I have classified these men according to the country of their birth and the lines in which they distinguished themselves, as appears in the table at the end of this chapter. For simplicity let us employ only the five major groups there given: (1) religion, philanthropy, philosophy, and education; (2) science of all sorts, both natural and mathematical, together with invention, and engineering; (3) history and economics; (4) literature and art; and (5) politics, government, and war. In addition to these there are a number of people whom we have omitted, but who are included in the Encyclopædia Britannica because of their success in business, because they were freaks, or because they inherited high positions but were otherwise insignificant and would not have been included simply on their own merits.

We do not yet know how far the real character of a nation is expressed in its men of genius. It seems, however, that such men to a large degree not only guide the activity of a nation, but are themselves more or less controlled in their choice of work by certain definite tendencies belonging to their country and epoch. For example, the Swedes, as a whole, or at least the middle and upper classes, are reported to be unusually scientific in their tendencies. A Swedish geologist and a Swedish geographer who have worked in other countries have both told me that Sweden is a peculiarly easy country in which to work. When the investigator crosses a farmer's fence he has only to explain what he is doing in order not only to be given permission to go where he will, but to be urged to investigate further. Except in the busy seasons, a farmer will often put off his own work and tramp around with the scientist, eager to help and eager to learn. He seems to do so for sheer love of scientific knowledge. Even in Norway this is by no means so true as in Sweden, and in Denmark much less so. On the other hand among the Swedes the interest in religion is remarkably slight. The vast majority of the educated people look upon religion as something for the ignorant, but of no real use in the world. Even among the lower classes there is an extraordinary degree of indifference. With this, though perhaps as neither cause nor effect, there goes what seems almost to be a real deficiency in the qualities of sympathy and affection. In view of these facts it is not surprising to find that twenty-six of the ninety-one eminent Swedes mentioned in the Encyclopædia Britannica were scientists. This is a larger proportion than in any other European country except Switzerland. On the other hand, only a single Swede was sufficiently famous as a leader in religion and philanthropy to be included in Britannica. This is a much smaller percentage than in any other European country for which we have data enough to draw any reliable conclusions. Thus it would seem that both among the Swedish people as a whole and among the leaders there is a tendency toward a type of mind that is cold, clear, irreligious, and relatively unsympathetic, but at the same time highly scientific.

Among the 107 modern Swiss in our five Britannica groups, 40 per cent are scientists as against 29 per cent in Sweden, 17 per cent in Germany, 13 per cent in France, and only 2 per cent in Spain. Switzerland likewise has had a strong religious tendency, for 15 per cent of her leaders have been eminent in the domain of religion and philanthropy, which is a larger percentage than in any other country. The reader may suggest that the religious prominence of Switzerland is due to the fact that people who were persecuted in other lands have fled to the free republic. This does not alter our conclusion, however, for we are dealing with men born since 1600 who did not become prominent until religious persecution was almost at an end. We are also classifying these men according to the country of their birth, not of their residence. Perhaps the fact that 15 per cent of the Swiss leaders and only 4.5 per cent of the French leaders have belonged to the religious and philanthrophic group may be due partly to the migration of French Huguenots to Switzerland during times of persecution. An investigation of this matter would be most interesting.

Scotland with 14 per cent of its eminent men as religious leaders, and 20.2 per cent as scientists, closely resembles Switzerland, while even in England and Germany the same tendency is seen in the fact that the percentage of the leaders who have been eminent in religion amounts to 11.8 per cent in England, and 11.1 in Germany, whereas for Europe as a whole the percentage is 9.2, for Italy 6.8, for France 4.5, and for Sweden only 1.1. It seems generally to be accepted that among the rank and file of the people of all these countries the tendencies are much the same as among the leaders. Hence we believe that the character of the leaders reflects the general character of the people, although it is equally true that the character of the leaders in turn guides the progress of the people as a whole.

If it be true that the lines of endeavor in which the leaders of a country distinguish themselves are an index of the character of the people, we may gain some idea of the geographical distribution of different types of ability from Figures 6 to 10. These show the percentage of eminent men in each country who have distinguished themselves in the various lines. For some countries, of course, the total number of eminent men is too small to be significant. Nevertheless, eighteen European countries have more than twenty-five men among the eminent persons born since 1600, and listed in Britannica, while in fifteen countries the number rises above sixty. In each map there are four

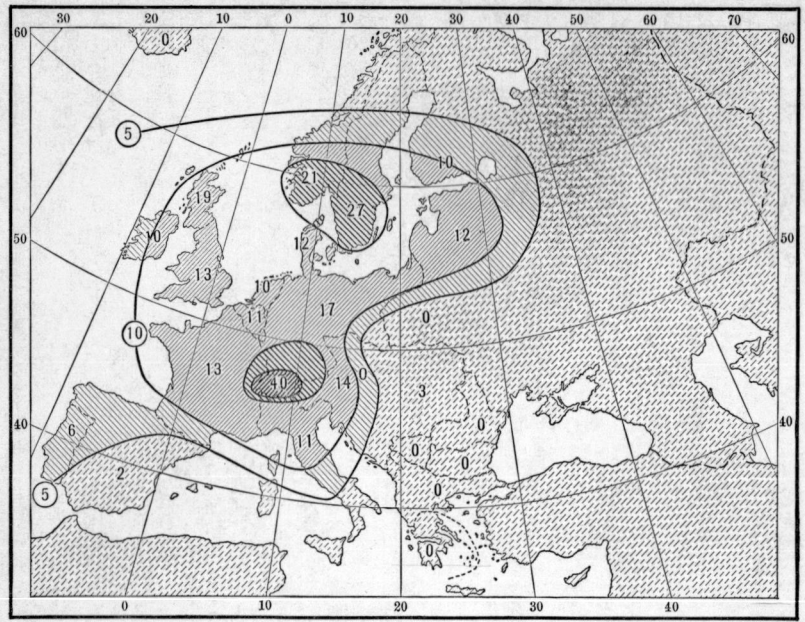

FIG. 6. PERCENTAGE OF SCIENTISTS AMONG EMINENT EUROPEANS.

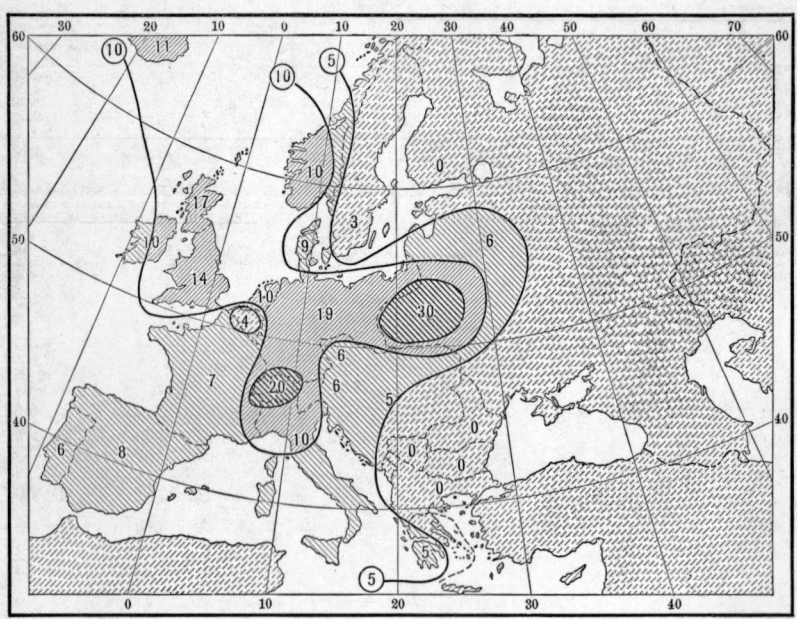

FIG. 7. PERCENTAGE OF EMINENT EUROPEANS ENGAGED IN RELIGIOUS, EDUCATIONAL, AND PHILOSOPHICAL WORK.

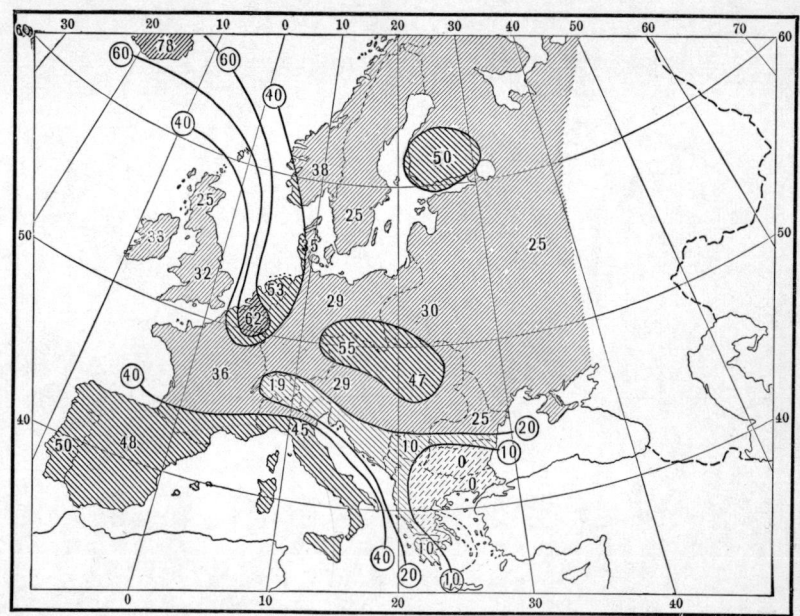

FIG. 8. PERCENTAGE OF EMINENT EUROPEANS ENGAGED IN ART AND LITERATURE.

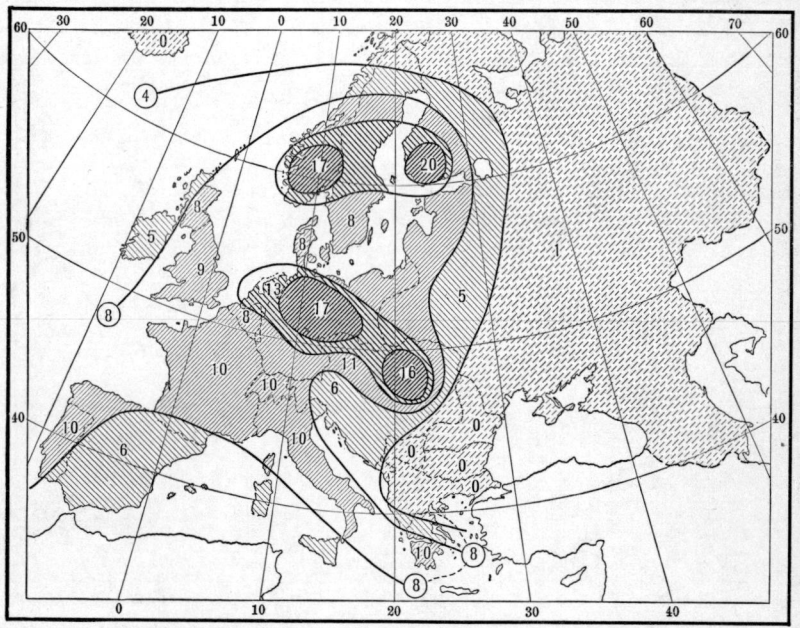

FIG. 9. PERCENTAGE OF HISTORIANS AMONG EMINENT EUROPEANS.

types of shading, to which a fifth exceptionally dark type is added in some cases to emphasize peculiar conditions in the Balkans, Switzerland, and elsewhere. The lightest shading indicates regions much below the average in the special lines of eminence shown in the map in question; the next shading denotes regions somewhat below the average; the third somewhat above the average; and the fourth and fifth much above the average. Figure 6 shows the degree to which scientists have been produced among the eminent men in each country. In Switzerland, as we have seen, 40 per cent of the eminent men have been of this type, and in Sweden and Norway the percentages are 27 and 21 respectively. Scotland and Germany likewise have produced a high percentage of scientists. In the rest of Europe the percentage has either been about the average, as in France and England, or less than the average, especially in the east and south. Of course a high percentage does not necessarily mean a large number of scientists, but merely that among the eminent men, whether they be few or many, the scientists hold a relatively important position. We shall consider the absolute number of eminent men a little later.

Now compare Figure 6 with Figure 7, showing the leaders in religion, education, and philosophy. Here again Switzerland, Germany, and Scotland stand high, but Poland outranks them all, while England is also prominent. But note the change in Scandinavia; Sweden falls into the lowest group. She is neither religious nor philosophical in her contribution to human progress. In art and literature, Figure 8, we come to quite a different type of distribution. Highest of all, not in actual numbers, of course, but in proportion to the total number of great men, stand Belgium, Austria, and the Netherlands, with Portugal, Finland, Spain, Hungary, and Italy following not far behind. England, Scotland, and Sweden are among the countries which have specialized relatively little in art and to a less degree in literature. Germany, Poland, Russia, and Austria fall into the same group. Here, as in each of the two preceding maps, the countries of southeastern Europe show no appreciable development.

Turning to history, Figure 9, we find a curiously spotted distribution. Men of this craft have been relatively most numerous in Finland, but Norway, Germany, and Hungary have likewise stood high, while the Netherlands, France, Italy, Portugal, and Austria stand above the average. In this group for the first time we find a southeastern country, Greece, rising above the average. Ireland, on the other hand, like Russia and Spain, seems to have taken little interest in recording its own history. Lastly we turn to Figure 10, which shows in what countries men have had the greatest tendency to distinguish themselves in war and politics rather than in more peaceful and more productive pursuits. Scandinavia, central Europe, and Portugal have had singularly few leaders of this type. In England and Scotland the average has been about like that for Europe as a whole, while in France, Denmark, Spain, Bohemia, and Hungary the number has risen a little higher. But note eastern and southern Europe. In Russia, Poland, and all the Balkan countries, including Greece, at least half of the eminent men have put forth their efforts in these two lines. In Servia nine out of the ten who are included in Britannica belong in this group, and eight of them were political leaders.

Let us now see if we can discover any reasons why the various countries have distinguished themselves in different lines. The task is extremely difficult and we are not sure of the value of our analysis. One thing stands out clearly. In the largest countries, or rather in those with the most persons in Britannica, the relative numbers of eminent men in each of the various branches of achievement show the smallest departures from the average for the rest of Europe as a whole. England is pre-eminent in this respect. This is partly, but probably not wholly, because the number of Englishmen mentioned in Britannica, 2,790 in our selected categories, is far larger than the number from any other country. Among the English, however, it may be that the able men have had such varied tastes that the different lines of endeavor are unusually well balanced. Scotland, with 604 people in our five main groups, seems to indicate pretty clearly that the differences from country to country are not accidental. The number of eminent Scotch is large enough so that accidents which might cause one group or another to predominate without any real cause are largely eliminated. Yet Scotland stands well above the average in men who have achieved fame either in religion or in the sciences. It is almost exactly average in literature, and in war and politics combined, while it falls distinctly below the average in historians, and has had only half the average percentage of artists. Inasmuch as the Scotch born since 1600 have had practically every opportunity that is open to the English or any other people, and have had no special handicaps as to the lines to which they could devote themselves, it seems as though it must be their racial character which leads them forward in science and religion, and keeps them back in history and art.

This seems the more probable when we consider Ireland. That country has 292 eminent men in our five groups. In spite of certain disadvantages, it is difficult to see any reason save their own character which should cause the Irish to have only two-thirds of the average proportion of scientists and half the average percentage of historians, whereas they have shown about the average tendency toward religion, art, and literature. Their strong tendency toward war and especially politics may be due to circumstances quite as much as to racial character, although it is hard to determine which is cause and which effect.

The high development of art and literature in certain countries is particularly interesting in its relation to racial character. In Belgium, for example, 48.7 per cent of the seventy-eight men in our five groups were artists, and 14.1 per cent distinguished themselves in literature. In Bohemia the figures are almost the same, though a trifle lower, while in Holland 39.5 per cent were artists and 13.7 per cent literary men. In Portugal, on the contrary, there has been a relatively strong development of literature, 57.7 per cent of the twenty-six men in Britannica belonging to that group, while only 3.8 have taken the lead in art. If we interpret Figure 8 in terms of both art and literature, we may say that in Europe the relative tendency toward literature has been particularly strong in three places, namely the far southwest, including both Spain and Portugal; the far north, including Iceland and Finland; and the south central part, Hungary. It would be hard to find three places that are racially more distinct, for the people of the southwest are primarily of the dark, long-headed Mediterranean type, those of Hungary are broad-headed intruders of Alpine stock, while in Finland we have an upper class of Nordic tendencies, especially in the west, and a lower class with broad-headed Alpine or Mongoloid characteristics. From the point of view of art Belgium, Bohemia, Holland, and Italy are the leaders, that is, they are countries where the proportion of artists compared with other eminent men has been greatest. In these countries the Belgians are relatively broad-headed, although with considerable Nordic admixture in the northern parts; the Dutch are more Nordic than the Belgians; the Italians are broad-headed Alpines in the north and long-headed Mediterraneans in the south; and the Bohemians are primarily broad-headed Alpines.

Perhaps there is some guiding principle in all this, but I cannot find it. Race, as we have just seen, does not seem to explain the presence of the artistic and literary temperaments. The size of the different countries does not seem to be of any particular significance, for Spain and Italy rank as large countries, while Belgium is small. Nor does there seem to be any environmental condition which would cause one part of Europe to tend to produce literary men and another artists. The distribution of people who attain eminence in religion and philosophy is equally puzzling. What common factor causes the Nordic Scotch and the Alpine Swiss to have a very strong religious tendency, while the Nordic Swedes and the Alpine Belgians have a minimum interest in religion? In most of the other maps the conditions seem equally puzzling. The only exception is the map of persons who distinguished themselves in war and politics, Figure 10. Here it appears clearly that in eastern Europe these have been the two branches in which there has been by far the most opportunity to rise to distinction. This may be explained partly on the basis of the backwardness of those countries, for in a backward country it is the warrior and the political leader who come to the front. It is also due in part to the fact that eastern Europe, even in the last three centuries, has felt the rough tide of invasion from the deserts of Asia. Only in 1683 was the Turkish conquest of Europe finally checked at Vienna, and the surgings due to the coming of the Turks and Tartars in Russia and the Balkans have scarcely died away even now. In such disturbed and backward countries the scientist, the religious leader, and the artist have little opportunity. But this explanation of the prevalence of warriors and politicians in eastern Europe does not help us in regard to the center. Switzerland, Holland, Belgium, Norway, and Portugal, being small countries and more or less protected by nature or by political agreement, may not have afforded much scope for political and military genius. Possibly the fact that Germany long consisted of a great number of little states had the same effect there. But why have Denmark, Sweden, Bohemia, Austria, and Ireland had such a large percentage of soldiers and political leaders among their great men? Ireland's percentage vies with that of Austria, and is thus above that of any country of Europe save Russia, Poland, and the states of the Balkan Peninsula. So far as political leaders are concerned, Ireland has had almost the same number as Scotland, 81 against 85, although it has but 292 men in our five groups compared with 604 in Scotland. We are again and again told that Ireland has been suppressed politically. Has such suppression bred political leaders? I think not, for the great majority of Irish political leaders since 1600 have not been representatives of little Ireland, but of imperial Britain. Moreover, in free America the Irishman tends to be a politician quite as much as in unhappy Ireland. Political genius seems to be born in him just as it is in the Turk, and just as a scientific temperament is born in the Swiss and Swedes, the artistic temperament in the Belgians, and literary proclivities in the Spaniards. Or would it be fairer to conclude that a series of historic accidents has given each country an impetus in one direction or another—an impetus which persists for generation after generation?

I am well aware that the preceding paragraphs fail to give any clear idea of the distribution of racial character. This is because no definite laws appear to be detectable in the distribution of types of eminence. One country certainly differs from another, but not according to any definite plan that can yet be recognized. If racial tendencies play a part in determining the lines in which genius manifests itself, as presumably is the case, those tendencies do not seem to manifest themselves as one would expect on the basis of the distribution of Nordic, Alpine, or Mediterranean blood. Nor do they conform to the distribution of racial kinship as denoted by the form of the head. So far as environment is concerned, it is equally hard to see any relationship, except in the tendencies toward war and politics, in the backward countries of the east and southeast. Perhaps historic development is the key to the problem, but that key would apparently have to be turned many times before it would unlock the door. The fact probably is that in its higher manifestations the peculiar qualities which we call racial character depend upon such complex and varied causes that for the present their analysis is beyond our power.

The character of a race, as of an individual, depends upon the energy, strength, concentration, and steadiness with which it pursues its aims quite as much as upon the elusive factors which determine what those aims shall be. In discussing that phase of our subject I shall recapitulate certain things which I have said in *Civilization and Climate*. The European ideal of civilization may be stated somewhat as follows: (Civilization is the stage of development which results when men display to the highest degree the power of initiative, the ability to dominate nature, the capacity for formulating new ideas and for carrying them into effect, the power of self-control, high standards of honesty and morality, the power to lead and to control other races, the capacity for disseminating ideas, the ability to express themselves in high types of art and literature, and other similar qualities. These qualities find expression in high ideals, respect for law, inventiveness, ability to develop philosophical systems, great works of art and literature, stability and honesty of government, a highly developed system of education, the capacity to dominate the less civilized parts of the world, and the ability to carry out

THE CHARACTER OF MODERN EUROPE 229

far-reaching enterprises covering long periods of time and great areas of the earth's surface.

On the basis of approximately this definition fifty-four geographers, anthropologists, historians, travellers, and others have classified the various regions of the world. These men represent fifteen countries, and the opinions of each of the following groups have received equal weight in the final classification: twenty-five Americans, seven British, six Germanic Europeans, six other Europeans (chiefly Latins), and five Asiatics. They rated the various countries on a scale in which 10 stood for the nearest approach to the ideal as expressed in the preceding definition, while 1 stood for the greatest divergence from that ideal. The average opinion of the five groups into which the fifty-four men were divided is shown in Figure 11. By giving America and Asia equal weight and by dividing Europeans into three groups animated by different ideals and different sympathies, we are able largely to eliminate the effect of racial prejudice. Fortunately the classification was made before the World War, so that it represents a far truer estimate of racial character than will again be possible until the hatreds and antipathies of the war have disappeared. The close agreement in the opinions of all five groups makes it almost certain that any other cosmopolitan group of eminent men would come to approximately the same conclusion. Hence Figure 11 appears to be the best available estimate of the geographical distribution of civilization and progress.

Let us compare this map with another, Figure 12, showing the number of persons mentioned in the Encyclopædia Britannica compared with the population of various parts of Europe in 1800. The estimates of the population at that time, as given in the table at the end of this chapter, are merely approximations in many cases; indeed, in certain instances, such as Greece and Turkey, they are little more than guesses controlled by the known facts of later times. Nevertheless they give a fairly true picture of the actual conditions in 1800 A.D. The reason for choosing that year is that it gives a fairly close approximation to the relative number of people in the various countries from

1600 to 1850 A. D., that is during the period when the men in our list were born. Previous to 1800 the data as to population are very unreliable. At that time, however, the rapid increase of population due to the introduction of machinery, railroads, steamships, and manufacturing had not begun. Hence the data for that year are fairly representative of the conditions during the two preceding centuries.

In the Encyclopædia Britannica the British Isles are of course represented by an unduly large number of names, and the same is perhaps true of France to a less degree. Moreover, a large country with a language in which many books are published may have some advantage over a small country. I doubt, however, whether this last source of error is significant, for in Figure 12 the numbers for Holland, Denmark, Switzerland, and Iceland compare very favorably with those for Germany and France. The mere fact of distance and of the degree to which the language of a country differs from English may also have something to do with the number of people who are included in the Encyclopædia. It is doubtful, however, whether even the fullest allowance for this would materially alter the appearance of Figure 12. "Ideas are light baggage" (Semple), and the editors of Britannica have for a century been on the *qui vive* to find new names appropriate for their book. In encyclopædias in other languages it is surprising to find that aside from persons living in the country where the book is published there are rarely any names which are not included in Britannica. Hence it appears that if the undue preponderance of the British Isles were toned down, and if the peripheral parts of the continent were given somewhat greater weight, we should have a fairly reliable map of the general distribution of eminent Europeans who have really influenced the world's progress. But such changes in Figure 12 would not materially alter its general appearance. The main feature of the map would still be an area around the North Sea where great men have been produced in relatively large numbers during the last three centuries. From this center the percentage of eminent men decreases in all directions except

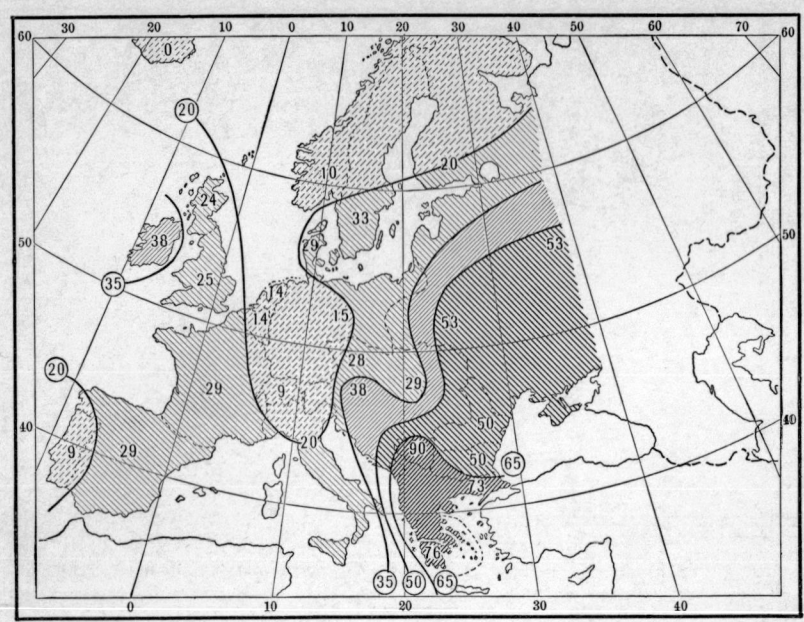

FIG. 10. PERCENTAGE OF EMINENT EUROPEANS ENGAGED IN WAR AND POLITICS.

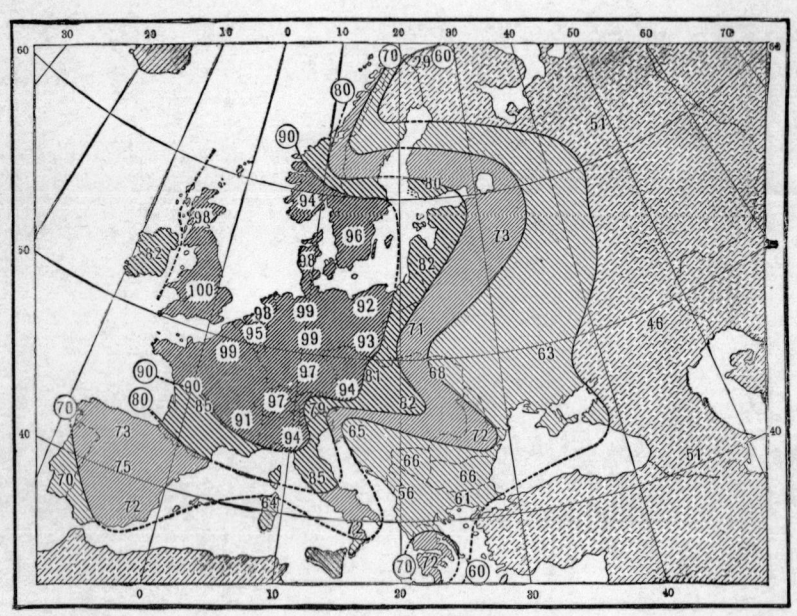

FIG. 11. DISTRIBUTION OF CIVILIZATION IN EUROPE.

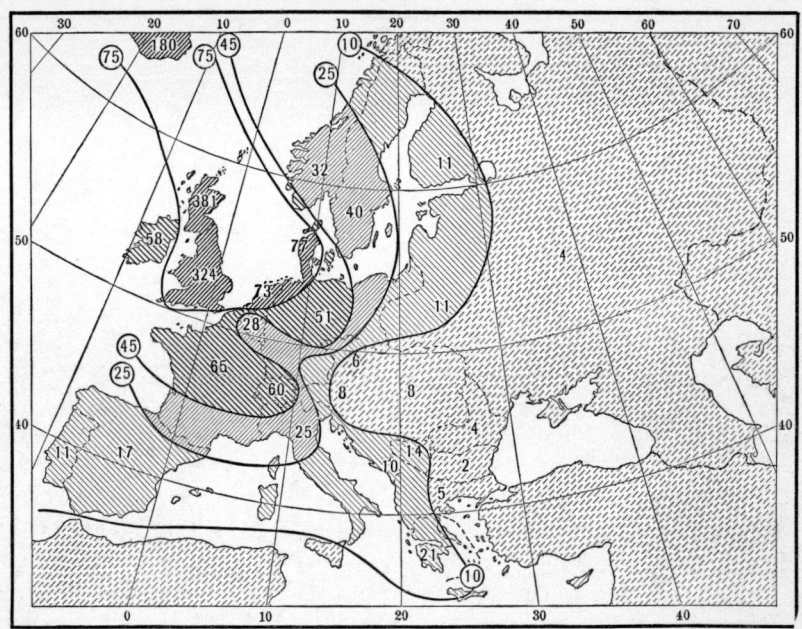

FIG. 12. EMINENT EUROPEANS BORN SINCE 1600 A. D. PER 10,000 OF ESTIMATED POPULATION IN 1800 A. D.

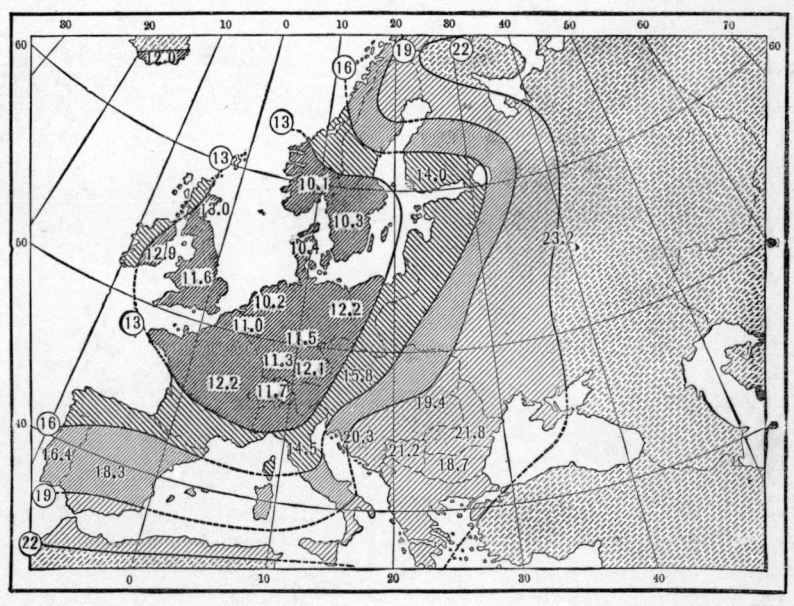

FIG. 13. DISTRIBUTION OF HEALTH IN EUROPE.

THE CHARACTER OF MODERN EUROPE

toward Iceland, where highly exceptional conditions prevail, as we shall see later.

Now compare the map of great men, Figure 12, with the map of civilization, Figure 11. The general aspect of the two is quite similar, and becomes still more so when allowance is made for the fact that we are following an English publication. In other words, it seems that the number of men of genius in proportion to the population agrees fairly closely with the relative position of the various parts of Europe in the scale of civilization. Note in each case the peculiar way in which there is not only an area of dark shading around the North Sea, but a projection toward the east in the Italian region and another in the Baltic region. Note also how Greece rises above the level of its neighbors in both maps. In one respect, however, Figure 12 contains a peculiar feature which is contrary to Figure 11. This is the belt of unexpectedly low achievement, which includes Belgium, Bohemia, Austria, and Hungary. Yet even in Figure 11 a hint of it can be seen in the relatively low position of Belgium, Bohemia, and the Austrian Alps. I shall not attempt to explain this, but it may possibly be significant that in general the central European belt of relatively few men of eminence coincides more or less closely with the belt where the broad-headed Alpine type of head is most common, as appears in Taylor's map, Figure 3, page 77. On the other hand, it must not be overlooked that in Switzerland, where the people are broad-headed, the proportion of eminent men is high.

The maps of the distribution of civilization and of men of eminence may both be regarded as illustrating the distribution of character. They do not indeed represent all phases of character, but they are based on the factors which are most vital in promoting human progress. In spite of the peculiar belt of low achievement in Belgium and Bohemia, the resemblance of the two maps is so great that we apparently must infer that they are related. But which is cause and which is effect, or are both the effects of some other cause? The answer is perhaps found in Figure 13, representing the distribution of health, and in

Figure 14, representing the distribution of climatic energy. The map of health is based on the average mortality statistics of the various countries of Europe, for it is generally agreed that the death-rate is the best available measure of health. The years 1909–1913 have been used in order to get normal conditions, such as prevailed before the World War. All the data have been reduced to what is known as a standard population, so that the differences in the death-rate due to a larger or smaller percentage of children or old people in the various countries are eliminated. Moreover, the deaths of infants have been eliminated, since the methods of recording these vary greatly from country to country. The deaths of people over seventy years of age have likewise been omitted, but these are too few to produce any appreciable effect. Even so the map is not perfect, for the mortality data in eastern and southern Europe are not very accurate, many deaths being unrecorded. But if these unrecorded deaths were included, they would simply increase the contrast between the more healthful countries, which are heavily shaded, and the less healthful, which are lightly shaded.

The map of climatic energy (Figure 14) is based on the way in which people's work, both physical and mental, varies from day to day and season to season among people of European origin in the United States. The amount of work done by pieceworkers in factories and the marks obtained by students on days with different kinds of weather give a fairly accurate measure of people's energy during a day or month with any given temperature, humidity, and degree of storminess. Records of health show variations almost identical with those of work. On this basis, knowing the average monthly conditions of the weather in each part of Europe, it has been possible to construct a map of climatic energy.

The similarity of the maps of civilization, genius, health, and climatic energy is so clear that it speaks for itself. In each map there is the same dark area around the North Sea, the same tendency toward a diminution in intensity in all directions, except that the maps of genius and health do not diminish toward Iceland. In all there are likewise projections in the regions of Italy

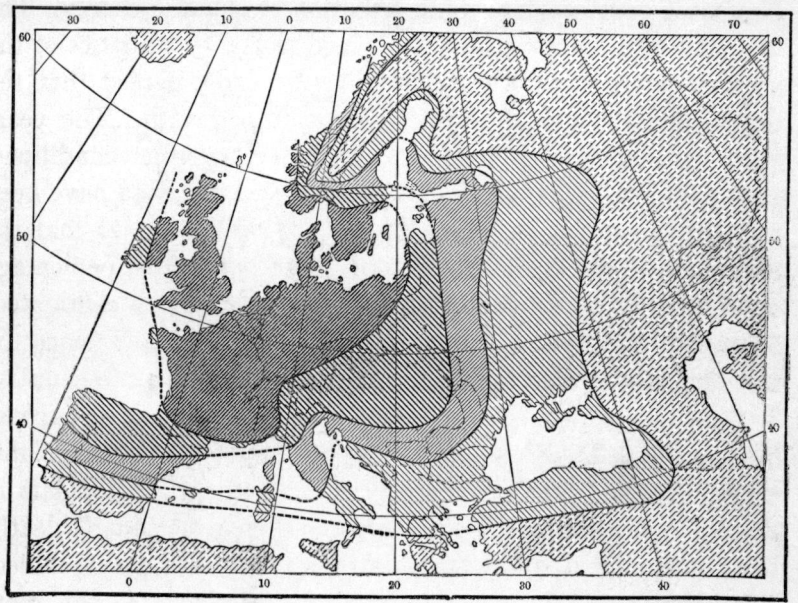

FIG. 14. DISTRIBUTION OF CLIMATIC ENERGY IN EUROPE.

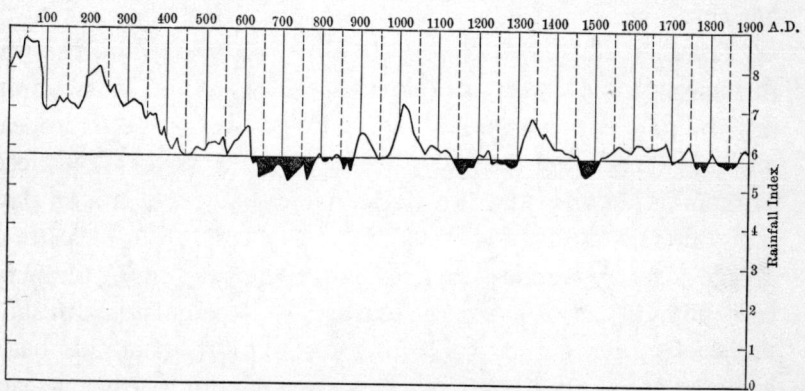

FIG. 15. APPROXIMATE VARIATIONS OF RAINFALL IN CALIFORNIA DURING THE CHRISTIAN ERA.

NOTE: This curve differs somewhat from those that I have generally published. It is taken from "Maya Civilization and Climate," *Proc. 2d Int. Cong. Americanists*, Washington, 1917. It has been adjusted to conform to the levels of the strands of Owens Lake. The part from 600 A. D. onward agrees closely with the curve of the Big Trees as corrected both by the author's method as described in "The Climatic Factor," Washington, 1915, and by the method of Doctor Antevs described in "Postglacial Climatic Changes," publication No. 352 of the Carnegie Institution of Washington, 1924. The part before 600 A. D. awaits further revision, but that does not materially affect this book.

THE CHARACTER OF MODERN EUROPE 233

and the Baltic Sea. That the four maps are intimately related can scarcely be doubted. The relation can be of only one kind. No one of the other three maps can have any effect on the map of climatic energy. That depends on nature, not man. Therefore it must be the foundation. Apparently climate influences health and energy, and these in turn influence civilization. The reverse is also true, for the stage of civilization has a great effect upon health.

But how is genius related to the other factors? Assume, for the moment, that the people in all parts of Europe are endowed with exactly the same degree of hereditary ability, and that the state of progress is everywhere the same, politically, industrially, socially, and otherwise. Would the proportion of men who rise to eminence be everywhere the same? I doubt it. The regions around the North Sea would probably always excel eastern and southern Europe. This is mainly because on an average the men of genius in the North Sea countries would be more energetic than those of other regions because they would enjoy better health, even though the medical service were everywhere equally good. They would be continually stimulated by their cool, bracing climate, and would feel like working hard all the year, whereas their southern and eastern colleagues in either hot weather or cold would be subject to periods of depression which are a regular feature of the less-favored parts of Europe. Because of their strength and energy the men of genius in the North Sea region would cause civilization to advance and incidentally would improve the conditions of health more than would their comrades of the south and east, even though the degree of innate ability was the same in all countries. Thus from whatever point of view the matter is approached, we seem forced to conclude that the phase of racial character which expresses itself in differences in energy, initiative, and the power of achievement is closely correlated with differences in the physical environment. This by no means indicates that inherited traits do not likewise have a great effect upon these qualities. It merely means that no bodily trait such as head-form, complexion, or other supposed indication of racial affinity has thus far been found to display a geographical distribution like that of men of eminence, or like human energy, initiative, and the power of achievement. On the other hand, it cannot be too strongly emphasized that the *type* of activity which prevails in different parts of Europe seems to be strongly influenced by inheritance. An artist presumably retains the artistic temperament no matter where he is or how great his energy or lassitude. A typical Irishman is a good politician and the Jew a merchant wherever he may be. Thus racial inheritance is presumably a strong factor in directing men's lines of effort. It is likewise a strong factor in determining the proportion of men of genius produced in any country. On the other hand, the degree of energy with which each race pursues its chosen lines, and the percentage of its men of genius who use their capacities so fully that they become eminent, seems to depend largely on physical environment.

Thus at least two of the fundamental reasons why Europe leads the world seem apparent. First, during the glacial period the ancestors of the present European races presumably suffered an extremely severe process of natural selection by reason of the alternate expansion of the ice-sheets on the one hand and of the deserts of Asia and Africa on the other. To this perhaps is due the superiority of the Mediterranean-Caspian blend as set forth by Dixon. Second, ever since the last glacial epoch had sufficiently passed away, some part of Europe or of the neighboring regions of western Asia and north Africa has enjoyed a climate unsurpassed in the qualities that give health and energy. As the climate has changed and as man has learned more fully to protect himself against the cold, civilization has advanced northwestward almost as if Scandinavia, Scotland, and Iceland were its goal. At every stage people of unusually good inheritance have been naturally selected to colonize those regions and have experienced the stimulus of an unusually good climate. Thus they have acquired another asset of enormous value, namely, the power to profit to an unparalleled degree by the knowledge, the institutions, and the ideals handed down from former generations.

CLASSIFICATION OF EMINENT EUROPEANS BORN SINCE 1600 WHOSE BIOGRAPHIES APPEAR IN THE ENCYCLOPÆDIA BRITANNICA

	Estimated Population in 1800 A.D. on Basis of Data of Levasseur and Bo-dio in Bull'n Inst. Internat. de Statistique, vol. 12, pt. 2, 1902.	Religion and Philanthropy	Group 1 Philosophy and Education	Group 2 Natural Sciences	Group 2 Math. Sciences and Invention	Group 3 History and Economics	Group 4 Literature	Group 4 The Arts	Group 5 Politics Revolution	Group 5 War and Adventure	Inheritance Without Other Special Merit	Business	Freaks	Total
Austria	16,000,000	5	1	16	1	8	19	17	23	25	10			125
Belgium	3,800,000	3		5	4	6	11	38	10	1		1		79
Bohemia	3,100,000		1			2	2	8	3	2				18
Bulgaria	1,000,000													2
Denmark	900,000	4	2	4	4	6	18	6	17	3	5			69
England	9,300,000	329	87	203	197	261	547	408	493	265	93	108	25	3,016
Finland	900,000			1			5		2					10
France	28,000,000	80	53	104	128	189	373	288	308	214	36	17	18	1,808
Germany	24,000,000	132	94	113	93	218	201	155	98	81	16	15	5	1,221
Greece	1,000,000 (?)	1	4		4	2	2		10	6				21
Holland	1,700,000	8	1	8		17	17	49	8	9				124
Hungary	9,500,000	3	2	2	15	12	25		10	4				73
Iceland	50,000						6							9
Ireland	5,200,000	27	15	14	31	17	63	38	17	35	4		3	302
Italy	16,000,000	26		13		39	60	118	62	18	10	3	1	396
Montenegro	100,000													1
Norway	900,000	2	1	3	3	5	6	5	1	2	1			29
Poland	4,000,000	3				2	5	8	1	11	1			43
Portugal	2,900,000	1		2		3	15	1	12	1	6			32
Rumania	3,000,000	1					2		2		2			12
Russia	35,000,000	7	2	6	8	4	25	10	6	19	3	2		140
Scotland	1,700,000	85	26	62	60	53	116	47	55	70	20	24	3	648
Servia	700,000													10
Spain	10,000,000	9	4	13	2	11	55	27	32	18	9	2		170
Sweden	2,350,000	1	2	13	13	7	17	7	24	7	4			95
Switzerland	1,800,000	16	6	26	17	11	14	7	5	5	1		1	108
Turkey	3,000,000 (?)								8	1	6			15
Totals		744	301	596	580	875	1,605	1,248	1,372	798	228	177	52	8,576
Percentage of Total		8.79	3.04	6.42	6.57	9.62	18.09	14.50	15.43	8.98	2.57	1.99	0.57	

CHAPTER XVI

THE CONTRAST BETWEEN GREEKS AND IRISH

HAVING finished our general survey of Europe, we are ready to turn to certain specific cases where natural selection may help to explain the character and history of nations. Among the few that we can here consider, Greece and Ireland are especially puzzling and present a remarkable contrast. In Greece we are puzzled as to why there was so sudden and magnificent an outburst of genius in a limited area. It is true that from 700 to 400 B. C. Greece seems to have enjoyed an unusually favorable climate, together with comparative freedom from malaria and other diseases. Nevertheless, the extraordinary concentration of genius in Greece as a whole, and especially in Athens, still remains unique and unexplained. In the same way the increasingly troubled and desperate condition of Ireland does not seem in harmony with the physical environment, nor is it adequately explained by the common views as to land tenure, absentee landlordism, and lack of home rule.

In spite of the wealth of legend, the early history of Greece is not clear. For example, there has been a violent discussion as to what part the Pelasgians played in old Greece, as to the reality of the Achæan invasion, and as to the connection between the Achæans and the culture of Mycenæ. After long controversy, however, the doubts as to the Pelasgians seem to be clearing away, while the work of such men as Ridgeway and Leaf is fast leading to conviction as to the Achæans. Hence without further explanation I shall recount the early history of Greece substantially as given by these latest authorities.

The earliest-known inhabitants of Greece appear to have been a people known as Pelasgians, or, as some prefer to say, Ægeans. They were apparently a Mediterranean race, and were probably short, dark, and straight-haired. Presumably they came from the south or east by sea. We do not know what contributions they made to human progress during the early days, when they were still under the stimulating and selective influence of migration and a new country. In later times, however, they seem to have settled down into a rather inert peasantry which did little worthy of note. They appear to have been largely serfs or helots, as in Sparta. In Homer the chiefs speak to them in the roughest and rudest manner, and do not hesitate to beat them on small provocation. In later times they continue to be despised, especially in Sparta, while in Attica we are told by Hecatæus that they were expelled by the Athenians. They were least important in the very places where later races displayed the greatest vigor and made the greatest progress.

Yet in the long run the Pelasgians appear to have gotten the better of each of the succeeding races. As Myers puts it in his lecture on *Greek Lands and the Greek People:*

> Broadly speaking, the history of Man in the Ægean, as in the Mediterranean world of which it is the microcosm, has been the resultant of two groups of forces. On the one hand, it has been the history of the attempts of non-Mediterranean men to penetrate from elsewhere into the Mediterranean coast lands, and of the earlier occupants to keep them out, almost always unsuccessfully. On the other, it has been the story of the attempts of the successful invader to acclimatize himself to Mediterranean nature, to learn her ways, and through conformity with them, to conquer her and survive. And these attempts also, in general, have failed. No type of non-Mediterranean invader has ever yet learned so quickly how to live under Mediterranean conditions, as to escape extinction in the process.
>
> There is clearly something in the physique of "Mediterranean" man which fits him in a peculiar way for life on the Mediterranean seaboard; for, once established there, in days of which we still know nothing, he has succeeded, in a remarkable way, in maintaining himself against all rivals within very rigid limits. A recent piece of work by Mr. Hawes illustrates well the rigidity of this physical control. Crete, during the centuries from the thirteenth to the seventeenth A. D., was a political appendage of Venice, and, like all Venetian dependencies, endured a copious inflow of settlers and functionaries from Venice itself and its home-colonies on the Dalmatian coast, a markedly "Alpine" region. Many Cretan families retain Venetian surnames today, particularly in certain provinces. It was therefore natural to suppose that if physical types could be implanted permanently by colonization extending over three or four centuries, these fam-

ilies would retain physical traits characteristic of the region from which their ancestors came. But, province by province, the Venetian-named Cretans today are of precisely the same physique as their Greek-named neighbors; in the more "Alpine" provinces they are "Alpine," in the "Mediterranean" districts they are "Mediterranean." Either, therefore, well-marked physical types of men are so unstable as to be modified by local conditions within a dozen generations, or they are so sensitive to a strange environment as to die out like exotic seedlings among indigenous weeds. Between these two alternatives we can hardly hesitate to choose, when we remember the deadly selection which is exerted in the South by infantile diseases; for it is believed that hardly one in three of the children who are born in Greek lands lives to its first birthday. With an infant mortality like this, and with a further steady drain on adults from malaria and other regional diseases, it is not surprising to learn that intruders must be very persistently reinforced from the outside, if they are to maintain their race alive under Mediterranean conditions. . . .

Outside the Mediterranean limits, however, the Mediterranean man is apparently as powerless to establish his race, in competition with the men of the Alpine and Anatolian highlands, as they are to displace him within sight of the sea which is his home. We reach, therefore, this notable conclusion, that in these apparently favored lands there is yet a physical control so efficient as to make acclimatization exceedingly difficult and slow; so that, though exotic types of man make their way from time to time either tumultuously or in persistent driblets into the Mediterranean world, their independent existence is destined to be brief; and after a very few centuries, their presence is difficult to detect. The physical evidence, however, is already sufficient to show that slow infiltration of foreign types has taken place nevertheless; that it has had its maxima at ascertainable periods; and that the types which alone show any ability to acclimatize or to amalgamate with the indigenous "Mediterranean race" are of the "Alpine" group, not the blond giants of "Northern" stock.

It may be that Myers goes somewhat too far in his conclusion as to the rapidity with which all traces of invading stocks disappear. Nevertheless, there seems to be overwhelming evidence that his general proposition is correct. Hence it would seem that the fundamental fact as to the Greeks is that the basis of the population consists of a Mediterranean race, the Pelasgians, or Ægeans. These people have made little or no contribution to history, and appear to have changed very little throughout the five thousand years during which we have more or less evidence of their existence. Their importance lies not in their relation to the brilliant epochs of Greece, but in the fact that after immigra-

THE CONTRAST BETWEEN GREEKS AND IRISH 239

tion has given rise to brilliant progress the Pelasgians have invariably ousted the invaders. They have not done this by war or violence, but by intermarriage and the absorption or dilution of the new blood, or else through natural selection. They have been able to endure the long, hot summers and the malaria and other diseases of Greece, while the invaders have been less resistant. Hence we may think of Greece as a country whose history would have been relatively placid, uneventful, and of little importance to the rest of the world if it had been inhabited only by Pelasgians. But because it received invaders and because migrations so regularly pick out the especially able types, Greece has risen to greatness at several epochs.

The first such rise was due to the coming of the Minoans or Cretans. The Cretans, it will be remembered, developed a wonderful civilization which dawned some three thousand or more years before Christ, and was in its prime a thousand or fifteen hundred years later. The Minoan palaces with their labyrinths, which gave rise to the tale of Theseus and the Minotaur, have become widely familiar in recent years. So, too, have the remarkably modern clothing of the early Cretans and their high ability in commerce and the arts. The record of the early relations between Cretans and Greeks consists mainly of ruins upon the Ægean Islands and at the heads of many bays which open southward on the Greek mainland. Whether the Cretans first came as merchants, pirates, or warriors, or as colonists escaping from trouble at home, no one yet knows. But in Homer, where first we have mention of the Minoans, they are leaders, even after Greece has been conquered by the Achæans. "Colonists on a large scale," as Leaf well says (*Homer and History*) "go as conquerors—they are conscious of their own worth, and confident in the future before them." In other words, if the difficulties which they overcome are great enough, such colonists are usually a selected people, who hand down their ability by inheritance. That such was the case with the Minoans appears from the fact that they imposed upon the Pelasgians a rule which apparently lasted many centuries. Moreover, they not only introduced

Minoan art, but developed it along lines of their own until it became the famous Mycenæan art which prevailed in Argos and other parts of Greece at the epoch of the Trojan War about 1200 B. C. Homer gives minute descriptions of the products of this first outburst of genius in Greece, an outburst due largely, it would seem, to the coming of selected invaders from Crete. Yet in the days of Homer, Mycenæan art was deteriorating. The Minoans had reached a splendid climax, but through intermarriage with the Pelasgians or for some other reason were beginning to decline.

The next step was further migrations, Achæan, Dorian, and Ionian. These probably began far north of Greece, perhaps in the fifteenth century B. C., and ended in the eighth or ninth. The three migrations were closely connected, and apparently involved only one race—the Greeks who gave Greece its glory—although different tribes were concerned. The fact that there is some doubt as to how distinct the Achæan migration may have been from the Dorian makes no difference in our general argument. The Achæans, to follow Leaf, were apparently the vanguard of a flood of blond incomers from the north, whose first wave overwhelmed Greece in the fourteenth century B. C., and passed on to Crete, where its ravages are seen in the destruction of the famous palace of Knossos, the most noteworthy of the labyrinths. In Greece itself, however, the conquerors did not destroy the old civilization. On the contrary, like the Norsemen in Normandy and their Norman descendants in Sicily, they appear merely to have seized the castles and strongholds and ousted many, though by no means all, of the old Minoan rulers. Then they assimilated the Minoan or Mycenæan art and even encouraged its practise by the survivors of the old dominant class.

Like all people who have migrated far or traveled widely, the Achæans, and likewise the Dorians who followed them, appear to have been a highly adaptable people, able easily to assume new customs and a new culture. Their long migration presumably from the grassy plains of the north, may have fostered this spirit of adaptability and of hospitality to new ideas, but it seems to be still more true that it was the innate possession of this spirit or temperament which made the Achæans willing and able to keep on migrating. Those of their original number who lacked this spirit presumably tended to stay in the old homes or to give up the onward movement long before the sea was reached in Greece.

In later times [as Leaf says] the people who most markedly of all the Greeks showed this readiness to imitate and adopt foreign manners, even after the distinction between Greek and barbarian was fully developed, were the Asiatic Ionians. Whatever the real origin of the Ionians may have been, it is hardly to be doubted that the Achæans had a large share in their formation. In this almost excessive adaptability we may fairly see an inherited type.

The Trojan War is an illuminating episode in the development of the Greece of classical times. The leaders were largely Achæans with a few Minoans, while the common people, although called Achæans, were presumably in large measure Pelasgians. The Greek expedition from the mainland across the Ægean may in some ways be counted as the beginning of the Ionian migration which later gave to the western shore of Asia Minor a thoroughly Greek character. Many of the greatest Greeks, including Sappho, Alcæus, and presumably Homer, came from that region. The real reason for the Trojan War is unknown. It may have been a conscious desire to conquer the lands across the Ægean and hold the control of the Hellespont, but it is doubtful whether the Greeks of that day thought in such broad terms or could get together so many independent princelings for any such endeavor. It seems more likely that an ambitious leader was able to crystallize a general state of unrest and commotion, and thus to lead an army to a region where there were possibilities of plunder and settlement. The reason for this unrest may have been the later heavings of the Achæan migration, or the first waves of the Dorian migration which broke in full force a little later.

There is a constant tendency for the historian to crystallize migrations into a single clear-cut event. As a matter of fact, primitive migrations usually consist of a long series of movements, a few of which involve the sudden inrush of a fairly large number of people, but most of which consist of the far slower infiltration of small groups. Viewed in this way the Achæan, Dorian, and Ionian migrations are all parts of a single movement whereby the Greeks finally gained control of Greece proper, the Ægean Islands, and the western shores of Asia Minor. First came the Achæan tribes, chiefly, we may suppose, by filtration, but perhaps with a sudden onrush at one time which took some of them across to Crete. The pressure of those who kept wandering down from the north presumably helped to induce some of those who had come earlier to attempt conquests east of the Ægean. Thus perhaps arose the Trojan War, which may be counted as the beginning of the Ionian migration. Two or three generations later another somewhat sudden and violent incursion of new tribes from the north is known as the Dorian invasion. This and the influx of new people which followed during the next two or three centuries made such commotion in Greece and disturbed the earlier comers so much that many of them, together with some of the old Minoan leaders, migrated across the sea to the Asiatic coast. This seaward movement is known as the Ionian migration. It produced famous Greek settlements like Chios, Ephesus, and Miletus. But in reality it was merely the outflow of the Achæans and Minoans who were displaced by the Dorians or retired in disgust because of the political and social confusion.

In this connection a passage from Thucydides is most illuminating. He is evidently talking about the centuries of migration and disturbance which we have just been describing. He not only states the general conditions clearly and forcibly, but indicates that the settlement of Attica was largely due to the Ionian migration, even though Attica lies west of the Ægean. Moreover, he gives an important hint as to one reason for the greatness of Athens.

The country that is now called Hellas was not regularly settled in ancient times. The people were migratory, and readily left their homes

THE CONTRAST BETWEEN GREEKS AND IRISH 243

whenever they were overpowered by numbers. There was no commerce and they could not safely hold intercourse with one another either by land or sea. The several tribes cultivated their own soil just enough to obtain a maintenance from it. But they had no accumulations of wealth, and did not plant the ground; for being without walls, they were never sure that an invader might not come and despoil them. Living in this manner and knowing that they could not anywhere obtain a bare subsistence, they were always ready to migrate; so that they had neither great cities nor any considerable resources. The richest districts were most constantly changing their inhabitants; for example the countries which are now called Thessaly and Bœotia, the greater part of the Peloponnesus, with the exception of Arcadia, and all the best parts of Hellas. For the productiveness of the land increased the power of individuals; this, in turn, was a source of quarrels by which communities were ruined, while at the same time they were more exposed to attacks from without. Certainly Attica, of which the soil was poor and thin, enjoyed a long freedom from civil strife, and therefore retained its original inhabitants. And a striking confirmation of my argument is afforded by the fact that Attica through immigration increased in population more than any other region. For the leading men of Hellas, when driven out of their own country by war or revolution, sought an asylum at Athens; and from the very earliest times, being admitted to rights of citizenship, so greatly increased the number of inhabitants that Attica became incapable of containing them, and was at last obliged to send out colonies to Ionia.

In ancient times both the Hellenes, and those barbarians whose homes were on the coast of the mainland or in islands, when they began to find their way to one another by sea had recourse to piracy. They were commanded by powerful chiefs, who took this means of increasing their wealth and providing for their poorer followers. They would fall upon the unwalled and straggling towns, or rather villages, which they plundered, and maintained themselves chiefly by the plunder of them; for, as yet, such an occupation was held to be honorable and not disgraceful. On land also neighboring communities plundered each other—all Hellenes carried weapons because their homes were undefended and intercourse unsafe; like the barbarians they went armed in their everyday life. And the continuance of this custom in certain parts of the country indicates that it once prevailed everywhere. (Bk. I, par. 5.)

Even in the age which followed the Trojan War, Hellas was still in process of ferment and settlement, and had no time for peaceful growth. The return of the Hellenes from Troy after their long absence led to many changes; quarrels, too, arose in nearly every city, and those who were expelled by them went and founded other cities. (Bk. I, par. 12.)

This quotation is interesting not only because it shows that during and after the period of migrations Greece long suffered from a state of chaotic confusion, but because it states that the people who were forced to migrate, or who chose to migrate, were the leaders. The common people, the old Pelasgians, remained in their homes; the Minoan, Achæan, and Dorian leaders were the ones who chiefly migrated. The case was like that which we have seen suggested in connection with Cambodia and which we shall see unmistakably in Norway. From among a people already highly selected by the nomadic life of the grasslands of Asia and by prolonged and strenuous migrations continued probably for hundreds of years, there was yet another selection whereby Ionia and especially Attica received a peculiarly large proportion of the ablest leaders, those who thought it well to seek new homes rather than yield to the invaders or endure an endless struggle.

It seems to me that this concentration of able people in Attica goes far toward explaining the preëminence of that region. If the Pelasgians had been expelled, as stated by Hecatæus, the "original inhabitants" mentioned by Thucydides must have been Achæans. Therefore Attica had no helots and was one of the few places, perhaps the only one of any size, where the population was almost purely Greek. Certain other conditions tended to intensify and perpetuate this group of unusually capable people. For example, the very poverty of Attica led the leading families to concentrate in Athens, which probably increased the chances that the upper classes would marry in their own group and not among the groups below them. Again, after the earlier period of colonization, that is the Ionian migration, Athens "took no prominent part in these later colonizing movements, preferring to concentrate her resources at home." (Edwards.) Moreover, her *citizens*, as distinguished from the lower classes, kept out of commerce and devoted themselves largely to affairs of state. Commerce was carried on by slaves and by the *metics*, or aliens. This again helped to prevent the upper classes from wandering to other regions. Much the same was true in Sparta, where also the display of genius was unusually great.

Another important condition was that in Attica, as in Sparta, the cleavage between the aristocracy and the lower classes was unusually sharp. The Athenian serf, as Abbott puts it in his *History of Greece*, "was degraded to a position in comparison to which the Helot of Sparta was happy." Hence intermarriage was less likely than in places where the separation was not so great. Thus the less competent Greeks were excluded from contributing their weaknesses to later generations of the dominant classes. This was the more true because the Athenians were very proud of their descent and very jealous of any admission of the children of women of the lower classes into the upper classes, no matter who the father might be. "To you, men of Athens, who are citizens by descent," says Demosthenes, "it is fitting that no sum of money however great should be preferred to respectable birth," but for freemen who have no pride of birth the case is quite otherwise. A further limitation against the loss of the pure Athenian inheritance is found in the law that an heiress must marry her eldest paternal uncle or eldest cousin who is the son of one of her father's brothers. Again, intermarriage between families of different states or cities continued down to the early fifth century, but at that time Athens became too exclusive to permit this. Here again is another of the many evidences that the Athenians of the upper classes guarded their inheritance with peculiar care. If Athens were unique in this respect, we might conclude that her care had nothing to do with her greatness. But inasmuch as we find many other cases, like the Parsis of India, where selection seems to give unusual ability and the ability persists as long as the inheritance remains unimpaired, it seems as though the greatness of Athens might be connected with the fact that her people were almost purely Greeks, not Pelasgians; they were the sifted remnants of a long and hard migration; they were to a large degree the leading families among the early migrants from other parts of Greece; and they preserved their inheritance unmixed to a peculiar degree.

In spite of their inheritance, Greece and Athens fell into decay. No doubt this arose partly and perhaps largely from political and social causes unconnected with inheritance. Yet there was apparently a change of inheritance. We have already quoted from Myers a widely held opinion that the Mediterranean climate and its diseases, especially malaria, tend strongly to [bre]ed out the non-Mediterranean types, especially the blond [no]rthern races. It seems quite clear that the upper classes of ancient Greece were a fair-haired people, for otherwise it is almost impossible to explain their ideals of beauty as disclosed in literature and in the blond shades in which their statues were painted. But to-day the blond element has practically disappeared from Greece. I am not one of those who thinks that all the world's ability is concentrated in any one race, but in this case it happens that such a race has experienced a peculiarly strong natural selection. The process by which the blond race died out seems to be indicated in some statistics which have been collected by Mayor in *A Companion to Greek Studies*. Judging by the statements of Thucydides and the figures as to the distribution of grain in 445 B. C., the number of Athenian citizens at the beginning of the Peloponnesian War (431 B. C.) was about 35,000 in the three upper classes, and 20,000 Thetes of the lower class, together with 10,000 or 15,000 metics or alien residents who did not enjoy citizenship. A century and a quarter later the census of Demetrius Phalerus (317–307 B. C.) shows a reduction to 21,000 citizens of all classes, but the number of metics, 10,000, does not seem to have fallen nearly so much. Part of this great loss among the citizens was due to the plague in 430 B. C., part to the Peloponnesian War, and part to other causes. At Sparta in early days there are said to have been 9,000 lots or portions for fully privileged citizens, and 30,000 for lower citizens. Herodotus says there were 8,000 full citizens at the time of the Persian War (490 B. C.), but Xenophon indicates only about 1,500 at the time of the battle of Leuctra (371 B. C.); Aristotle, before 300 B. C., gives the number as 1,000, while in the days of Cleomenes (240–220 B. C.) there were said to be 700. In the first century after Christ the whole of Greece is reported to have been able to produce only 3,000 heavy-armed hoplites.

THE CONTRAST BETWEEN GREEKS AND IRISH 247

Whatever may be the exact figures, it is obvious that from about 400 B. C. onward there was an ominous and increasingly rapid decline in the Greek population. So far as figures are available they suggest that the upper classes suffered most. The present Greeks suggest the same thing: they are almost universally brunets of the Mediterranean type; the blond type, which seems to have predominated in the upper class of ancient times, is practically gone. Elsewhere I have shown that one reason for this may be an apparent change of climate. During the great age of Greece the climate appears to have been stormier, a trifle cooler, and decidedly more bracing than now. Malaria likewise appears to have been relatively scarce and to have done little harm. These conditions were presumably an important element in allowing the selected Greek migrants to rise to such heights of achievement. When the climate changed for the worse, especially from about 300 to 200 B. C., a great adverse selection seems to have taken place whereby the fair Greeks of the intellectual classes died out, thus leaving only the old Pelasgian stock as progenitors of the modern Greeks. In other words, Greece apparently agrees with many other parts of the world in showing that rigid selection produces a competent race, but the degree of achievement of that race and the permanence of its virility depend largely upon the amount of mixture with other races and the kind of climate.

Now let us cross Europe and glance for a moment at Ireland. Because that island lies on the extreme margin of Europe, one might expect that it would have been a terminal area into which many migrations would have poured, thus giving it a peculiarly selected and competent population. But this has been the case to only a moderate degree. At an early date Ireland received Mediterranean migrants who were presumably quite capable when they first came, but whose capacity may have diminished because, like the Pelasgian Greeks, they have not been subjected to the stronger types of natural selection. In later times other migrants have come to Ireland; but only in the case of the Scotch-Irish of Ulster have they become a permanent and distinct element of the population. It seems to me that much of the difference between Ireland and England is attributable to two geographical facts: first, Ireland lies farther from continental Europe than does England and is separated from the continent by the larger island; second, although Ireland has a fairly good climate, it is too damp, the crops are limited, famines occur now and again, and the difficulties of getting a living are much greater than in the neighboring island. In neither case is the difference between Ireland and England great, but both are important.

England has been the final resting-place of large groups of competent migrants—Kelts, Romans, Angles, Saxons, Danes, Normans, Huguenots from France, and various others. These newcomers have in most cases been much more competent than the average of the countries or races from which they have come. Therefore England has been greatly stimulated. In many cases these same migrants have spilled over into Ireland, but generally that has been all. Because Ireland is more remote than England and because the force of migration has been broken by England, and also because Ireland is on the whole less desirable than England, the Emerald Isle has received little more than the splashings of the migrations.

But something else has likewise happened. Ireland's cloudiness, dampness, famines, and remoteness from the centres of Europe have constantly tended to drive away the most able of her sons. The chances for a career have been far greater in England or France than in Ireland. Life in those other countries has been pleasanter, and the rewards of success much greater. So the young leaders have made the easy journey across the water, while the less competent peasants have stayed by the "stuff" at home. In later years the same process has taken place on a still greater scale. For several decades or generations during the late eighteenth and early nineteenth centuries, a comparatively dry period and the absence of summer frosts, such as that which ruined the potato crop in 1739, seem to have given Ireland excellent crops most of the time. Hence the population increased

THE CONTRAST BETWEEN GREEKS AND IRISH 249

rapidly. In 1845, in spite of six seasons of dearth which almost led to local famines between 1831 and 1842, the population reached a maximum of 8,300,000. Then came a series of damp, rainy years with such complete failure of the potato crop that 200,000 to 300,000 people died of starvation and fever. The British Government provided work for over 700,000 people at one time, but this was not enough. Then food was distributed in enormous quantities, and over 3,000,000 people were at one period supplied with rations. Nevertheless such great discontent arose that in 1848 a rebellion was attempted.

Perhaps the most important result of the over-population of Ireland coupled with the distress due to famine was that it set the Irish to talking about migration. The talk was not idle, for a rapid migration to America began in 1846. In five years the population had diminished to 6,600,000, or about 20 per cent. During succeeding years the climatic conditions were fairly dry and favorable and migration declined rapidly, as is shown in the following table. In the eighties, however, another prolonged wet period with poor harvests again brought home to Ireland its great handicaps of over-population and a poorly organized social system. People again flocked to America. This Irish migration is one of the most wholesale known to recent history.

POPULATION IN IRELAND

Date	Population	Decrease in Population	Percentage of Decrease
1841	8,200,000	....	....
1851	6,600,000	1,630,000	20
1861	5,800,000	770,000	12
1871	5,400,000	390,000	7
1881	5,100,000	240,000	4
1891	4,700,000	470,000	9
1901	4,500,000	240,000	5
1911	4,400,000	80,000	3

There can be little question that the migration was selective. Doubtless many competent Irishmen still live in the Old Country. But there is equally little doubt that on the whole it was the more energetic, wide-awake, adaptable, and industrious Irish who left their old homes and fared forth across the sea. That is perhaps one of the chief reasons why Ireland has had only about one-sixth as many eminent men in proportion to the population as have its two English-speaking neighbors, England and Scotland, as appears in Figure 12. For hundreds of years England has been getting the cream of Ireland, and during the past century America has scooped off most of the top milk. On the blue liquid which still remains some cream may rise, but it would be contrary to the lesson of history to expect that a country which has been so depleted shall thrive as does a country into which fresh, creamy milk has just been poured. The prospects would be sad, indeed, were it not that a race like the Irish possesses so highly mixed an inheritance that high types continue to appear now and again in spite of the constant removal of such types by migration. With proper selection there is little doubt that Ireland could again possess a large body of competent leaders.

England, like Ireland, has suffered from migration, as have most of the countries of Europe. But England has had a great safeguard in her aristocracy and her system of primogeniture. Gibes at the British aristocracy are common, but in proportion to its numbers that aristocracy has probably produced more men of genius than almost any similar group except the Athenians, the Jews, the Icelanders, and a few others. To-day, in spite of all that may be said in comic operas and novels, I would rely on a hundred average British aristocrats with more certainty than upon an average hundred picked at random from almost any other group of men. These aristocrats, simply because they belong to a ruling class, have largely remained in England. Many younger sons have indeed gone to new lands, but, thanks to primogeniture, the oldest sons, and the heirs when those sons have died, have almost invariably made England their permanent home. Thus, although England has been sadly drained of her middle classes, a fact which perhaps aggravates her present difficulties, she still has a large, strong body of gentry who inherit not only a social tradition of public service, but a physiological capacity for honest, painstaking, and often brilliant leadership. That is her salvation. It is Ireland's misfortune to have lost practically all of that type together with a large share of her middle class.

CHAPTER XVII

THE DISPERSAL OF THE NORTHMEN

IF the reader reviews the line of reasoning pursued thus far in this book he may be surprised, as I have been, at the concentration of advantages around the North Sea and southern Scandinavia. The northern location of those regions and the prevalence of westerly winds from the Atlantic keep the summers cool and invigorating. In winter, however, those same winds come from an ocean which is warmed by the Gulf Stream and Atlantic Drift, so that its waters are not much colder in winter than in summer. Hence harmful extremes of temperature are rare. Again, in this northwestern section of Europe cyclonic storms are especially abundant. Therefore, not only is there plenty of rain and atmospheric moisture at all seasons, but stimulating variations of the weather are frequent. These conditions of climate produce two highly desirable results: First, they make it possible for people to get a good living by agriculture, but do not make this easy unless people are highly industrious and have a strong spirit of thrift. Second, they tend to make the inhabitants healthy and active to a degree almost unparalleled elsewhere.

The North Sea region and especially Scandinavia have still other advantages. They were among the last parts of the world to be freed from ice; hence their cooler parts, at least, have been inhabited only a few thousand years. Again, this part of the world is remote from regions whence it could be repopulated after the last ice age. Accordingly its inhabitants had to migrate far and undergo strong natural selection before settling in their final homes. Inasmuch as most of the Scandinavian and North Sea people reached their homes quite late, their inheritance has as yet been relatively little impaired by adverse selection or by the mere cessation of the selective action of migration. Moreover, the earliest of the selected migrants found in their new homes no inefficient aborigines. Even the latest invaders found a population which was presumably fairly competent, and which did not usually tend to replace the newcomers because of its numbers or because it was better adapted to the climate.

Another significant fact arises from the climatic vicissitudes of the glacial period, together with the relation of Europe to the African and Asiatic deserts. Because of these conditions the races which invaded northwestern Europe had presumably been subjected to an almost unprecedented degree of racial mixture and natural selection before they experienced the final renewal of these processes during the repeopling of Europe.

As a result of these advantages the people of northwestern Europe have been able to assimilate and utilize new ideas and inventions while still in the freshness and vigor of racial youth. Hence it has been their privilege to rise especially high in such matters as the use of bronze and iron, the Christian religion, representative government, and many other modern ideas and institutions. From such conditions of environment and selection, then, arises in considerable measure the present undoubted supremacy of the Nordics. From this also arises the groundless fiction that practically all progress everywhere is due to Nordic infusion.

Having seen how the Nordics reached their high racial level, let us inquire what lessons may be learned from the history of one of their most typical branches, the Northmen or Norse who lived in Denmark, southern Norway, and southern Sweden in the eighth century. At the end of that century they appear to have been living in relative quiet for some three centuries, without any special migrations or wars other than the ceaseless local struggles which are almost universal among energetic barbarians. Then something happened, and the Norse suddenly began migrating in all directions. In their open boats they swarmed along every coast, plundering, ravaging, pillaging, burning, enslaving, and killing. At first they merely landed on the coast, commonly using an island as a base, and making a swift dash upon the inhabitants. Then they made camps where they stayed a few weeks while part of them made forays up the rivers. Next they spent the winter and made more extensive raids, harassing the country for hundreds of miles. Finally they established themselves permanently. In Ireland they founded Dublin and there set up a kingdom which lasted about three hundred years, or from about 851 A. D. to the Norman invasion from England in 1155. In Scotland and northern England they did likewise, and finally in 1015 Canute became king of all southeastern Britain. At the mouths of the Elbe, Rhine, Somme, Seine, Loire, and Garonne the Norsemen also established themselves more or less permanently. In northern France they actually forced the French to cede to them the territory of Normandy to which they gave their name. Farther south these Norse Vikings made predatory raids around Spain, upon the African shore, and along the Mediterranean coast to the mouth of the Rhone and North Italy. In another direction, eastward from their home, they migrated into Russia under Rurik and his brothers. Rus they were called, and thereby Russia received its name after the Norse had established a kingdom at old Novgorod. From there wandering bands went down the Dnieper and reached Constantinople, where for generations the Byzantine emperors had a famous Varangian Guard of Norse soldiers. Thus Europe was almost encircled by these wild, bold, barbarous adventurers. One other migration took place at this time, westward to newly discovered Iceland. This, unlike all the others, was peaceful, for there were no inhabitants in Iceland aside from a few Irish monks.

The degree of permanence in the Norse settlements varied greatly, as did the extent to which the Norse retained their racial inheritance. In Iceland practically all the settlers were Norse, and they have remained unmixed for a thousand years. In Normandy the Norse exterminated many of the former inhabitants and took a considerable area for their own possession. Nevertheless, being a highly adaptable people they adopted the French language, religion, and customs, and in due time became so amalgamated with the surrounding people that they are now almost indistinguishable, as is also the case farther north in Belgium and

THE DISPERSAL OF THE NORTHMEN

Holland. This was easy and natural, because many of the people of northern France were closely allied to the Norse in race. Nevertheless, before the Norman Norse had wholly lost their racial individuality they sent out two sets of migrants who have played an important part in history. One set established a Norman Kingdom in Sicily and South Italy; the other followed William the Conqueror to England.

In England itself, two centuries more or less before the Norman conquest, the Norse—partly Danes and partly Norwegians—had settled in such numbers that they not only established a kingdom, but imposed their speech upon certain small areas, for example, in Yorkshire and Lincolnshire. Nevertheless, in Britain, as in France, the Norse ultimately adopted the language, religion, and customs of their fosterland, and amalgamated with the former inhabitants, especially the upper classes, who were their close racial kin. Since England received a second important Norse addition when the Normans came in 1066, and since it had no large outward migrations like those from Normandy, it owes more of its blood and its character to the Norse than does any other country except those of Scandinavia and Iceland.

In Ireland along the north and east coasts the Norse were an important element for a while, but like every other upper class they tended to move away from that country. Many migrated to Iceland within a generation or two after coming to Ireland, while others at later dates went to England and the continent. Thus the Norse inheritance has played a relatively minor part in Ireland. In Russia much the same is probably true. Although the Rus may not have migrated from Russia so quickly as many of their kinsmen did from Ireland, they were absorbed in the great mass of Slavs around them and soon lost their individuality. Elsewhere except on the Swedish border of Finland, the Norse appear to have been a more or less ephemeral element.

But note one important fact. Wherever the Norse went they were dominant. This is not strange, for we have already seen how rigid had been the selection by which the pioneer qualities had been evolved in them. Moreover, it was the best of the

Norse who went on Viking raids. Even before the great period of migration it was the common thing for young Norsemen to go "a-viking," much as the eighteenth-century young man in England took a year of travel. But it was only the young men of relatively good birth and position who could afford to do this. Even the retainers who accompanied them were "trained warriors, chosen for their high spirit" (Enc. Brit., vol. IX, p. 468). When Canute's great army went to England in 1015 it was a picked host which contained "not a thrall or a freedman." Many of that particular army returned to Denmark, but the other Viking hordes which settled in foreign lands were also largely of the upper classes.

When the Norse first began to spread abroad over the coasts of Europe, their homeland contained three chief classes of people, Thralls, Karls, and Jarls. The first or typical Thrall is thus described in the Elder Edda known as Rigsmal:

> "Edda a child brought forth, whose swarthy skin
>   With water then was sprinkled, and its name was Thrall.
> And as it grew, it throve, but on its hands
>   Were fingers thick, a shriveled skin, and knotted knuckles.
> A hideous face it had, a curving back,
>   And sharp, protruding heels. As it gained strength,
> It proved its might by binding bast in heavy loads;
>   And carrying faggots home unwearied all day long."

The Thralls, we are told, "erected fences, manured the fields, tended swine, kept goats, and dug turf." Perhaps they represent the bleached Proto-Negroid element which Dixon believes that he finds among the Nordics.

The Karl stood higher than the Thrall. The Rigsmal does not tell much of his personal appearance except that when the first Karl was born his eyes twinkled and his mother swathed his ruddy head in linen. When he grew up,

> "He learned to tend the oxen, make a plow,
>   Build barns and houses, fashion carts, and turn the furrows."

His occupation was not very different from that of the Thrall, but he owned land, and in those days that was everything. The

Karls perhaps were the Mediterranean and Alpine elements of the population.

Highest in the social scale were the Jarls or Earls, whose description suggests that they were pure Nordics, or possibly of Dixon's Caspian race. Of the first Jarl's noble parents we are told:

> "The husband sat and twisted strings and bent his bow,
> And arrow shafts prepared; but his good wife
> Glanced down at her fair arms, smoothed out her flowing veil,
> Fastened her sleeves and made her headdress straight.
> A clasp was on her breast, holding her ample robe.
> Her sark of heavenly blue seemed dull, for her bright brow
> And neck and breast were whiter than the driven snow."

When this fair wife brought forth a boy,

> "In silk they wrapped him; then with water sprinkled,
> And named him Jarl. Light was his hair, bright were his cheeks,
> And his young eyes gleamed keen as any serpent's.
> In his own home he grew, learning to hold the shield,
> To draw the twanging bow string, bend the bow,
> And shape the arrow with a smooth and deadly shaft.
> He likewise learned the javelin to hurl,
> The spear to brandish, the fleet horse to ride apace,
> Dogs, too, he managed, and he learned to wield
> The sword, and boldly swim through fierce and boisterous waves."

A little later he was taught the ancient runes, given a new name, and told that he was now of age to possess his own lands and the dwellings of his fathers. Then his nobles went forth to a hall where dwelt a great chief.

> "There they found a slender maiden,
> Fair and most elegant. Erna her name.
> Her they demanded and soon led her home.
> So Jarl espoused her."

It was the Jarls who went a-viking, taking with them the best Karls and only a few chosen Thralls. Thus it was the ablest of the Norse who spread themselves abroad over the coasts of

Europe. After the first preliminary raids they took their wives with them, although many, of course, married the women of the conquered countries. They were a prolific race, and the most successful leader had the most children. Harold Fairhair, a king of Norway of whom we shall shortly hear more, divided his kingdom among "about twenty" sons at his death in 933 A. D.

Almost everyone who has studied the Norse migrations with any care is puzzled to account for the comparatively sudden way in which the Norse in all three of the Scandinavian countries began to move outward at the same time. Two political events have been suggested as the cause, namely, Charlemagne's conquest of the Saxons and advance into Denmark, and Harold Fairhair's taxation in Norway. It appears to me, however, that movements so widespread and so prolonged cannot have been due to these causes alone, although they unquestionably were important. Charlemagne's war on the Saxons lasted from 772 to 785, and immediately preceded the first great Norse raid on England, 787 A. D. But it is scarcely probable that the effect of Charlemagne's intermittent and relatively unimportant war on the extreme southern edge of the Norse territory could last a century and a quarter and could lead to a widespread and prolonged disturbance in all three of the Scandinavian countries.

Harold Fairhair's reign (860–933) did not begin until the Viking raids had been in progress nearly a century, nor can his taxation have had much, if any, effect on countries other than Norway. Nevertheless, it is interesting as illustrating how a political cause may combine with other causes to produce migration and natural selection. It is likewise important because it had much to do with determining the kind of selection which differentiated the Icelanders from the Normans.

The whole social organization of the ancient Scandinavian communities rested upon the possession of land. If a man had no land he was inevitably a Thrall. If he held land but owed rent or tax to any other, even the community, he was a Karl. If he held his land in his own right, free and untaxed, he was a Jarl or Earl. Now it happened that in the ninth century there was a

THE DISPERSAL OF THE NORTHMEN 259

great movement in Europe toward kingship. Charlemagne is the outstanding example, for it was he who made the kingship real in central Europe. In England in the ninth century the West Saxon Ethelbert likewise secured the supremacy of the whole island. In Denmark Gorm established a monarchy. In Sweden Eirik Eymundsson, king of Upsala, was strengthening the tottering "overkingship" which then existed into a regular sovereignty. In Norway this consolidation of small independent states into a strong monarchy was carried out by Harald Harfagr (Harold Fairhair). He subdued the petty kings of Norway and after a hard struggle made himself sole ruler over the whole country. The new king needed revenue. One of his early acts was to appropriate to himself the lands which had formerly been the common property of the whole community. Thus the Karl who had formerly paid rent to the community became a tenant of the king. Harold also levied a poll tax, but this was no innovation and raised no complaint. But when the king levied a tax on the land of Jarls, the free proprietors who from time immemorial had lived on their own estates free from all impost, then indeed, as Coneybeare says, "there arose a universal cry of indignation which it is not difficult to sympathize with. Not only was an invasion thus attempted of the most cherished, nay, sacred rights of those who were free born throughout the land, but it swept away the only distinction which raised the allodial lords above those who were tenants either to the state or to private freeholders. For in those early times it was the *free* possession of land that conferred social and political distinction; and the loss of such property necessarily entailed loss of condition. Long after our present social system had grown up, did the same idea linger in the minds of the old landed gentry of England."

One result of Harold's taxation was to increase the migration from Norway which had already been in progress for a century. Many of the proud Jarls who might not have left Norway as raiders now went as settlers; some to England, but most to Iceland. So great was this migration that Harold Fairhair, fearing that his realm would be left barren, imposed a tax on all who wished to leave the country.

In addition to political causes, such as Charlemagne's wars and Harold's taxation, three others may be suggested as reasons for the outburst of the Norse at the end of the eighth century. One is the inherent character of the Norse, their curiosity, boldness, initiative, physical vigor, and ability to fend for themselves. But there is no reason to suppose that this was any more marked in 900 A. D. when viking raids were at their height than in 600 or 700, when the raids were of a mild character and involved merely a little plundering instead of wholesale destruction of towns and permanent occupation of distant territories. Another possibility is over-population. Although no exact data are available, it may well be that the population of the Norse territories had been increasing, and had reached such a point that some relief was necessary. It hardly seems probable, however, that a mere gradual increase would cause so sudden an outburst unless supplemented by some other event which would serve to bring matters to a crisis.

It appears to me that such an event is probably found in vagaries of climate, and hence in changes in the capacity of the land to support population. Unfortunately, no direct evidence is available as yet from Scandinavia, but there is considerable of an indirect kind. The rings of growth of the great Sequoia trees of California, as I have explained in *The Climatic Factor*, give an approximate record of climatic changes which is fairly reliable for two thousand years. Archæological and historical evidence shows that the California curve of tree growth can be used as a key to the climatic changes in other parts of the world. Only in certain places, to be sure, do the changes appear to have been of the same type as those of California, while in others they seem to have been of a directly opposite type, and in still others intermediate. Now in Iceland, Ireland, England, and Scandinavia there is considerable evidence that since about the twelfth century the main fluctuations of climate have been essentially the same as in California, although they may have varied considerably in details. The evidence is particularly strong in respect to a period of diminished storminess and rainfall in the twelfth and thirteenth centuries, and another of increased rainfall in the fourteenth, as I have explained in *Climatic Changes*. As to earlier times the data are very scanty, but there seems to be reason to think that the same degree of agreement has prevailed.

If the California curve of climatic changes really applies in a general way to Scandinavia, it may help to explain the raids of the Vikings. From the time of Christ until the middle of the seventh century, in spite of certain distinct exceptions, there was presumably a general tendency for the climate to become less stormy, drier, and more mild, as may be seen in Figure 15. Scandinavia lies so far north and has so many storms that such a tendency is highly advantageous, for it increases the productivity of the soil, and makes it easier to get a living. Thus from before the middle of the seventh century until after the middle of the eighth, Scandinavia probably enjoyed one of the most favorable climatic periods in its whole history. Presumably there was a general condition of prosperity, and the population increased until there was little or no room for more, according to the standards of production which then prevailed.

In the latter part of the eighth century a change apparently set in and the climate became more stormy, so that there was more danger that the crops would fail. Whether the change was of the same general magnitude as is indicated in Figure 15, we do not yet know. Even if it were no larger than appears in the diagram, it might cause grave difficulties if the country were near the limit of population, and especially if there were a few especially wet years when the crops did not ripen well. The hypothesis, then, which I would advance is that the combination of Charlemagne's wars upon the Saxons and of a series of poor crops together with the normal increase in population led to a strained situation. In accordance with the innate character of the Norse, this suddenly relieved itself in a series of unusually severe raids. The first of any magnitude reached England in 787, France in 789, and Ireland in 795. When once the Norse had undertaken such distant raids and had discovered how exciting and profitable they were, and how easily they could be carried out because of the fear which the Vikings inspired, they continued them. Thus they ravaged the Frisian coast of Holland in 808, harried Ireland unmercifully in succeeding years, raided the French coast in 841, and appeared in Russia in 859.

All the Norse movements thus far described were mainly plundering expeditions. Not till after 866 did real settlement of the Danes begin in England. Then followed a period of about sixty years, 870 to 930, when the Norse from Denmark settled in considerable numbers in England; those from Norway, together with a considerable number from Ireland, went to Iceland; while many from Sweden went to Russia. The maximum migration especially to Iceland, took place in the forty years from 880 to 920. Now this, it will be noticed, was the time of Harold Fairhair's taxation, but that applied only to Norway. It was also a time of unfavorable climate, if we may judge by Figure 15. Therefore it seems reasonable to suppose that at this time, just as had happened about a century earlier, economic distress due to unfavorable seasons combined with political exigencies to cause migration in addition to the raids that had previously taken place. By 940 A. D. the last movements of this particular period seem to have died away in some minor Danish raids in England. Then for a generation or two the Norsemen left the rest of the world in peace. But in the last decades of the tenth century the political and economic causes again combined. Judging by the conditions in California, it would seem that the period from 990 to 1010 must have been very unfavorable in Scandinavia. If that is so, it may help to explain why the Norse again began making raids in England in 991. At the same time political events were moving in such a way as to culminate in the conquest of England by Canute and his picked army in 1015.

In this dispersal of the Norsemen, as in their earlier migration into Scandinavia, we seem to find that many causes, physical, social, and political, have probably combined. All appear to have exerted a more or less selective effect, so that the Norse

THE DISPERSAL OF THE NORTHMEN

who went out from their homes during the period from 797 to 1015 A. D. were not only a picked group, but were divided into two distinct types whose fortunes we shall follow in the next chapter.

CHAPTER XVIII

WARLIKE NORMANS AND PEACEFUL ICELANDERS

The Norse who migrated to other lands fall into two general classes. One class migrated primarily because they were warriors in search of plunder. Only incidentally did they become colonists. Such, in the main, were the earlier raiders upon the coasts of England, Scotland, Ireland, the Low Countries, France, and countries even more remote. Their most typical settlement was Normandy, which came into their possession because the French feared their further depredations. The other class of Norse migrants went out primarily as colonists. Some of this type doubtless went to Normandy, more to Ireland, and many to England, but the greatest number, or at least the most compact and homogeneous group, went to Iceland. Of course there is no sharp line of cleavage between the two groups, but the difference is real and important. One group, intent on plunder, selects a goal that is rich, prosperous, and not too well defended. The other, determined to escape from political and economic trouble at home, selects a remote goal with scanty resources, but where there will be no need of fighting. Distinctly different types are attracted in each case. How far this difference explains the obvious differences in the history of the two groups we cannot tell, for many other factors enter into the matter. Nevertheless, it will be instructive to compare the history of the Normans with that of the Icelanders.

One of the most conspicuous facts about the Normans is their conquests. Almost everyone knows that in 1066 William the Conqueror, a Norman at the head of an army whose leaders were largely Normans, and whose rank and file were probably one third Norman, came to England and made himself king. He and his army settled in England, and other Normans followed. They at first exterminated great numbers of the Saxons, but later amalgamated with them and were soon almost indistinguishable from their Danish and Norwegian kinsmen who had come to England in the earlier Norse migrations and who had already become more or less mixed with the older Saxon aristocracy. Less familiar, but no less important from our point of view, is the Norman conquest of the two Sicilies, that is, South Italy and Sicily. This began in 1017 and in a generation or two gave rise to a kingdom "second to none in Europe for wealth and magnificence. . . . This Norman conquest of Sicily forms the most romantic episode in mediæval Italian history." (Enc. Brit.)

In the articles on Normandy and Sicily in the Encyclopædia Britannica the historian Freeman gives a vivid picture of the achievements and character of the Normans. When they went to England they had already abandoned their Norse religion, language, and national traditions, and had adopted those of their Gallic neighbors. They apparently had no sense of kinship with the Nordic Saxons, and it is even doubtful whether they knew that their name meant Northmen. The war-song which Taillefer chanted as they marched to battle was not a Viking saga, but the song of Roland and the peers of Charlemagne. They had adopted the art of their French home so completely that though they developed the Romanesque style more fully, there is no evidence that their coming caused any recognizable break in English architecture. Yet they brought much to England in the way of language and of the spirit of feudalism. Even in these respects, however, the steady progress of England along lines already established was scarcely interrupted. This is not strange, for the Angles, Saxons, Danes, Norse, and Normans all appear to have been closely related branches of a single great stock. Hence both the Viking and the Norman invasions of England merely represent an accretion of new and able people to strengthen the old stock. Since the newcomers experienced no great change of environment either physically or socially they simply mingled with their more or less warlike kindred and helped to give the England of to-day its strength. Much of the old Nordic inheritance still survives and is perhaps one great reason why England has been the greatest of colonizing nations and likewise has never lost a war except when it fought with a chosen band of its own children in America. Even now it is especially the Norman type both in body and spirit which surges forth to the ends of the earth. It was perhaps the latest Norse warriors whom Gibbs describes in *Now It Can Be Told*. During the Great War, as he watched the soldiers in the old cathedral at Amiens, the Australian soldier boys with "their clean-cut hatchet faces, sun-baked, tanned by rain and wind, their simple blue-gray eyes, the fine strong grace of their bodies, as they stood at ease in this place of history, struck me as being wonderfully like all that one imagines of those English knights and squires—Norman-English —who rode through France with the Black Prince. It is as though Australia had bred back to the old strain."

The spirit which in modern times makes the Australian and New Zealand soldiers rank so high caused the Normans to conquer southern Italy and Sicily nine hundred years earlier, and made them extremely vigorous and enterprising as Crusaders. The career of the Normans in Sicily throws an interesting light on their character. They did not adopt either of the two languages of the island, Greek or Arabic, but in other respects they adopted Sicilian habits instead of imposing their own. Since most of the Sicilians were Christians, there was no need of change in this respect, but one would at least expect that the newcomers would impose some traces of themselves upon literature, art, and architecture. But when the Normans came to Sicily the Saracens had already been in possession for two hundred and thirty years. Their advance in civilization was so great and their style of architecture so good that the Norman kings simply studied and promoted Saracen literature, and always called in Saracens for any special work of engineering or architecture. The Normans, as Freeman puts it, came into the inheritance of the two most civilized nations of the time, the Greeks and Arabs, and allowed the two to flourish side by side. Sicily's most brilliant period began with the coming of the Normans. Yet the Normans did nothing new; they simply maintained order, and let the Saracens build their palaces and churches. So completely was this the case that if we judged solely by the architecture, we should never know that the Normans had set foot in Sicily.

As to the character of the Norse we are told by Keary that they were filled with an immense joy in battle, a kind of mad fury which made them proof against everything, including even superstition. They had need for such courage, for more than most men they attempted every kind of adventure. They went to sea in little ships that would now be considered quite unseaworthy; they also made long expeditions inland, using horses and the other paraphernalia of warfare on land. Cruelty and faithlessness were almost as characteristic of them as was boldness, at least among that portion of the Vikings who became famous as raiders. Nevertheless, these same men had an uncommon capacity for government and for peaceful organization. That is why Sicily was so prosperous, while Keary calls Normandy the best-governed part of France in the eleventh century. In this connection Freeman points out that even in their most unbridled acts the Norse and the Normans had a peculiarly high respect for the forms of law. In Iceland we shall see this fully exemplified. In England it led William the Conqueror and his followers to make the most elaborate attempts to find legal justification for dispossessing the former owners of the soil. Among all the qualities of the Norse none is more noticeable than their adaptability. It was this which made them adopt Saracenic art in Sicily, the French language and the Christian religion in Normandy, and British customs and modes of thought and speech after they reached England. This same quality is emphasized by MacDougall in his discussion of the Nordic race, as we have already seen; it is one of the chief reasons why Japan rivals many European nations in no small number of ways.

The quality of adaptability upon which Freeman lays so much stress has also come to our attention as pre-eminently characteristic of the Greeks. The reason is probably the same; namely, that only the most adaptable people can or will persist to the end of a long and arduous migration. Leaf seems to be right when he points out in *Homer and History* that "the Norman invasion of Italy is in fact a type of a whole class of invasions—the conquest of a decaying civilization by a comparatively small number of warlike adventurers." He might have added that the Normans did not necessarily lack originality as Freeman seems to imply. It is more likely that they were so busy adopting and adapting to their own use the new things that they found, and were so engrossed in the business of governing, that they had no time for new inventions. It took the Greeks and the Khmers at least a century or two before they began to display the great originality for which they are renowned. But in Sicily the Normans were swamped before they reached that stage. It was their misfortune to be a relatively small band, and to migrate to a country farther south than Greece. They went to an unfavorable climate at a time when it appears to have been deteriorating to perhaps its worst level, and when malaria is known to have been a most terrible scourge. They appear to have intermarried freely with the Christian part of the population. All these conditions presumably tended to cause the richly endowed inheritance of the Normans to be extinguished within a few generations. Their fate was like that of the Greeks, the Khmers, and many other similar invaders. It was remarkable chiefly because it came upon them so soon. A small minority of uncommon ability was quickly swamped in a large majority, presumably because the selective processes which had sorted out the ability ceased to be effective. Constructive evolution through migration was superseded by racial mixture and destructive selection through climate and disease.

Now turn to Iceland. There we are confronted by a slightly different initial selection and by an almost diametrically different later evolution. Here is what Coneybeare says about the original movement to Iceland:

> It was indeed a migration *en masse*, and that, too, not of the meanest of the population; ... on the contrary, the noblest and worthiest of the land, the most peaceably disposed, and the most cultivated, formed the bulk of the migrant host. While the most warlike of the nation sailed southward and founded a new Norman kingdom in Gaul, the richest land

owners eventually settled in Iceland; which fact, coupled with the necessity imposed upon them of trading with the British Isles for the supply of many of their wants, accounts for the peaceable and even mercantile spirit which characterized the Icelanders.

It must not be supposed, however, that the Norse who migrated to Iceland were peaceable according to our standards. They are called by Bryce a people "whose chief occupation was to kill one another." But in those days all men who amounted to much had to fight. The Jarl, by his very position, was a fighter, and the Karls who followed them were doubtless in many cases chosen for their ability to fight. Nevertheless, Coneybeare is almost certainly right in saying that the Norse who went to Iceland were more peaceful and more inclined to mercantile pursuits than were those who harried the countries farther south. Thus the original Icelanders were a highly selected group of people. In the first place, the Norse as a whole appear to be one of the world's competent races. Second, the Jarls were Jarls because their ancestors had shown unusual ability to fight, to organize, and to hold their conquests, and they themselves maintained their position largely by virtue of an inheritance of ability. Third, from among this most gifted group in an uncommonly able race there were selected first those who loved freedom more than the favor of kings, and second those who were inclined to try their chances in a new and difficult land rather than gain a living by fighting and plundering. Thus it was the best of the best who founded Iceland, and they created there what Bryce calls "an almost unique community whose culture and creative power flourished independently of any favoring material conditions and indeed under conditions in the highest degree unfavorable."

The migrants to Iceland included a fair sprinkling of Swedes and Goths, and some from the Norse settlements in Scotland and Ireland, but apparently not a single Dane. Among the others, as among those from Norway, there was an unusually large percentage of high-born Jarls. For example, Queen Aud, the widow of Olaf the White, the Norse king of Dublin, was preceded and followed by a large number of her kinsmen and relations, many of whom like herself were Christians. Some of the other migrants were also Christians, but the great majority were heathen. Many of the Norse chiefs had lived for a long time in foreign lands. Thus while there was remarkable uniformity of race, ability, and social condition, there was some diversity of ideas. This uniformity deserves notice, for it is unique. Other migrations, such as that of the Puritans, have perhaps comprised an equally uniform body of people, but in most such cases there has soon been much diversity through further immigration or through mixture with previous inhabitants of the new country. In Iceland nothing of the kind took place. By 930 A. D. the population of that island appears to have amounted to 30,000 to 50,000, a number which probably was not much exceeded down to the census of 1823. Practically no immigrants have gone to the island since the first migration. Thus for a thousand years Iceland has been the home of a peculiarly homogeneous people.

In almost every respect the new Icelandic home of the Norse was less favorable than their old home. Both Norway and Iceland are highly mountainous, with only a few plains here and there, chiefly in the south. Both regions likewise have depressed coasts where the ocean penetrates far into innumerable picturesque fiords. Deep sheltered harbors, steep cliffs, sparkling waterfalls, grim canyons, and cold glaciers are also numerous, but give to neither country any special advantage. But the Icelandic mountains are a much greater handicap than those of Norway in one respect, for they are volcanic. Time and again Iceland has suffered terribly from eruptions and earthquakes, especially in the last part of the thirteenth century. In 1362 Oeraefa Jökull, the loftiest mountain in Iceland, 6,424 feet high, swept forty farms, together with their inhabitants and livestock, bodily into the ocean. In 1783 an eruption of Laki spread so much dust over the island that it occasioned the loss of 53 per cent of all the cattle, 77 per cent of the horses, and 82 per cent of the sheep which form the main source of livelihood. After that the island was visited by a famine, which destroyed 9,500 people, or one-fifth of the total population.

WARLIKE NORMANS AND PEACEFUL ICELANDERS 271

Sometimes the volcanoes cause destruction in more indirect ways. For example, many of the Icelandic volcanoes during their periods of quiescence are covered with ice and snow. When an outbreak occurs these melt and give rise to sudden floods. Katla has caused serious damage in this way by converting several cultivated districts into barren wastes. Earthquakes, likewise, are a frequent menace. In the southern lowland in 1784 some 92 farmsteads were totally destroyed by an earthquake, and 372 farmsteads and 11 churches were seriously damaged. Again in 1896 a series of earthquakes destroyed 161 farmsteads and damaged 155 others.

Although earthquakes and volcanoes figure largely among the disasters chronicled by the historian, they seem to exert no serious check upon progress. They are temporary accidents, and if people are energetic, the material damage that they do is usually soon repaired, as in Japan, southern Italy, and California. In the famine and distress which follow such disasters, the more provident, industrious, and intelligent inhabitants, to be sure, are least likely to die, but such conditions last so short a time and recur so infrequently that they probably have little effect upon racial character. Yet in one important respect they have probably had an important effect on Iceland. They have joined with other factors in rendering the island relatively unattractive, thus helping to prevent immigration and thereby preserving the purity of the original Icelandic blood. At the same time they have presumably encouraged some Icelanders to migrate back to the old country. During the Middle Ages many able Icelanders went to Norway, and some of them became bards and advisers at the courts of Scandinavian kings. We are told that among 230 bards practically all were Icelanders. Most of these, however, returned to Iceland. Nevertheless, some remained in Norway and Denmark. The memory of the volcanoes can scarcely have been an incentive to go back to Iceland. At any rate, we know that in modern days volcanic eruptions have played a not inconsiderable part in stimulating migration to Canada.

Another marked feature of the geographical environment of

Iceland is isolation. Distances of six hundred miles from the Norwegian coast and five hundred from the nearest part of Scotland may not be great for ocean liners, but they were very great for the small vessels of more primitive times. Only a bold, adventurous spirit would brave a journey between Europe and Iceland in one of the low, half-open boats of the early Norse; for the voyage lies across one of the stormiest seas. Iceland's isolation presumably kept the timid and weak at home when the island was settled. In later days it has probably been much more effective than the volcanoes and earthquakes in preventing either immigration or emigration, and thus has been a factor in preserving the original inheritance and making the Icelanders one of the most homogeneous peoples on the face of the earth. How homogeneous they are is illustrated by the fact that even as late as 1920 there were only 710 foreigners in Iceland out of 95,000 people, and of the foreigners 507 were Danes and Norse. The Icelanders are equally homogeneous in religion, for in 1920 only 463 were dissenters from the Lutheran Church, which has long been the established form of religion.

Mineral resources are another of the respects in which Iceland, according to common standards, is even worse off than Norway. There are, to be sure, some deposits of the peculiar pure form of calcite, called Iceland spar; and sulphur from the volcanic deposits was exported to England as early as the thirteenth century; but there is no coal, no petroleum, no iron, no copper, no gold or silver; none, in fact, of the mineral resources whose importance in our day is so much emphasized. This may have been a disadvantage in early times, and it has probably retarded modern Iceland somewhat in the domain of manufacturing. Even now the island sends most of its abundant wool to Europe instead of making it into cloth at home. But the countries that have developed manufacturing in the last century are likely to pay dearly for the privilege. They have herded their people into great and unhealthful cities; they have in some cases introduced vast numbers of immigrants whose racial inheritance is none too high and who as individuals appear to be much below the average of their respective races. To-day in far northern Spitzbergen the presence of coal is drawing a group of laborers who are ignorant, rough, and unprincipled. They appear to be mostly men who do not succeed well at home but who have some love of adventure and so take a chance in the far north. If Iceland had to receive ten or twenty thousand such men to work coal mines, her racial inheritance would at once be greatly altered for the worse. But she can easily get coal from England. Moreover, she has numerous waterfalls which are already beginning to be developed and which in time will probably supply her cheaply with power, light, and heat, without introducing a great number of low-grade miners.

The greatest disadvantage of Iceland compared with southern Norway, whence came most of the Norse settlers, is the climate. In no other country has a genuine center of civilization grown up in such high latitudes. This does not mean that Iceland is really cold, for its shores are washed by the relatively warm Atlantic Drift, so that the average January temperature at Reykjavik, the capital, is about 30° F., or essentially the same as that of New York City and central Denmark. In summer, however, Iceland is cool. Even in the warmest places the average temperature of July is only about 50° F., or the same as that of the northern coast of Alaska, the mouth of the Lena River, and the southern tip of Tierra del Fuego. Elsewhere such summer temperatures are associated with people like the Eskimos, the Lapps of northern Europe, and the degraded Indians of Tierra del Fuego. Only among the Incas of Peru does any people ever seem to have risen to a high degree of civilization or maintained such a civilization for any length of time where the summer temperature was so low. Cuzco, the Inca capital, 11,400 feet above the sea, has an average temperature of only about 52° F. in November, the warmest month. Its coldest month, July, however, averages 46° F., so that the growing season is far longer than in Iceland. The Icelanders, like the Incas, present the remarkable spectacle of a race where a unique inheritance seems, for a while at least, to have overcome the almost incalculable disadvantages of a cold summer.

The real importance of temperature to Iceland is seen in its effect on agriculture and forests. Agriculture in the sense of raising crops is almost impossible. Potatoes, turnips, and cabbages can indeed be raised in fair quantities in the southern part of the island, but there are only a few hundred gardens all told. The raising of grain is practically impossible, although occasionally an attempt is made to raise rye or barley. Forests are likewise lacking. There are, to be sure, many trees on the island, but most of them are little birches three to ten feet high. Only in a few areas do the birches reach a height of twenty feet. Willows, too, are fairly abundant, but are generally only seven to ten feet high. The largest trees are a few mountain ash, some of which attain a height of thirty feet. None of the trees are large enough to be of any appreciable value for lumber, and even as fuel they are of small use.

One reason why vegetation and agriculture are so hampered in Iceland is the extreme variations in the temperature of the same month from year to year. If the winds blow from any direction between southwest and east they come from relatively warm parts of the Atlantic where the Gulf Stream, after ceasing to move as a distinct current, spreads out and forms the Atlantic Drift. If, on the other hand, the winds blow from the west and north they cross an arctic current and may bring great fields of polar ice against the Icelandic coasts. Thus from year to year there may be enormous differences in the temperature of the same month. In 1856 at Stykkisholm the average temperature of the month of March was 40.1° F., for the winds blew from the southwest; but in 1866 when the winds blew from the north or northwest the same month averaged 12.4° F., while in 1881 the month of March averaged only 7.6° F. In few parts of the world are there such great differences in the same month from year to year. If the ice closes in and cold winds blow in summer, the season may be so wet and cool that the hay cannot be dried. The grass may grow abundantly under such circumstances and generally does, and the cattle and sheep grow fat, but when winter comes there is little for them to eat. An unusually warm and rainy summer may produce the same result, as in 1870. Even though the grass is cut and tended most carefully it rots. In 1759 conditions such as these caused a famine in which 10,000 people died. This was followed in 1762 by a disease which caused 280,000 sheep to die or be slaughtered.

Because of the climate the only important resources of Iceland are grass and fish, and these determine how perhaps nine-tenths of the people live. A fine, dense growth of grass covers all the lower parts of the island, especially in the south. It is rich and nutritious and admirably adapted to animals. Horses and cattle thrive fairly well and sheep admirably. In 1921 Iceland had 49,300 horses, or as many as Iowa in proportion to the population; its 23,700 cattle were proportionally rather less numerous than in most parts of the United States, but its 554,000 sheep, or six for every inhabitant, gave it relatively more than in any State except Nevada and Wyoming. In proportion to the population Iceland has more horses than any other country except Argentina and Canada. It has relatively about as many cattle as Austria, Germany, and Great Britain, while its sheep are proportionally more numerous than those of any countries except New Zealand, Uruguay, Australia, and Argentina. Few countries are forced to depend upon animals more completely than Iceland.

The truth of this last statement becomes more apparent when it is realized that aside from domestic animals the other great resource of Iceland is fish. According to the Census of 1910 the population of Iceland depends upon the following sources for a living:

Agriculture	43,400	Commerce and transport	3,940
Fishing	15,900	Immaterial production	2,602
Day laborers and domestic servants	10,103	Relieved by public assistance	1,660
		Pensioners and capitalists	902
Industry	6,031	Profession not stated	644

Among the agriculturists, however, a great many follow the sea part of the time, so that fishing is much more important than would appear from this table. In 1915 the exports of fish products amounted to about 70,000 tons or three tons for every man over twenty years of age.

Such, then, was the unfavorable environment in which the relatively peaceable, thoughtful, and mercantile group of Norse settlers found themselves after their migration to Iceland. According to all the ordinary standards, the volcanoes and earthquakes of Iceland, its isolation, its lack of minerals, its cold summers and consequent lack of practically all resources except animals and fish, and worst of all its severe and recurrent famines, make that island a far less favorable place than the old Norwegian home from which it was settled. Compared with Normandy and England it is hard to find any respects in which the Icelandic environment is superior. Even Sicily, in spite of its malaria, would be counted much better than Iceland by the vast majority of mankind. Yet the Norse migrants to Iceland, although greatly handicapped, have rivaled England in their achievements, and have far surpassed Sicily. What ground there is for this assertion we shall see in the next chapter.

CHAPTER XIX

THE PERSISTENCE OF A SELECTED INHERITANCE

ICELAND is an astonishing country. In spite of its physical disadvantages it has stood in the forefront of civilization for a thousand years. It may almost claim that in proportion to its population its contribution to human progress has been greater than that of any other region except ancient Greece and Palestine.

If this seems extravagant, consider the facts. A real though crude measure of the influence of a country is the extent to which the country and its doings are discussed. The Library of Yale University has no special interest in Iceland, but its catalogue contains approximately 326 cards under that heading. About half deal with the geography, history, and general description of the island, 120 are devoted to the Icelandic language, and 50 to Icelandic literature. Ireland, with a population approximately forty-five times that of Iceland, has only about 1,440 cards, and no section devoted to literature. Again, Mexico is almost twenty times as large as Iceland and has about one hundred and sixty times as many people. It speaks a language which is more easily understood by Americans than Icelandic, and which is known to literally hundreds of thousands, while Icelandic is known only to scores. Yet that country, so large, so populous, so near, so rich, and so intimately connected with the commercial progress of the United States, has only about 1,200 cards in a great representative library—one-fortieth as many as Iceland in proportion to the population.

Another rough measure of the importance of a country is its men of eminence. The Encyclopædia Britannica, as we have seen, furnishes perhaps as good a summary of the world's opinion as can be found in any single source. It gives accounts of nine Icelanders, born since 1600, while many others are discussed in connection with the golden Icelandic age when the Sagas were written. In proportion to the population in 1800 this gives Iceland three times as many eminent authors as Ireland, France, or Switzerland, three and one-half times as many as Germany, six and one-half times as many as Belgium, and twenty-three times as many as Austria. Iceland's representation in the Encyclopædia Britannica during the last three centuries is proportionally greater than that of any other country outside England and Scotland.

A third measure of the importance of a country is the opinion of men of sound judgment. Lord Bryce perhaps sums up such opinions as well as any one. Here is what he says:

> Iceland is a country of quite exceptional and peculiar interest, not only in its physical but also in its historical aspects. The Icelanders are the smallest in number of the civilized nations of the world. Down till our own days the island has never had a population exceeding seventy thousand, yet it is a Nation, with a language, a national character, a body of traditions that are all its own. Of all the civilized countries it is the most wild and barren, nine-tenths of it a desert of snow mountains, glaciers, and vast fields of rugged lava, poured forth from its volcanoes. Yet the people of this remote isle, placed in an inhospitable Arctic wilderness, cut off from the nearest parts of Europe by a stormy sea, is, and has been from the beginning of its national life more than a thousand years ago, an intellectually cultivated people which has produced a literature both in prose and in poetry that stands among the primitive literatures next after that of ancient Greece if one regards both its quantity and its quality. Nowhere else, except in Greece, was so much produced that attained, in times of primitive simplicity, so high a level of excellence both in imaginative power and in brilliance of expression.
>
> Not less remarkable is the early political history of the island. During nearly four centuries it was the only independent republic in the world, and a republic absolutely unique in what one may call its constitution, for the government was nothing but a system of law courts, administering a most elaborate system of laws, the enforcement of which was for the most part left to those who were parties to the lawsuits.
>
> In our own time Iceland has for the student of political institutions a new interest. After many years of a bloodless constitutional struggle between its people and the Danish Crown, Denmark conceded to Iceland a local legislature, and an autonomy under that legislature which has greatly improved the relations between the two countries and furnished another argument to those who hold that peace and progress are best secured by the application of the principles of liberty and self-government.

PERSISTENCE OF A SELECTED INHERITANCE 279

One more important fact needs to be emphasized in order to appreciate Iceland. Not only did that country stand remarkably high nearly a thousand years ago, but it has sturdily maintained an enviable place among the nations in spite of periods of depression. In the sixteenth century, for example, printing was introduced into Iceland and soon became common. In Europe the first book was printed in Mainz in 1455. During the next half century the art of printing spread slowly to Italy in 1465, France 1469, England 1476, and Denmark and Sweden 1482–1483. In 1490 a book was printed in Constantinople, but even in Scotland 1507 was the date of the earliest book. All these are European countries, easy of access, and with populations numbering millions or at least many hundred thousands. Iceland was far away across a stormy sea, and had a population of only perhaps 50,000. It had been through a period of terrible depression and was only weakly in touch with the rest of the world. Yet in 1530 a printing-press was introduced, and between 1540 and 1600 at least forty-six books were published. In Norway it was not till 1651 that the first book was printed.

That the old ability has not been lost is evident from the fact that even during the last century Iceland had four native sons sufficiently eminent in literature to find a place in the Encyclopædia Britannica and others of the same sort, such as the sculptor Thorwaldsen and the explorer Stefansson, who were born of Icelandic parents in other countries. Nor are its great men Iceland's only claim to fame. Iceland to-day is highly progressive politically and socially. For example, it is one of the few countries which have a well-managed system of old-age pensions. Judging by the figures for 1908–1913, which are the last for a normal period, the death-rate in Iceland, 14.8 per thousand inhabitants, averages lower than in any country in Europe except England, the Netherlands, and the three Scandinavian countries. Such advanced countries as France, Germany, and Switzerland have distinctly higher death-rates than Iceland. Still more significant is the fact that in Iceland the reduction in the death-rate between the decades 1876–1885 and 1906–1915 was greater than during the corresponding period in all but one of the seventeen regions of Europe for which data are available. In Iceland the death-rate in the later period, when reduced to what is called a standard population, was only 58 per cent of the rate in the earlier period; in Saxony the corresponding figure was 56, but in Prussia it was 69, in France and England 75, and in Ireland 84.

Another evidence of the progressiveness of modern Iceland is its large number of learned institutions. Remember that we are talking about an island whose poor and widely scattered inhabitants are about as numerous as those of the three northern counties of New Hampshire, and are scattered over an area one-fourth larger than the combined area of Vermont, New Hampshire, Massachusetts, Rhode Island, and Connecticut. Remember that the island is as hilly as those three New Hampshire counties, which contain the White Mountains, and that its largest city is about the size of their largest city, Berlin. Remember, too, that in July the temperature even in the warmest places is only about 2° F. above that of the top of Mt. Washington. Yet Iceland has a university with schools of theology, medicine, law, and philosophy. At the university Denmark, France, and Germany, at least until the World War, maintained lecturers to give instruction in their respective languages and literatures. Iceland has also a normal school, a school of navigation, and two agricultural schools, as well as all sorts of scientific societies, a Bible Society, a national picture-gallery, and an archæological society which has published dozens of volumes.

Still another evidence of Iceland's present position is the literacy of its people. Only a few countries like New Zealand and Finland have so small a percentage of illiteracy. For a long time legal marriage was forbidden to a girl until she could read and write. Schools are scarce because the population is extremely scattered; often a community consists of only a few widely separated families. Yet so highly do the Icelanders appreciate literacy that almost every mother is a schoolmistress and well-nigh every child is taught at home. It is significant that the age when attendance at school is required is from ten to fourteen years.

PERSISTENCE OF A SELECTED INHERITANCE

In many places the younger children cannot go to school because of their isolation. Nevertheless education is compulsory. So interested are the Icelanders in their books that the favorite recreation is reading aloud during the long winter evenings. A traveler reports a boy of twelve who was studying botany from a Latin text-book while he was tending his flock of a thousand sheep. In another place a girl of fourteen had the duty of carrying the milk to the creamery every day. She carried it on horseback, two cans tied together and slung over the back of each horse. She tied her horses together in a string, nose to tail, as is the custom, and rode ahead on her own horse, reading as she went.

Books are a necessity in every family, even in the isolated northern island of Grimsey,

whose sorely isolated people look out upon the boundless boreal seas beyond the Arctic circle. They form the northernmost indigenous little community of our Germanic race—living by fishing in the most frigid of waters, and by capturing, at the peril of their lives, the stormbirds which build their nests on the almost inaccessible cliffs bounding the eastern shores of their tiny home. Dwelling on a dozen sterile farms, they maintain, with difficulty, three or four score of sheep, and half a dozen cattle and ponies, whose existence, like that of their owners, is one of perennial hunger. Yet they have a little church and an intelligent pastor, and a much-read island-library of a few hundred volumes (many of them on birds and fishes). (Mimir.)

Another extraordinary fact about Iceland is the high moral standards which seem to prevail. According to *Hygeia* (December, 1923), a journal published by the American Medical Association, syphilis has never been common in Iceland.

Block states that syphilis was not observed in Iceland until 1753, when it was recognized among certain employees in a manufacturing establishment at Reykjavik. It appears that even then this disease did not gain a foothold on the island, and soon it disappeared altogether. Finsen, who practised medicine in Iceland for nine years in the twenties of the last century, stated that the disease was introduced for a second time in 1824, but that he saw only five cases and these were confined to foreigners. Finsen remarks that it is strange that notwithstanding the fact that Iceland is visited annually by the men from hundreds of trading and fishing boats, the disease has not spread among the natives. If this is true, there can be but one explanation, and that is that sexual promiscuity does not prevail among the natives of Iceland as it does in Europe and America.

In reply to this it may be said that about thirteen per cent of the Icelandic births are illegitimate. In all probability, however, this merely means that in about one case out of fifteen or so, young couples consummate marriage without waiting for the wedding ceremony. This is a common practice where clergymen are scarce and the people are isolated. It does not mean immorality.

Perhaps the most surprising fact about Iceland is that this high development of culture and ability exists among a people whose mode of life is not merely humble, but arduous, monotonous, and in certain respects almost repulsive. Most of the Icelanders, as we have seen, are of necessity either farmers or fishermen. The majority dwell in one-story houses with enormously thick walls of stones stuffed with moss and turf. To get into the house one must pass through a long, low, dark passage, where the newcomer is almost sure to bump his head. This leads into the combined kitchen and living-room, which is often also a sleeping-room. Only in that room is there a fire. Ventilation is almost lacking, for the windows are very small and are rarely or never opened during eight months of the year. The stench is intolerable to those who are not used to it, for not only do the Icelanders rarely bathe, as is usually the custom in such cool climates; but aside from black bread made of imported rye or barley, their food is largely fish which are often half-decayed, milk which is usually sour, and butter which is purposely allowed to become rancid. Their woolen clothes are often wet with rain, their boots are foul from walking among the sheep and cattle. The combination of smells is almost indescribable, especially when it is mingled with smoke from the fire and from poorly trimmed kerosene lamps.

The work of the Icelanders is as humble as their dwellings. In February the fishermen begin to go out on the stormy sea, and a large share of the men from the farms go with them. In little boats manned by four to ten men they brave the storms and fogs, catching many kinds of fish, but especially the cod, which is one of their main sources of food. Often the weather is so bad that the fish rot even when properly cleaned and salted. In May the farmers usually go back to the farms. One of their first tasks is to cut the peaty turf, which is almost the only fuel. But by July a much more weighty task is before them, for the winter's supply of hay must be cut and cured. The cutting in itself is a heavy task, but the part which occasions trouble is the drying, for Iceland is a rainy, foggy country. All too soon the short summer nears its close. While the warm weather lasts the farmers must collect the sheep which have been wandering wild among the mountains. The wool must be plucked off, for it has become loose as the new growth begins to make the animals ready for the ensuing winter. To shear the sheep would remove the long, coarse hair which is needed as a protection against the rain. During this same busy season the houses must be repaired, the cracks must be stuffed once more, the turf which has been slowly drying during the cool summer must be brought home, and the precious hay-fields must be manured as soon as the hay is removed. Perhaps the women make a gay excursion into the interior to gather the lichen known as Iceland moss, but their chief work is to care for the animals, milk them, make butter, and tend the home. Supplies, too, must be laid by against the winter, when snow-storms may hold the people at home for weeks at a time. That may mean a long trip on horseback to the nearest town, where imported food and manufactured goods are for sale. Aside from the weekly church services and the occasional weddings, such trips are almost the only break in a life that is peculiarly monotonous.

Then comes the early winter, for Iceland has neither spring nor autumn. That is the most demoralizing part of the year, for there is little to do, and the tendency is to sit idle day after day. Perhaps that is why the Icelanders have obtained among some travelers the reputation of being lazy. This is far from the truth, for though the Icelander is physically slow and is not quickly aroused, he is wonderfully persistent and competent when once he bestirs himself. The long winters when there is little or nothing to do except feed the animals are often called the cause of the great love of books among the Icelanders. They are a condition rather than a cause, for the cause lies in Icelandic character. But the long winter evenings certainly give an opportunity for the expression of that love, and the habit of reading aloud is almost universal. Perhaps the love of books would wane if the Icelanders could enjoy themselves out of doors at almost all seasons as freely as do the people of Italy. While one of the family reads, the rest are usually busy making ropes and saddles, knitting stockings and mittens, or embroidering bed-covers.

In spite of the seemingly unfavorable conditions under which the Icelanders live, they appear still to retain much of their old ability. Yet they do not seem to be quite the same as their ancestors. They are more peaceful, slower, more cautious, less vivacious, and perhaps less ready to take the initiative, although by no means backward in attempting new things when once they are persuaded to try them. And possibly they are more honest. At least the habits of robbery and piracy which stand out among their ancestors are no longer noticeable, while almost every traveler is impressed with the extraordinary honesty of the Icelanders of to-day. They are indeed well able to drive a sharp bargain, and do not hesitate to charge a good price to the supercilious foreigner who looks down upon them and makes disagreeable comments upon their poor food and ill-smelling houses. But let a foreigner go among them in a spirit of appreciation and goodwill, and they not only give him their best ungrudgingly, but guard him from discomfort wherever possible. Russell, for example, in his book called *Iceland*, reports that he lost a riding-belt, but did not think it worth while to go back and hunt for it. Three weeks later it was handed to him, having been passed from hand to hand and from farm to farm along his route.

According to the hypothesis of this book, the original cause of Iceland's greatness was the repeated selection to which its people were subjected. When the last selection had brought to the island a peculiarly homogeneous, sober-minded, and competent group of people, there ensued within a century or two a wonderful outburst of genius which produced the famous sagas and sent Icelandic bards and advisers on more or less temporary pilgrimages to the courts of northern Europe. In our own day, a thousand years later, Iceland still has the right to boast of her people's character and achievements. A selected inheritance, when isolated, protected, and kept up to the mark by further selection, seems to be able to persist indefinitely.

CHAPTER XX

THE DIRECT EFFECT OF ENVIRONMENT ON CHARACTER

In trying to explain the character of a race, we must constantly remind ourselves that character as it displays itself before our eyes is the complex result of three great factors. One of these, as we have seen again and again, is inheritance; a second is physical environment, with its multitudinous effects upon nutrition, health, occupations, and modes of life; and the third is social environment and the vast mass of ideas, habits, inventions, and discoveries which are handed down from one generation to another. Even in a country like Iceland, where the importance of inheritance is especially clear, the other factors have also been highly important and may even have changed the biological inheritance to a certain extent.

An example of the effect of environment is the fact that because of the dangers of seafaring, the death-rate among the young men of Iceland is very high. Although only about a sixth of the islanders are classified as fishermen, a large part of the farmers also engage in this occupation part of the time. Figure 16 shows the effect of this occupation, together with that of the dangers due to herding sheep and cattle in a rough and foggy land full of bogs and cliffs. The solid curved line indicates the number of deaths among Icelandic men and boys for every hundred deaths among women and girls. In infancy about 125 boys die for every 100 girls, partly because more boys than girls are born and partly because boys are less resistant. By the age of four or five, the deaths among the two sexes are practically equal, while at the age of ten there are more deaths among girls than boys. Then the normal sequence of events is completely shattered, for in spite of the dangers to young women incident to the bearing of children, the deaths among young men enormously exceed those among their wives and sisters. The dotted line in Figure

EFFECT OF ENVIRONMENT ON CHARACTER

16 shows the conditions in Switzerland, where the men are engaged in ordinary farming, and in relatively healthful industrial occupations. There the number of deaths among men is less than among women, up to the age of thirty. This is the normal condition in most parts of the world. If Iceland were a normal country it would presumably have about 95 deaths among young men between the ages of fifteen and twenty-four for every 100 deaths among young women. As a matter of fact, for the forty years from 1876 to 1915 the ratio of deaths at those ages was 169 for young men, compared with 100 for young women. Put in another way, this means that as soon as the boys become old enough to go out in the fishing-boats upon the stormy Atlantic or to tend the sheep upon the remoter mountains they begin to be killed in large numbers. In no other country where accurate records are available is there any such enormous killing off of the young men in early manhood. Even in Norway, where seafaring is more prevalent than in almost any country except Iceland, the excess of deaths of young men over young women at the ages from fifteen to twenty-four is only 25 per cent, as appears in the dash line in Figure 16.

What effect has all this on racial character? It must have some effect, and it may have much. Although the excessive death-rate of men compared with women continues to old age, it is greatest among the young lads when they first go to sea. Many boys who in a few years would be married and become the fathers of the next generation are killed, and many young fathers are killed before their families are complete. This ruthless slaughter of the young men does not happen occasionally, like the destruction due to a volcanic eruption. It happens every year, and it has happened every year for ten centuries. It is a steady, inexorable, selective factor which affects every generation and practically every part of the Icelandic population. No demonstration is needed to show that in general the careless, the rash, the slow-witted, those who do not get on well with their fellow boatmen, and those who lose their heads in an emergency are the ones most likely to lose their lives.

The total number killed in any one year is not large, but think what happens when a process of selection like this goes on for a thousand years. Every year takes from among the people a certain group who have the qualities described above. Remember that those who are thus picked out are chiefly young men who in many cases are not yet fathers. Thus the fathers of the next generation tend constantly to be those whose lives have been spared because they are cautious, careful, level-headed, and at the same time quick, alert, and strong, able to co-operate with their fellows, resourceful, observant of winds and waves, and able quickly to draw correct conclusions in the face of danger and death. Read a book like Stefansson's *The Friendly Arctic*. There one sees a concrete picture of those qualities and their results. In the whole realm of exploration few books exemplify a greater capacity for careful, accurate observation, for caution combined with daring, and for correct inferences from natural phenomena. Not all the Icelanders have these qualities to such a degree as Stefansson, but on the whole they are the qualities which stand out in Icelandic character. Doubtless something of this character had already become part of the Norse racial inheritance before the migration to Iceland, but it seems probable that these qualities have steadily become more and more ingrained through the natural selection due to a fisherman's life in those far northern waters. The shepherd's life too has had a similar effect. When a boy is lost in the mist and is in danger of falling over a precipice or stumbling into a bottomless bog, he needs not merely a sturdy body but a clear, cool head, and the faculty of observing carefully and then calmly deducing accurate conclusions. If he has those qualities, he may survive and pass them on to his sons and daughters. If he has not, he perishes and his weaknesses perish with him.

The climate of Iceland probably plays a direct part in determining Icelandic character, as well as an indirect part through inheritance. On the whole the climate appears to be healthful. This is presumably one of the reasons why the death-rate in Iceland has for many years been nearly as low as that of the most advanced countries of western Europe, in spite of uncleanness and poor sanitation. The climate itself may also be largely responsible for the lack of cleanliness. So far as I am aware, this lack prevails among every people who live in a cool, moist climate where the water is always cold and where animals are the chief means of support. This is true of the present Norse as well as the Lapps, of the mountain Swiss as well as the Tibetans, and of Eskimos, Kirghiz, Tunguses, and a host of others. The cleanest people in the world are the inhabitants of warm, moist countries, where the state of culture requires clothing, and where there is plenty of water. The Malays of Asia and the Mayas of Yucatan are extraordinarily clean because in their climate frequent bathing is a pleasure and also a necessity if the people would be free from insects and other afflictions. Their case is not at all like that of the Icelanders at the time when Christianity was introduced. That conversion, it might be noted incidentally, was a good example of the sane way in which Iceland has always done things. After a group of missionaries had preached for some time the Icelanders met in their Althing, or general assembly, to consider whether the new religion should be adopted. The matter was finally left to the Speaker of the Law, as the head of the Althing was called. He deliberated by himself for two days. Then he decided that Christianity was the religion of the future. Therefore Christianity was adopted, but there was no compulsion. Those who wished to worship the old gods could still do so. The amusing part of the story is that one great objection to the new religion was that it required baptism. The people of northern Iceland objected especially, because they did not want to be plunged into cold water. They were not used to bathing. Finally it was agreed that they should be baptized in the hot springs at Reykir, and everything was harmonious. To-day the descendants of these early Christians rarely learn to swim because the Icelandic waters are too cold, especially in the north. This doubtless tends to increase the death-rate among the fishermen.

The climate of Iceland is not only healthful but stimulating.

Among European races physical activity appears to be greatest when the temperature averages not far from 65° F., whereas mental activity seems to be greatest at a lower temperature, averaging perhaps 40°. The ideal is a climate like that of England, which ranges from the physical optimum in summer to the mental optimum in winter. While Iceland cannot rival such a climate, its temperature fluctuates about the average which seems to be best for mental activity. Another highly important element in stimulating mental activity seems to be variability. Here again Iceland is much favored. It lies close to the path where storms are most numerous. One of the things which are especially noticeable in Iceland is the clear skies following rain or snow. Since most of the people live near the coast, they rarely experience extremely low temperatures, such as prevail in continental North America; yet they pass through a great number of stimulating changes. In this respect Iceland is far better off than Sicily or than the cool home of the Incas in Peru, where the temperature resembles that of Iceland, but the changes from day to day are negligible.

One of the most interesting facts about Iceland is that the climate seems to have fluctuated back and forth near a certain level above which the conditions are quite favorable, while below they are repressive. When the climate has become repressive, an actual change in Icelandic character seems almost to have occurred. The racial inheritance has presumably remained unchanged, but the stress of the environment has inhibited it from expressing itself in progress. Let us briefly review the history of Iceland with this in mind. When the Norse Jarls with their attendant Karls migrated from Norway the climate, as we have seen, was probably unusually cool and rainy (Figure 15, page 232). After 930 A. D., however, the climate in California appears for a generation or two to have returned part way toward the conditions of the seventh century, which were apparently favorable for Iceland and Norway. In Iceland it is probable that the mitigation of climate was greater than appears in Figure 15, for grain seems to have been cultivated, and that is now almost impossible. It was during this favorable epoch, although by no means because of it, that the Icelandic constitution took form. After an early period of discord and broils the leading islanders, being an unusually sensible set of men, agreed that some central government was necessary. They sent one of their number, Ulflgot, to Norway to report on the system there in vogue. The result of his report was the establishment of the Althing. This national assembly is highly peculiar. Its members were chiefly the heads of the various local districts. Its chief was called Speaker of the Law. The Althing met once a year in the latter half of June on a green, grassy plateau almost surrounded by the rocky cliffs bordering a small lake. There they decided what the existing law was, and occasionally framed new laws. One of the Speaker's main duties was to recite the whole body of law, section by section, completing it in three years; if any one asked the law it was also his duty to give the needed information. Neither the Althing nor the Speaker had any authority to enforce the law. They decided the cases that were presented, but enforcement was left in the hands of the people. It is marvelous that the law was so well respected. Many times, indeed, the legal decisions were enforced by violence, or even ignored. Usually, however, a defendant who was declared to be in the wrong yielded simply through the force of public opinion. Until Iceland passed under the rule of Norway in 1262–1264 this same system continued with only slight modifications. The island was ruled by law, not by a government. In a weakened form the Althing continued to sit until 1800, and was revived in a modified form in 1843. So great was the respect of the Icelanders for the law that even the ghosts of dead men were believed to respect it. One of the old sagas, as quoted by Bryce, contains a most instructive story.

A chief named Thorodd, living at Fró á in Breidifjörd, on the west side of Iceland, had just before Yule-tide been wrecked and drowned with his boat-companions in the fjord. The boat was washed ashore, but the bodies were not recovered. Thereupon his wife Thurio and his eldest son Kjartan bade the neighbors to the funeral feast; but on the first night of the feast, as soon as the fire was lighted in the hall, Thorodd and his companions entered, dripping wet, and took their seats round it. The guests welcomed them: it was held that those would fare well with Ran (the goddess of the deep sea) who attended their own funeral banquet. The ghosts, however, refused to acknowledge any greetings, and remained seated in silence till the fire had burnt out, when they rose and left. Next night they returned at the same time and behaved in the same way, and did so, not only every night while the feast lasted, but even afterwards. The servants at last refused to enter the fire-hall, and no cooking could be done, for when a fire was lit in another room, Thorodd and his companions went there instead. At last Kjartan had a second fire lit in the hall, leaving the big one to the ghosts, so the cooking could now be done. But men died in the house, and Thurio herself fell ill, so Kjartan sought counsel of his uncle Snorri, an eminent lawyer and the leading Godi of Western Iceland. By Snorri's advice Kjarten and seven others with him went to the hall door and formally summoned Thorodd and his companions for trespassing within the house and causing men's deaths. Then they named a Door-Court (Dyradómr) and set forth the suits, following all the regular procedure as at a Thing-Court. Verdicts were delivered, the cases summed up, and judgment given; and when the judgment word was given on each ghost, each rose and quitted the hall, and was never seen thereafter.

Ghosts have given much trouble in many countries, but it is only the Icelanders who have dealt with them by an action of ejectment.

Another extraordinary example of respect for law occurred in 1006 A. D. That year there occurred an indecisive duel between the poet and viking Gunnlaug Ormstunga (Snake's Tongue, as he was called from his satirical powers) and another poet, Hrafn. Gunnlaug had been betrothed to Helga the Fair, one of the most famous Icelandic heroines, but had been detained by King Ethelred II and had returned to Iceland to find her married to Hrafn. So the two men fought for Helga. As the duel was indecisive Gunnlaug wished to renew the fight and Hrafn was willing. But next day the Althing passed a law prohibiting formal duels. The duellers obeyed the law, but went to Norway to avoid it. There, years later, they fought again, and killed each other.

After the establishment of the law Iceland was fairly prosperous for a few years, but by 1000 A. D. a period of decay had set in. Then, according to Burton:

A mighty change came over the island mind when Olafr Tryggvason (Olaf I, Trustyson) induced the Althing to accept Christianity as the national religion. The old pagan creed had become age-decrepit. . . . The "great Sire of gods and men" was dying or dead, a gloomy fate which equally awaits superhuman and human nature. The decline and fall of Odinism only repeated the religious histories of Palestine, Egypt, and India; of Greece and of Rome, whose maximum of effeteness has ever been at the period of the Christian invasion.

How great a degeneration there really was at this time it is difficult to determine. It is interesting to note that what Burton calls "the maximum of effeteness" came at a time when Figure 15 suggests that the climate may have been quite unfavorable. This does not mean that it was worse than that of to-day, for as to that we are not certain. In fact the tale of Burnt Njal seems to speak of grain-fields about 1000 A. D., although now there is practically no such thing in the whole island. But perhaps these were merely hay-fields. At any rate it seems probable that about 1000 A. D. the climate was colder and stormier than in the preceding and succeeding centuries. Whether this caused degeneration we cannot say, but it may well be that then, as at a later time, discontent and poverty were so great that the people were ready for a change.

After the introduction of Christianity there followed a wonderful period of two or three hundred years during which Iceland rose to its highest level. This was the age of the great sagas, when Iceland almost rivaled Greece in its literary ability and in the versatility and vigor of its writing. During this halcyon period the first school was established in the middle of the eleventh century; a little later hospitals were endowed and about a half dozen monasteries and nunneries were established. The people, according to Burton, became "less crafty and cruel." It can scarcely be doubted that Christianity was an important cause of this great outburst of intellectual activity, economic prosperity, and moral improvement. There may also be much truth in the idea that this was the normal flowering of Iceland's racial genius. Perchance also, as some would say, we are dealing here with a mere historic accident.

In addition to all this it may be pointed out that the golden age of Iceland not only followed the introduction of Christianity but was also a time when the climate was becoming more and more favorable, as is suggested in Figure 15. According to Pettersson and others the temperature in the twelfth century was such that large trees grew in several parts of the island and were used for building houses. Such areas are specifically named in the *Landnamabok*, which gives a minute account of Iceland comparable to the account of England in the *Domesday Book*. Fruit-trees were raised, and grain was cultivated. Communication between Iceland and Europe was relatively easy, for there was little ice even on the north coast. Presumably because of the higher temperature, prosperity was the rule. The change in character noted by Burton may mean merely that being more prosperous the Icelanders were in better temper, less prone to fight with one another, more ready to be friends with the rest of the world, and hence "less crafty and cruel."

Toward the end of this prosperous period Iceland was annexed to Norway in 1262-1264. This was partly due to the fact that many Icelanders were at the Norwegian court, a condition which may have been accentuated because the favorable climate not only made it easy to travel across the ocean, but helped to provide products for export. After the annexation a flood of troubles overwhelmed the island. First there was a series of terrible eruptions and earthquakes. Then epidemics began, one of the worst being in 1306, while the worst of all was the Black Death in 1348 and again in 1402. During the fourteenth and fifteenth centuries the Icelanders were reduced to a "low dead level of poor peasant proprietors careless of all save how to live by as little labour as possible, and pay as few taxes as they could to their foreign rulers." (Powell.) This depression is commonly ascribed to the union with Norway, but an interesting sidelight is thrown on this conclusion by the fact that during this same period "all spirit seemed to have died with the commonwealth; even shepherding and such agriculture as there had been sank to a lower stage; wagons, ploughs, and carts went out of use and knowledge; architecture in timber became a lost art, and the fine carved and painted halls of the heathen days were replaced by turf-walled barns half sunk in the earth; the large decked luggers of the old days gave way to small undecked fishing-boats. ... The early falling off of the Norse trade threatened to deprive the island of the means of existence; for the great epidemics and eruptions of the fourteenth century had gravely attacked its pastoral wealth and ruined much of its pasture and fishery." (Powell.)

All this is merely the normal picture of what must inevitably happen in Iceland if the climate becomes cold and stormy. In Figure 15 it appears that in the fourteenth century such a climatic change actually occurred in California; abundant local evidence indicates a similar occurrence in Iceland. The ancient records tell us that ice began to crowd against the north shore so that communication was very difficult. After the Black Death there followed a winter so long and cold that it destroyed nearly all the cattle and nearly ruined the country. In Norway the summer rains were so frequent and the summers were so cold that the crops failed year after year. In England similar conditions produced the greatest of all English famines from 1308 to 1321, with others in 1351 and 1369. In Iceland the same thing seems to have happened. We have already seen how a slight change in the direction of the wind and in the icepack may greatly alter the temperature. Even as late as 1866 and 1882 Iceland suffered severely because cold summers and the consequent impossibility of curing hay led to the destruction of vast numbers of sheep and cattle.

From 1300 onward, the trade of Iceland markedly declined and changed. The first noteworthy change was a decline in commerce with Norway; then there was a transfer of trade from Norway to England, and an increase in the importance of fish compared with wool as an Icelandic export. Finally by 1440 there was almost no commerce with Norway, and very little with England. In those days there was no salt, no cloth, no bread in Iceland, nor any liquor except milk and water. The change from architecture in wood to the building of turf houses, the giving up of the use of the plough and the cart, the change from large-decked luggers to small undecked fishing-boats, the decline in trade with impoverished Norway, the substitution of fish for wool, and the final abject poverty and absence of foreign trade in Iceland—all these are what would be expected during a prolonged period of cold summers. Doubtless Norse oppression helped at first to impoverish Iceland, but that was scarcely the reason why Iceland fell to the lowest ebb in its history. The very isolation of the island, and the ice and storms which then prevailed, must have tended to prevent intercourse between Iceland and Norway. Moreover, Norway herself in the fourteenth and fifteenth centuries was in dire straits from repeated failures of the crops, and political confusion was her lot partly on that account. She had to leave Iceland to its own misery. All things considered, it appears that the climatic stress of the fourteenth and fifteenth centuries was one of the chief factors in the contrast between the twelfth century with its great intellectual activity and the fourteenth with its collapse of civilization. The inherent character of the Icelanders probably did not change, but a repressive environment held them down.

When this time of climatic stress came to an end there was a new awakening in Iceland. By 1475 A. D. trade had begun to revive; in 1518 some three hundred and sixty British merchants were in Hafna Fjord alone; and trade continued brisk during the first half of the sixteenth century, although it declined in the second half. It would be unwarranted to say that the more favorable climatic conditions which are indicated in Figure 15 at the end of the fifteenth century were the cause of the renewal of relations with Europe and of the coming of the Icelandic Reformation. Nevertheless it is a fact that when the climate appears once more to have become relatively favorable, though perhaps not so good as in the earlier days, Iceland once more revived and took up the march of progress. Printing was introduced, the Bible was translated, and printed books began to be common.

Then came another change.

The progress of the seventeenth century is principally marked by adverse events, physical evils, and the rapacious violence of men who united to waste this miserable island; while the wretched inhabitants, long unaccustomed to the use of arms, could offer no resistance even to a small band of pirates. The oppression they suffered from these marauders was extreme, no part of the coast being for a moment secure from their attacks. It is a melancholy fact, that the majority of them were French or English, as if the two most powerful and civilized of the European nations had combined to oppress the poorest and most helpless, and to visit on their descendants the evils which had been endured from the ancient Northmen. In 1627 some Algerine corsairs, too, who found their way to that remote region of the ocean, spread universal dismay round the whole coast.

The last part of the seventeenth and the whole of the eighteenth century are described as a period not only of physical disasters but of increasing superstition. From 1660 to 1690 Iceland like other parts of the world became excited over witches, and sixteen persons were burnt alive. In 1707 smallpox killed 18,000 people, a third of the population. This may almost be called a direct result of isolation, for when a people is long isolated it loses the power to resist epidemics. In the middle of the eighteenth century the seasons, as we have seen, were so inclement that in 1759 a famine swept away 10,000 people, while vast numbers of sheep perished in 1762. A volcanic outbreak in 1765 and the terrible eruption of Skaptaa in 1783 spread destruction far and wide. After the eruption,

The noxious vapors that for many months infected the air were equally pernicious to man and beast, and covered the whole island with a dense fog which obscured the sun, and was perceptible even in England and Holland. The steam rising from the crater, or exhaled from the boiling waters, was condensed in the cooler regions of the atmosphere, and descended in floods that deluged the fields and consolidated the ashes into a thick black crust. A fall of snow in the middle of June, and frequent showers of hailstones of unusual magnitude, accompanied with tremendous thunderstorms tearing up huge fragments of rock and rolling them down into the plains, completed the scene of desolation. The grass and other plants withered, and became so brittle that the weight of a man's foot reduced them to powder; and even where the pastures seemed to have recovered, the cattle refused to touch them, dying of actual starvation in the midst of the most luxuriant herbage. Small unknown insects covered many of the fields, whilst other portions of the soil formerly the most fertile were changed by the ashes into marshy wastes overgrown with moss and equiseta. A disease resembling scurvy in its most malignant type attacked both men and cattle, occasioned in the former no doubt by the want of food, and the miserable, often disgusting, nature of that which alone they could obtain. Many lived on the bodies of those animals which had perished from hunger or disease, whilst others had recourse to boiled skins, or substances still more nauseous and unwholesome. The numerous earthquakes, with the ashes and other matter thrown into the sea, caused the fish to desert many parts of the coast, whilst the fishermen seldom daring to leave the land, enveloped in thick clouds during most of the summer, were thus deprived of their usual stock of winter provisions. We cannot better conclude this frightful catalogue of evils than by the following summary of the numbers of men and cattle more or less immediately destroyed by it in two years. The most moderate calculation makes these amount to 1,300 human beings, 19,488 horses, 6,801 horned cattle, and 129,937 sheep. . . . The smallpox also added once more its fatal influence, and in a few years 11,000 individuals fell victims to these combined attacks. The destruction of the fisheries on the southern coasts was an evil of a more lasting character, and one from which the country was long in recovering." (Nicoll.)

The last great recovery of Iceland has come since 1850. This apparently has been helped a little by an amelioration of climate, but it is largely the result of modern means of communication. Steam has partly broken down the isolation of the island. Thus contact with other countries has given a new impetus to literature as well as commerce.

The history of Iceland may be summed up as the resultant of three main factors—race, climate, and isolation. A thousand years ago Iceland began its historical career with a picked group of immigrants. Only in a few places like New England and New Zealand has the population of a country been so rigidly selected. Nowhere else, so far as we are aware, has such a body of people been so nearly isolated for a thousand years, as in a great experiment, and thus allowed to show what happens to a picked racial stock when left almost unmixed and undepleted. Does this experiment suggest that races develop to a certain point and then inevitably decay? Does it indicate that racial mixture is necessary in order to maintain the vitality and power of a race? Does it show that a race must in the long run conform to the type of physical environment in which its lot is cast? Does Icelandic history support the view that the introduction of new political and social institutions is the main basis of human progress as many historians seem to believe?

The answer to all these questions seems to be negative. The one clear lesson of Iceland appears to be that no one theory accounts for the history and character of a people. From first to last three great facts stand out in Icelandic history. The first is the persistence of racial character. The earliest Icelanders were above the average of their race or almost any other race in intellectual ability: the modern Icelanders still stand above the average of their race. But there has been a change, for the second great fact is that the Icelander to-day is different from his ancestors. The difference is hard to define, but it seems to lie in greater caution, greater deliberation, greater seriousness, less activity, less initiative, perhaps greater power of self-preservation, and probably a stronger tendency toward purely intellectual as opposed to physical activity. The change may be partly the result of a complete mingling of Jarls and Karls so that the Icelanders are extraordinarily uniform in inheritance and hence in character. In larger measure, however, it is probably the result of natural selection whereby the more daring, adventurous, and rash young men have been killed off in every generation. The more nervous, vivacious types seem likewise to have been weeded out, perhaps because they are less able to endure anxiety and uncertainty.

The third great fact is that in Icelandic history we seem to see a constant repetition of expansion and repression in response to a changing environment. When the Icelandic climate has been mild, life has been comparatively easy, and men's minds have been free from the constant repression and discouragement which are inevitable among a thoughtful people who are continually on the verge of starvation and economic ruin. A people with lower mentality can endure such conditions with much less mental strain than can so alert a people as the Icelanders. During mild epochs the people have also been better nourished than during the cold periods, for in the worst periods they are almost restricted to fish and Iceland moss, whereas in the good periods the islands themselves furnish plenty of milk and meat and even a little grain and fruit, while an abundance of flour, salt, and other necessities can be and have been imported. Thus during the periods of good climate the diet of the people has presumably caused their health and energy to be better than in the periods of repressive climate, while at the same time the direct effect of the climate in stimulating health and mental activity appears to have been not inconsiderable. In addition to all this, intercourse with the outside world has been relatively easy during the periods of good climate and of expansion, and difficult during the periods of poor climate. This is not merely nor even primarily a question of storms and ice and fog. The main factor in restricting communication with the rest of the world has apparently been poverty. When people are too poor to keep their boats in repair, and when they have nothing to sell to the rest of the world and no means of paying for what they would like to buy, commerce must decay, no matter how smooth the seas may be. Thus the environment co-operates with natural selection, and the effects of both are profoundly modified by intercourse with other peoples and the introduction of new ideas. Racial character, or that which we commonly suppose to be racial character, arises from extremely complex causes.

CHAPTER XXI

THE SELECTION OF MODERN AMERICANS

Much of the story of Iceland is repeated in that of the early white settlers in America. To tell the American story in such detail as that of the northern island is not feasible in this book. We can merely show the relation of two noteworthy examples to our general theme. We shall discuss the early New Englanders and their descendants farther west in the prairies and California. We might with almost equal profit consider the French and English in Canada, the plantation owners and negroes of the southern United States, or even the Spaniards and Portuguese of Latin America. In each case we should find the same great principles at work, but with an almost infinite variety of absorbing details.

For reasons which are but dimly understood, the sixteenth century saw in England a great awakening. It expressed itself in intellectual leaders like Shakespeare and Bacon, in a ruler like Elizabeth, in explorers like Drake, and in men like Raleigh, who was soldier, explorer, historian, poet, and gentleman. Among men of a certain type the awakening took the form of questions as to moral duties, the Bible, and man's relation to God and to spiritual authority. In some it led to strong convictions as to duty toward their fellows, and as to the worship of God according to the dictates of their own consciences, or according to their interpretation of the Bible.

In the first forty years of the next century this same spirit produced a period of constitutional conflict at home in England, and

was marked by an outburst of romantic activity that sent hundreds of Englishmen out into the western seas in search of adventure and profit. Coincident with the later days of these half-piratical expeditions and organized commercial enterprises were the migrations of those who, moved by impulses that were partly religious, partly political, and partly economic, sought independence of worship and permanent homes in the New World. Though differing widely in purposes and results, these journeyings into the unknown West were often closely related in origin, and were supported by groups of men, aristocrats, commoners, merchants, and adventurers, who were ready to promote any undertaking, whether commercial or religious, that promised a profitable return. It is difficult to grasp the full significance of the settlements of Virginia, Maryland, Massachusetts, and Saybrook, without a knowledge of the circumstances under which the colonies of Bermuda, Barbadoes, and Old Providence were established; for all represented in different forms and propositions the influences at work in the motherland which were arousing in men of all classes the spirit of adventure and revolt. No single motive governed the men who voyaged over seas during this romantic period. The zeal of the viking and the lust of the capitalist were inextricably interwoven with the hopes of the godly in the task of opening and occupying the great frontier which stretched westward from the maritime states of Europe. (Andrews.)

For our present purposes the most important of these British adventurers were the Puritans, especially the more daring among them who fared forth across the sea.

The English Puritans formed a veritable clan, intimately bound together by ties of blood, marriage, and neighbourhood, and they acted together in all that concerned colonisation on one hand and autocratic rule on the other. (Andrews.)

They attempted to found colonies not only in New England, but in and around the Caribbean Sea.

The founders of both wished to provide a refuge for the oppressed victims of Laud's ecclesiastical régime, each was to be a sanctuary where the Puritans might worship God after their own fashion, each was to be a society ordered according to the dictates of religion and governed with justice and equity, but upon the strictest Puritan pattern. (Newton.)

Before we follow these people across the seas, turn back a moment to their origin. The majority of them came from East Anglia, that is, from the part of eastern England north of London and the Thames and south of The Wash. Now this region, as Havelock Ellis discovered in his *Study of British Genius*, is one of three great foci of intellectual ability in England. In proportion to its population the East Anglian focus appears to have made the greatest contribution to British genius. Ethnologically it is

the most recent of the three. East Anglia is a region very open to invasion: Brythons, Romans, Angles, and Normans all seem to have come here in large numbers [this was the place where the Danish Viking landed in greatest numbers] and it differs from every other English district (except to some extent Kent, a country closely allied to it) in continuing to welcome foreigners—Dutch, Flemish, Walloon, French—all through mediæval times down to the revocation of the Edict of Nantes at the end of the seventeenth century. . . . East Anglia is productive of great statesmen and great ecclesiastics; it is also a land of great scholars. At the same time nearly half the British musical composers and more than a third of the painters have come from this region. It has no aptitude for abstract thinking, for metaphysics, but in concrete thinking, in the art of treating science philosophically, it is easily supreme. Its special characters seem to be its humanity, its patience, its grasp of detail, its deliberate flexibility combined with a profound love of liberty and independence. The characteristic English love of compromise is rooted in East Anglia. . . .

It may be noted that the founders of New England, both on the political and the religious side, were largely produced by East Anglia. The Washingtons came from the related county of Northamptonshire; the Emersons were from Suffolk; Winthrop, who, it has been said, more than any other man moulded Massachusetts, which moulded New England, belonged to Central Suffolk. (Ellis.)

Judging by these quotations, the Puritans were a selected migrant group derived from a long line of earlier migrations and selections. Only a certain type of strong-minded, thoughtful, self-controlled, and self-sacrificing people accepted the new doctrines and had the courage and determination to live up to them. Out of the whole English people they comprised a small percentage who were selected because they possessed certain highly valuable qualities. In the case of the Pilgrims who eventually came to Plymouth the first selection was followed by a second. Persecution and contempt weeded out the Puritans who were less steadfast, and then led some to migrate. One group went to Holland. A migration to a new country, as we have seen again and again, is never a light matter. This is especially true when the migrants leave home because they are persecuted, when they go to a new country without special invitation, and most of all when the journey is long and dangerous. From among the people who had the courage and temperament to become Puritans only a small and highly selected number went to Holland. Those who went were in general the ones with unusual courage and determination, strong convictions, and great initiative.

Again there occurred a third selection, and the Pilgrims were the result. When the Pilgrims left Holland, the ones who braved the dangers of the long, hard voyage were in general those most strongly imbued with the Puritan temperament and the spirit of adventure, and likewise highly endowed with physical vigor, initiative, and courage. The man who was conscious of physical weakness dared not go. The woman who was timid kept back her husband. The family with an ailing child was also deterred. Thus there occurred a rigorous selection on the basis not only of mental and moral character, but of health. As a rule it was the motor-minded, determined, conscientious, religious, adventurous, and physically strong who embarked in the *Mayflower*. Even yet the selection was not ended. Before the voyage was finished death had set its hand upon the little band. During the first winter 44 of the 102 who composed the original *Mayflower* passengers died of diseases which found an easy prey because of exposure, hunger, and weariness. At the end of a year, when new recruits arrived, half had succumbed. Here again it was the weak who perished, especially those who were weak in spirit as well as body and therefore gave up in discouragement.

The experiences of the Pilgrims are more or less typical of those of all the early Puritans who came to America. Throughout the whole period of migration, from the time the emigrants first left England until they had become comfortably established after some years in America, the women and children suffered most. For them a migration is always harder than for the men, partly because they are by nature more conservative, partly for purely physical reasons, and partly because the responsibility for the children and the pain of their birth and death fall on them far more than on the fathers. Hence we may be confident that in many a case the men of a family were ready to go, but the women deterred them. After the new land was reached it was still the women among whom the selection was greatest. During the first sad year at Plymouth thirteen of the eighteen married women were laid in unmarked graves, and most of them were young. In cases where families returned from America to England we may be certain that far more went at the instigation of the women than of the men, for that is human nature. To live in early America as well as to go there required a higher, stronger type of body, mind, and spirit among the women than among the men.

Those early Pilgrims and the Puritans who followed them for a century more or less may have been too stern and too oblivious of the value of joy and beauty, but they had marvelous qualities. Because they were practically all selected for certain religious and moral qualities plus physical strength, hardihood, independence, and the spirit of adventure and because like mated with like, their various characteristics, their racial characteristics we may almost say, were handed down for generations. They brought to America certain new ideas and those ideas helped to mold the society of early New England, but I believe that the selection of the migrants rather than the ideas was the main reason for the great progress of the American colonies.

Although the Pilgrims and the Puritans of New England were the most outstanding example of the process of selection in early America, something similar occurred on a different scale in all the colonies. The Huguenots of the South, the Quakers of Pennsylvania, and various other groups represent the same type of selection as the Puritans. In some of the Southern States, especially Virginia and South Carolina, there were many immigrants derived from the best families of England, younger sons who came to the New World because they were animated by a spirit of adventure and enterprise. Their case was very different from that of modern immigrants. It is one thing to go to a new country under the guidance of a paid agent in a great comfortable steamship and find a job ready at the other end, and quite another to go to a new country in a small sailing vessel and establish one's self in the face of naked savages. Thus to a greater or less degree the colonists along the whole Atlantic seaboard from Maine to Georgia were a selected people.

When a race is selected in this way certain results may be expected. The history of the American colonies illustrates them. As soon as the period of fierce struggle for existence was over, the high inheritance of the colonials began to assert itself in new institutions. It was not the institutions which made the people, as is so often implied and even asserted. It was the people who made the institutions. One of the things done by those selected Americans was to develop a remarkable system of universal education at public expense, a system which all the world is to-day copying. They also devised or at least remodelled the town meeting, and made it an astonishingly effective and educative institution of government.

When the time was ripe the Puritan race in conjunction with the Quaker race, the race of younger sons from farther South, and other sons of the same kind fought the Revolutionary War on the basis of an old English doctrine to which the Americans gave a deeper meaning: "No taxation without representation." In later days that doctrine has spread in every direction and is still spreading. The next step was to form a new kind of government. Among students of political history it is almost universally agreed that the American constitution is one of the masterpieces of human invention. It stands to-day unshaken and is being copied all over the world because its authors, descended from Southern aristocrats as well as Northern Puritans, were men of uncommon insight who were likewise versatile and adaptable, and therefore gifted with that strong quality which makes a man see the other side and concede something for the sake of a great result. This same versatility, together with a high degree of that curiosity which we have seen to be so strong a factor in producing pioneers, presumably had much to do with the great outburst of inventive genius which has sent American machinery, telephones, telegraphs, and automobiles all over the world.

Many other ideas and institutions were evolved in early America and to-day are still potent. For example, one of the greatest differences between Old England and New England in the seventeenth and eighteenth centuries was the position of women. The women of America had far greater freedom than their sisters in Europe. They occupied a more important place in society, they had a higher degree of education, they were accorded a greater freedom in choosing their husbands, and they became the teachers of both boys and girls not only at home but at school. To-day women are probably more free to follow whatever career they may choose in America than almost anywhere else. We are told again and again by foreigners that we spoil our women, partly by submitting to them and partly by pampering them. But Europe and the rest of the world are more or less emulating our example. All this, I believe, is in part the result of the great selection whereby the women who were the mothers of the first American generations were peculiarly strong both in body and mind. They were more highly selected than the men; thus they rose rapidly toward a position of equality with their husbands. A similar tendency to elevate the position of women seems to occur in almost every difficult migration. We have seen it among the Hakkas, the Norse, and the women who teach their children in the Icelandic homes; and now we see it even more clearly in America.

Another great social change which occurred in early America was the abolition of class distinctions. Here again it seems to me that the cart has often been put before the horse. Class distinctions were not abolished in America because some one introduced a new idea. On the contrary they came over to America in full force. The idea of abolishing them arose out of the actual conditions of life. Only in rare cases has so large a group of people displayed such great equality of ability. Of course there were weaklings and servants, and even in the North there were a few slaves before the Revolution. Nevertheless, in those days, in a general way, all men in New England were "free and equal." The framers of the constitution did not give expression merely to an empty aspiration when they penned those words. They put into writing a fact of observation which was approximately true. It was true because the people whom they represented were descended from a highly selected ancestry and were themselves of such high average quality that the vast majority were really fit to take some part in solving political and social problems of great complexity. To-day the same idea of universal equality is abroad in the world. It was expressed in that one of President Wilson's fourteen points which states that all peoples have the right to self-determination. The President believed what he said, perhaps because of his engrossing study of early American history. But what was true in early America may not be true to-day in the world as a whole, for other countries have not been through the same process of rigid natural selection.

After the American Revolution the United States entered upon a new phase of development. Because of the high average ability of her sons and daughters there ensued a wonderful growth in transportation, commerce, manufacturing, and science, a rapid expansion westward, and a truly marvelous utilization of the great resources of a new country. Here again there is often a curious misapprehension. Coal, iron, harbors, and broad wheat lands did not make America great. They were merely tools which were effective because they were in the hands of unusually competent people. How much would they have amounted to in the hands of the sub-normal Chinese whom we have seen in some of the famine areas? But gradually a great change, perhaps a great decline, begins to appear in America. Old ideals are shattered in many places and there is a more insistent search for mere pleasure and excitement. Is this the result of wealth and luxury? Partly, no doubt, but I believe that it is due still more to a definite and rapid change in the quality of the people. The idea that America was the refuge of the oppressed and the land of the free gradually gave place to the idea that America was the land of the dollar where the poor foreigner could live in comfort and the competent American could make a fortune, provided labor were cheap and abundant.

THE SELECTION OF MODERN AMERICANS

Thus there occurred a most significant change in the type of immigrants. Whereas at first the strongest incentives were the dictates of conscience and the love of adventure, the dominant motive later became the desire for personal prosperity. At the same time the growth of steam communication made it more and more easy for even the incompetent to reach America. With steamship agents all over Europe urging people to come to the land of promise and making it easy to come, the selection of immigrants became less and less stringent. Many able people still came, but the average steadily fell. How could it be otherwise when the factors which had formerly exercised such rigid selection became more and more inoperative?

In this connection a little table prepared by Clarke is worth reproducing. Among the thousand leading American men of letters born in the United States prior to 1851 the number who were born in each decade per million of the white population is as follows:

Before 1771	10	1811–1820	22
1771–1780	15	1821–1830	13
1781–1790	15	1831–1840	11
1791–1800	23	1841–1850	7
1801–1810	20		

This table furnishes a good example of the way in which the same facts are susceptible of diverse interpretations according to whether one places more emphasis on nature or nurture, on inheritance or environment. Clarke is inclined to see in the table an evidence of nothing except a change in the social environment. He believes that the pronounced way in which the births of eminent literary men reached a maximum from 1790 to 1820 is due to the opportunities which arose after the Revolutionary War. The decline from 1820 to 1850 is attributed to the fact that after the middle of the last century, that is, during the time when the men born between 1820 and 1850 were making their reputations, the temper of the American people was rapidly changing so that the demand for many of the higher types of literature greatly diminished. Still more important, as it seems to me, is the fact that during that period almost boundless new opportunities presented themselves for achievement along commercial, industrial, and scientific lines. Many men who in an earlier era might have become authors actually became railroad men, bankers, chemists, and the like.

My own work, as illustrated in *Civilization and Climate*, for example, has dealt so largely with the influence of environment that I am naturally inclined to accept an environmental explanation whenever possible. Nevertheless, I cannot help raising the question whether even in the present case inheritance may not play a part. In the year 1820 the white population of the United States numbered 7,867,000; that same year only 8,385 immigrants were admitted and very few had come for at least two generations. During the next thirty years a great change took place. In 1850 the white population numbered 19,553,000, of whom 2,245,000 were foreign immigrants. On the basis of the censuses from 1870 onward, for which alone the data are available, these foreign-born immigrants together with their American-born children are estimated as forming at least one-quarter of the white population in 1850. But it is well known that the immigrant stocks that have come to this country since 1820 have made relatively little contribution to American literature, even though they came from the same countries as the earlier colonial stocks. They apparently were not so highly selected as the earlier migrants or were selected on another basis. Hence the inclusion of about 5,000,000 immigrants and their children would automatically cause us to expect the figures in Clarke's table to drop from 22 in 1820 to about 17 in 1850.

During this same period another factor may also have diminished the inherent tendency of the Americans toward literary achievement. That influence was the decline in the birth-rate which is now so noticeable among the more competent parts of the American people. It first became evident in America at about the time we are discussing and among the very type of people from whom a large part of our literary men have originated. The decline in the size of families at this time is no figment of the imagination. Here are the figures as given in a most illuminating census bulletin entitled *A Century of Population Growth in the United States, 1790–1900.*

	Number of Children Under 16 Years of Age per Thousand of White Population in 1790	Decline in the Number of Such Children per Decade				Number of Such Children in 1900
		1790 to 1820	1820 to 1850	1850 to 1880	1880 to 1900	
New England........	470	9.0	28.3	16.3	9.0	291
Middle States........	494	3.7	26.7	15.7	11.0	326
Southern States......	502	2.0*	14.7	11.0	14.5	402

* Increase.

The great decline in the number of children in New England and the Middle States between 1820 and 1850 is especially remarkable. Its significance is increased by the fact that the presence of the immigrants with their large families would tend to keep the proportion of children at a high level. Nevertheless, in the three decades from 1820 to 1850 that proportion fell 19 per cent in New England, 16½ per cent in the Middle States, and 12 per cent in the country as a whole. Among the old colonial families it can scarcely have fallen less than 25 per cent. But the decline in the birth-rate was more rapid among the cultured people of the towns than among the farmers, and it was from the townspeople that the literary men were derived for the most part, as will soon appear more clearly. Hence on the basis of the birth-rate, that is, from the point of view of inheritance alone, we should expect that the proportion of eminent literary men per million of the white inhabitants would have fallen off not merely to 17, as already indicated on the basis of immigration, but to a decidedly smaller figure, perhaps as low as 12, by reason of the changing birth-rate. I confess that I am much surprised at this result, for when I began to study Clarke's table I supposed that the environmental causes of the decline in American literary achievements were stronger than the biological causes. Now, however, I am forced to conclude that while both sets of causes are operative, the biological cause, that is, the actual decline in the proportion of people carrying a high inheritance, is probably the more important of the two.

Since 1850 the further lowering of the inherited mental and moral fiber of the United States through unselected immigration has apparently accelerated the tendencies which have already come under our attention. Take, for example, the scientific men of the country. Among the one thousand most eminent scientists, as indicated by asterisks in the third edition of *American Men of Science*, Visher finds that the nationalities forming the bulk of our recent immigration do not furnish nearly so large a proportion as do those whence the earlier immigrants were derived. The first column of figures in the accompanying table shows the actual number of starred scientists born as natives in each country and living in the United States; the second column shows the number of men over twenty-one years of age who were born in each country and lived in the United States in 1900; while the third column shows the number of such foreign-born men per starred scientist. Immigration from the countries of Group I seems really to help American science, for they excel the standard of native white Americans among whom there is only one starred scientist for about 21,000 men over twenty-one years of age in 1900. Notice that four of the countries—Belgium, Holland, Scotland, and England—encircle the North Sea and lie in the center of the most favored part of Europe. Japan has a special advantage in this classification because immigration from that country is limited to the higher callings; Switzerland leads Europe in its general scientific aptitudes, as we saw in an earlier chapter; Canada is almost like part of the United States; while the eminent Russians are practically all Jews. Thus there is some special and obvious reason for the supremacy of each group.

Group II is composed largely of the countries of central Europe. Whatever may be the innate quality of the people of those countries, they certainly have not made a large contribution to American science, for in proportion to their numbers they

THE SELECTION OF MODERN AMERICANS

rank only one-fourth as high as Group I. Group III, on the other hand, contributes only one scientist where Group I contributes thirty-five. The showing of Poland, Ireland, and Italy with their large immigration to America in recent decades is truly "pitiful," as Visher well says.

Country of Birth	Number of Starred Scientists	Number of Men Over 21 Years of Age in U. S. in 1900	Number of Such Men per Starred Scientist
Japan	3	28,000*	9,300
Belgium	1	10,000*	10,000
Holland	4	46,000	10,500
Scotland	11	116,000	10,550
Canada	47	505,000	10,700
Switzerland	5	63,000	12,600
England and Wales	33	463,000	13,000
Russia (mostly Jews)	11	173,000	15,600
Total. Group I	115	1,404,000	12,200
Denmark	3	82,000	27,300
Bohemia	2	70,000	35,000
Austria	4	143,000	35,600
Hungary	2	73,000	36,500
Sweden	6	294,000	49,000
France	1	55,000	55,000
Germany	23	1,327,000	57,600
Norway	2	173,000	86,500
Total. Group II	43	2,217,000	51,500
Poland	1	183,000	183,000
Ireland	2	722,000	361,000
China	0	78,000	......
Mexico	0	45,000	......
Italy	0	253,000	......
Total. Group III	3	1,281,000	427,000

*Estimate: One-third of all persons born in these countries. Data for men over 21 years of age not given separately.

Facts like these might be multiplied indefinitely. They suggest that unselected immigration has already brought the United States to the point where over-population and the radical lowering of the standard of living seem to be imminent; it has presumably accelerated the tendency toward small families among the most competent; and it has apparently increased the tendency for young people who have a fine inheritance to marry those whose inheritance is not so good. Of course this danger has always existed, but it has been intensified by the widespread mingling of the older and newer types. I recognize that some of the recent immigrants are of as high a type as the early colonials, but I am talking about the average, and that is what determines the general quality of a people. Education may raise the individual enormously, but no amount of education will give the high inheritance which existed abundantly in the Puritans, the Quakers, the Huguenots, and the aristocratic colonial planters of the South, and which is relatively scarce in the later immigrants. What America and every other country needs is an abundance of fine minds in sturdy bodies and a diminution of the incompetents.

Many studies of eminent Americans leave little room for doubt that in spite of their relative decline in numbers the old American stocks whose ancestors were in the country before the Revolution still inherit a degree of ability which makes them dominant. The preponderance of these early stocks among men of eminence is evident from Figure 17, which is based on the 21,579 persons who are included in *Who's Who in America* for 1922–1923, and who were born in the United States. The map shows the number of eminent persons who were born in each State per million of the average white population at the time of their birth. Among 310 consecutive names selected by chance where the book happened to open, the numbers born in successive decades were as follows:

1826–1835	1	1866–1875	110
1836–1845	15	1876–1885	51
1846–1855	43	1886–1895	7
1856–1865	83	1896–1905	0

Inasmuch as very few persons in this edition of *Who's Who* were born before 1845 or after 1885, only the censuses of 1850, 1860, 1870, and 1880 have been employed in calculating the average population. Since about twice as many eminent people were born in the decades from 1856 to 1865 and 1866 to 1875 as in the decades before and after, the census figures for 1860 and

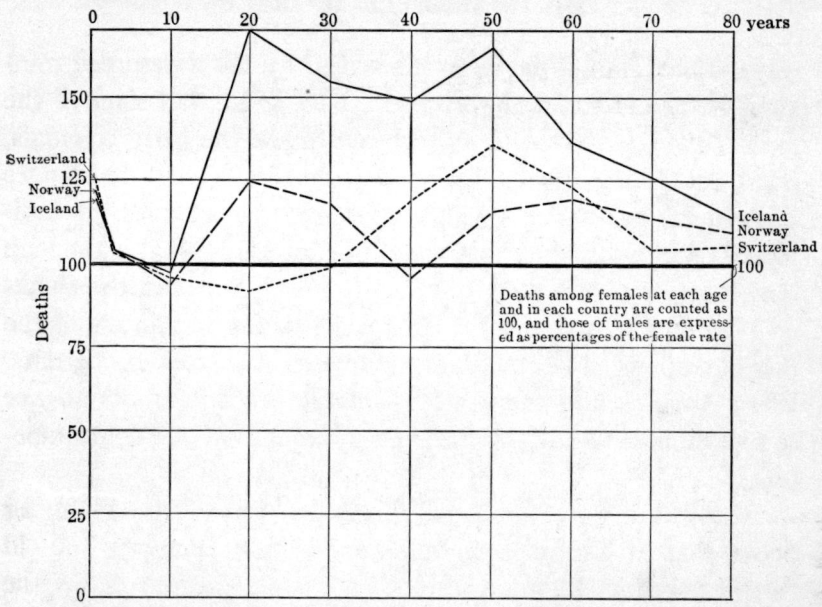

FIG. 16. DEATHS OF MEN COMPARED WITH WOMEN AT VARIOUS AGES IN ICELAND, NORWAY, AND SWITZERLAND, 1876–1915.

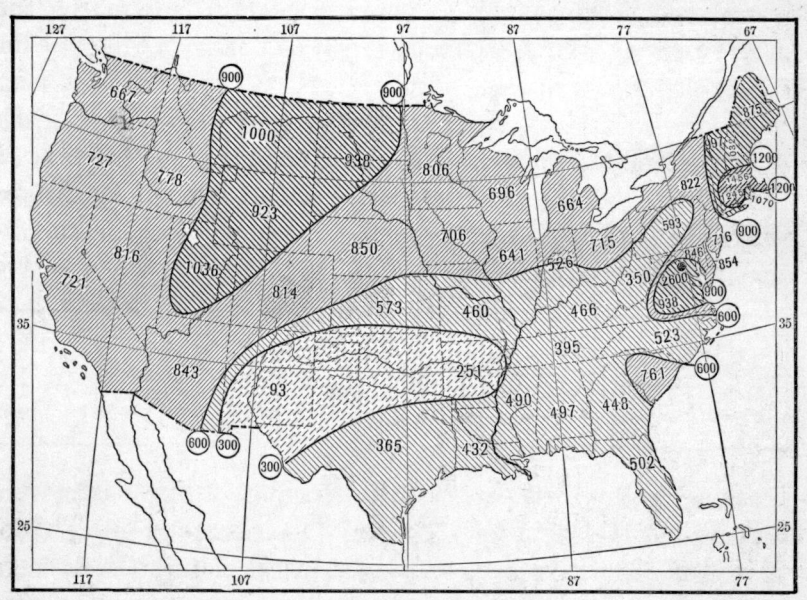

FIG. 17. PERSONS IN *WHO'S WHO* (1922–1923) BORN IN EACH STATE PER 1,000,000 OF MEAN WHITE POPULATION 1850–1880, GIVING DOUBLE WEIGHT TO 1860 AND 1870.

1870 have received double weight. In order not to put the South at a disadvantage the colored people have been omitted. It would have been equally fair to omit the foreign-born, which would have caused the figures for the Northeastern States to be appreciably higher than is now the case. Thus the map errs on the side of giving the Southeastern States an advantage and the Northeastern States a disadvantage.

In spite of this disadvantage the extraordinary way in which the New England States, especially Massachusetts, with 1,456 eminent persons per million population, and Connecticut with 1,245, surpass all other sections of the country is brought out most clearly by the dark shading. The other dark areas in Figure 17 likewise seem to point to the value of inheritance in producing leaders. The District of Columbia with a ratio of about 2,600 eminent sons per million of the population surpasses even Massachusetts. The reason is obviously that the population of Washington contains an extraordinarily large number of persons chosen for their ability, and naturally a considerable number of the children follow in their fathers' footsteps. In Virginia and again in South Carolina many more eminent men have been born than would be expected from the conditions in the surrounding States, and this again seems to find no satisfactory explanation except in inheritance. Moreover, in the Western States of Utah (1,036), Montana (1,000), the two Dakotas (938), and Wyoming (923), we seem to have an area where an unusually large proportion of able men were born between 1840 and 1885. At that time the population of those States was very small and consisted largely of pioneers. If the general hypothesis of this book is right, those pioneers produced an unusually large number of eminent sons because the difficulties of early settlement kept away the less efficient type of settler.

On the other hand, the extremely low position of New Mexico seems to be due largely to the fact that in the period under discussion the greater part of the people whom the census reckons as white in that State were actually Mexicans, often with a large Indian admixture.

It must be clearly understood that Figure 17 does not show where the eminent people of the United States now live. It does not even show where they were born, but merely how many were born in each part of the country in proportion to the total population. The actual proportion of eminent residents who were included in *Who's Who* in 1922 appears in Figure 18, which shows the number of eminent persons in *Who's Who* per million of the white population in 1920. The contrast between this map and Figure 17 indicates that many men and women of ability are moving away from their old homes and settling elsewhere. If these more prominent people are moving in this way, other people of more than average ability are presumably moving in the same way. Note that the District of Columbia, as might be expected, has a far greater proportion of eminent people than any other region, its ratio being about 4,450 per million inhabitants. Nevada appears to come next, but this is probably because that State has so small a population that the mere inclusion of certain officials and others who fall in fixed categories in *Who's Who* gives it an undue representation. Next to Washington as a place of real concentration of persons of ability comes New York City and its vicinity, thus giving a figure of 521 for New York State while Connecticut follows closely with 510. The rest of New England still holds its own fairly well, although Massachusetts, largely because of its great foreign population, is falling behind.

In the West, although eminent people were born in large proportions in the States from Utah to North Dakota, as shown in Figure 17, they were too few in number to give those States any specially large proportion of eminent people to-day, or else most of them have moved away. Colorado and California, on the other hand, perhaps by reason of their climate, appear to exert a strong attraction upon persons of ability, as do the other Southwestern States and Illinois to a less degree. Texas, Oklahoma, Arkansas, and Florida, on the contrary, seem to be places which people of unusual ability avoid as permanent homes, even though they may go there for business or pleasure.

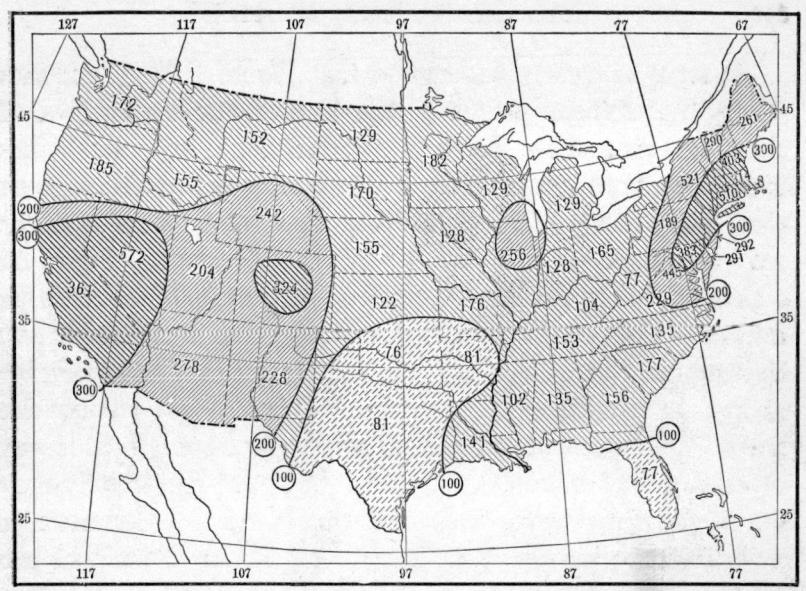

FIG. 18. RESIDENTS INCLUDED IN *WHO'S WHO* (1922–1923) PER 1,000,000 POPULATION IN 1923.

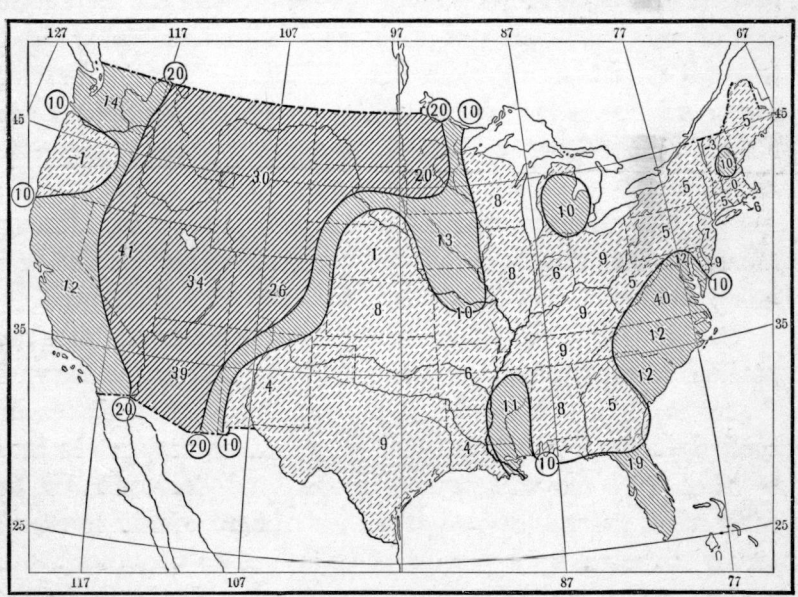

FIG. 19. INCREASE (OR DECREASE) (1912 TO 1922) IN PROPORTION OF EMINENT PERSONS BORN PER 100,000 OF POPULATION.

The figures in the map show the differences between the percentage of each State in 1912 and 1922, the percentage being reckoned on the basis of Massachusetts as 100 in both cases.

What we are mainly interested in, however, are the people who pass on to their children the kind of inheritance that leads to eminence. In order to determine not only where these people resided in the past but where they are moving to, I have prepared Figure 19. This represents a comparison between two maps of the same kind but different in date. One of these is Figure 17, showing the proportion of persons born in the various States according to *Who's Who* for 1922; the other is a similar map published in *Civilization and Climate* and showing the corresponding proportions in 1912. Since the two maps cannot be directly compared, the number of eminent persons born per million persons in Massachusetts has been reckoned as 100 in each case and corresponding values have been given to the other States. On this basis New York had a rank of 51 in 1912 and 56 in 1922; that is, as the birthplace of eminent persons New York is gaining on Massachusetts, although still far behind. The difference between the New York percentages in 1912 and 1922 amounts to 5, which is the figure given for that State in Figure 19. On the other hand, the percentage for Rhode Island has changed from 79 to 73, so that it appears on the map as − 6. Figure 19 indicates that during the course of ten years, that is, when the period from 1830 to 1875 is compared with the period from 1840 to 1885, the migration of strong stock out of Massachusetts and the immigration of weaker stocks from Europe caused that State to decline in its relative production of men of unusual ability when compared with all other States except Rhode Island, Vermont, and Oregon. The large area which is lightly shaded in Figure 19 gained only slightly if at all over Massachusetts, but in the heavily shaded areas of the Northwest and West aside from Oregon the gain was great. That was the part of the country into which settlers were then bringing their families. The Rocky Mountain section from Arizona to Montana, and also the Dakotas, were in the stages of development where the successful pioneers first begin to bring their wives and build up families. That apparently is one reason why the proportion of eminent sons increased so rapidly.

Opportunity, as well as birth, doubtless enters into this matter. The relative increase in the birth of children who attained to eminence in Virginia, the Carolinas, Florida, and Mississippi may be largely the result of the new opportunities which came after the Civil War. Likewise in the Rocky Mountain States the growth of opportunities for education, travel, and the like doubtless played a part. Other factors also need to be considered to explain such anomalies as the very low position of Oregon and the surprisingly high position of Virginia, but we cannot enter into their discussion in detail.

Let us turn back once more to the question of the Puritans and their descendants. On the basis of replies from 18,400 persons whose names appear in the 1922-1923 edition of *Who's Who in America*, Visher has made a study of the occupations of the fathers of the eminent people of America. The first column in the following table shows the total number of fathers reported as belonging to each main group of occupations. From this it is easy to calculate that on the basis of the census of 1870 only one unskilled laborer in approximately 48,000 was a father of an eminent son included in the persons for whom data are available in *Who's Who*. Among other professions, the number of sons for each 48,000 was approximately as appears in the second column of the table.

	Number of Fathers of Persons Reporting in *Who's Who*, 1922-1923	Number of Eminent Sons per Approximately 48,000 Men
Men of leisure	49	?
Unskilled laborers	94	1
Skilled laborers	1,165	30
Farmers	4,310	70
Business men	6,473	600
Professional men, except clergymen	4,265	1,035
Clergymen	2,036	2,400

These figures are very significant. It needs no demonstration to show that clergymen, as a body, especially those of fifty years ago, possess the Puritan spirit to a greater degree than almost any other part of the community. They are the spiritual, if not the physical descendants of the Puritans. As a matter of fact, however, the clergymen whose sons are most likely to attain eminence are the physical descendants of the Puritans, as we infer from the next table, where Visher gives details as to the various professions.

	Total Number Reporting Their Fathers as Belonging to Specified Professions	Eminent Sons per 48,000 Men
Engineers	24	298
Physicians	480	460
Methodist clergymen	150	495
Lawyers	625	923
Baptist clergymen	210	1,105
Sea captains and pilots	76	1,140
Presbyterian clergymen	570	4,325
Episcopal clergymen	270	5,565
Congregational clergymen	360	6,000

Notice how completely and decisively the Presbyterian, Episcopal, and Congregational clergymen excel all others. Among these three the Congregationalists stand decidedly in the lead. But Visher's data, so he informs me, suggest that the Unitarians rank even higher than the Congregationalists. They are not included in the table merely because their numbers are small. An almost identical result was reached by Clarke in his study of an earlier generation. Among a thousand leading American men of letters born before 1851 he finds that

in respect to absolute numbers the Congregational body stood far above its nearest competitor, the closely related Presbyterian Church. If relative numbers are considered, however, the Unitarian body apparently had the greatest proportion of literary persons born within its ranks, and the Congregationalists, Friends, and Universalists followed in order. All four had a relatively large number of men of letters born to their members. On the other hand, Methodist Episcopal, Baptist, and Roman Catholic families possessed relatively very few literati.

Clarke's explanation of these conditions is interesting as indicating the variety of factors which must be considered in any fair discussion of our problem.

The fact that there were born within the ranks of some denominations relatively more men of letters than in others is of interest, but standing by itself it cannot be considered particularly significant. Odin found that, in proportion to the numbers in each religious division, many more French men of letters had been brought up as Protestants than as Catholics. He thought that there had been a number of reasons for this superiority, but believed the most important to be that, on the whole, Protestant children received superior educational opportunities because of the superior wealth of their parents. Possibly both economic and educational factors may serve to explain the differences discovered in America. It is a well-known fact that, during the period studied, the Unitarians and Friends, for instance, were on the whole in comfortable circumstances, while the Roman Catholics were relatively poor. The resulting differences afford at least partial explanation of their differences in literary productivity. On the other hand, it must not be forgotten that Protestants enjoyed greater freedom of thought than Roman Catholics. This factor may be only less important than poverty and lack of education. Data are not now available, however, on which to base studies which would indicate the relative importance of these various factors. At present one can simply conclude that without question religious training has played some part in the production of American men of letters.

It seems to me that while Clarke is right in assigning great weight to environment and training, he attaches too little importance to inheritance. It must not be overlooked that environment and training depend largely upon inherited ability. Every one knows that people of unusual strength of character generally create for themselves and their children a relatively good environment, both social and economic, no matter what their origin may be. They are also more likely than others to give their children proper training, even though the so-called "advantages" may be lacking. Again, it seems to me that environment and training are not sufficient to explain the extraordinary differences disclosed by Visher's data. Consider once more the remarkable fact that among the families of the Congregational ministers who were in active life from 1850 to 1880 one out of every eight, on an average, produced a son or daughter of sufficient distinction to be included in *Who's Who*. At that period the son of a Congregational minister was 6,000 times as likely to attain eminence as was the son of an unskilled laborer, and approximately 86 times as likely as the son of a farmer. In a country like the United States it seems impossible that such great differences can arise unless inheritance as well as environment plays an important part.

The effect of inheritance seems to become still more clear when persons of a single profession are taken. Selecting the clergymen from the preceding table and adding an estimated number on the basis of the partial returns for sons of Universalist and Unitarian clergymen, we find the following number of eminent sons per 100 fathers:

Methodists	1.0	Episcopalians	11.7
Baptists	2.3	Congregationalists	12.5
Universalists	7.0	Unitarians	15.0
Presbyterians	9.0		

Here we have a group of people among whom the ideals are very similar. Aside from the Methodists and Baptists, whose economic condition and degree of education are somewhat lower than those of the others, the mode of life, methods of training, and general environment of all the rest are extremely similar. Yet we find highly significant differences in the production of men of notable ability.

Taken as a whole the extent to which the clergymen of the various denominations have been the fathers of eminent sons is almost directly in proportion to the degree to which the denominations have suffered persecution and selection. The Methodists are the latest of the sects here considered. Not until about 1740 did John Wesley establish what may be called a genuine Methodist church. While the church represented some new ideas, it suffered no special persecution and only a little ridicule and opposition. Therefore there was no great selection, such as took place among the Puritans. In the United States the growth of the Methodist church was even more gradual than in England, and never involved any appreciable opposition or contempt on the part of other sects like that which the Universalists and

Unitarians had to face. The Baptists, as we know them to-day, originated in England soon after 1600, when they separated from the Brownists, who ultimately became the Congregationalists. In America the Baptists were relatively few in number up to the Revolutionary War. Under the leadership of Roger Williams and Mrs. Anne Hutchinson they were bitterly persecuted by the Congregationalists. The main growth of the Baptists, however, did not come until after the Revolutionary War. Hence it involved no persecution, while such selection as took place both among the Baptists and Methodists was on the whole emotional rather than intellectual. This seems to be one of the important reasons why these two sects have produced a smaller proportion of eminent leaders than have some of the others. Both of these churches, by reason largely of that same emotional appeal, have grown far more rapidly than the denominations whose appeal is more intellectual. The degree of selection among their members has been correspondingly less rigid.

The Universalists are a sect of relatively recent origin derived in considerable measure from the Congregationalists. They have been conspicuous for their emphasis upon the guidance of each man's own inner light somewhat like the Quakers. When they severed their connection with their former churches a considerable degree of courage and perseverance was required, for they were frowned upon and ostracized as practically never happens in these later generations. The selection due to this cause, however, was probably less important than the selection due to the insistence of the Universalist church upon individual thought rather than upon emotion as among the Baptists and Methodists, or upon external authority as among the Roman Catholics.

Turning now to the Presbyterians, we find that in the early days in Great Britain they suffered some persecution. This did not result in a violent severing of old ties like that which marked the birth of the Congregationalists. On the Continent, however, the Presbyterians suffered such great hardships that many of the most determined among them fled to England. Thus the early Presbyterian churches contained not only a large Scotch element, which imparts strength almost everywhere, but also a considerable element of chosen people who had fled from other lands. From the very beginning the Presbyterian church appealed to the intellectual rather than the emotional side of man's nature, and therefore attracted a type of people different from the Methodists and Baptists.

Coming to the Episcopalians we find a communion which as a whole has not suffered any very strenuous persecution since the early period of conflict with Roman Catholicism which resulted in its establishment as the State church of England. Because of its official character both in England and in many of the early American colonies, the Episcopal church has always been aristocratic. Its clergymen, more than those of almost any other sect both in England and America, have been selected from among the upper classes. Among them, and presumably among their sons, the capacity to maintain a dignified place in society, using the word in its broader sense, has been quite as important as intellectual keenness and has presumably had an appreciable effect upon the type of leaders which the church has produced.

The Congregational church furnishes an example of a selection more rigid than that of the preceding churches. We have already seen that practically the whole of the early members of that church were people whose moral convictions lead them to separate themselves from the dominant Episcopalians and make the hard migration to the new world.

The original Unitarians were almost wholly Congregationalists. Their leaders were an unusually intellectual and thoughtful group who separated from the Congregational church because a century ago they had arrived at certain convictions which now have become the common property of most scientific men and of a large number of the more thoughtful people in all the more intellectual denominations. The emphasis of the Unitarians upon pure intellect and the repression of appeals to the emotions or to external authority have caused them to remain a relatively small body, but have at the same time acted as selective factors to keep within the denomination a high proportion of unusually intellectual people. Therefore it is not surprising that in proportion to their numbers the Unitarians and especially their clergymen have contributed a greater number of eminent leaders than has any other group of Americans for whom we have statistics. Here is a list of a few of the many Unitarian leaders. Those whose fathers were clergymen are indicated by asterisks:

> Charles Francis Adams, father and son, diplomatists and historians.
>
> Henry Adams, and his brother, Brooks Adams, historians.
>
> The fact that the six members of the Adams family mentioned in this list and in that which follows represent four generations of the same family is typical of many cases where inherited talent appears for generation after generation.
>
> Louisa M. Alcott, author.
>
> James Freeman Clarke, historian.
>
> Richard H. Dana, author of *Two Years Before the Mast.*
>
> Dorothy Dix, philanthropist and author.
>
> Charles W. Eliot, president of Harvard College.
>
> *H. H. Furness, Shakespearean scholar.
>
> *Edward Everett Hale, author and clergyman.
>
> Thomas W. Higginson, author, philanthropist, merchant.
>
> Helen Hunt Jackson, author.
>
> Sarah Orne Jewett, author.
>
> David Starr Jordan, college president.
>
> Henry C. Lodge, senator and author.
>
> Theodore Parker, theologian.
>
> John Randolph, statesman.
>
> Laura E. Richards, author.
>
> William H. Taft, President of the United States.

Impressive as this list may be, it by no means shows the full quality of the contribution of Unitarians to the life and thought of America. The University of New York contains what is known as the Hall of Fame. The sixty-three names thus far recorded there have been selected with great care by a jury of one hundred eminent people after prolonged and thoughtful study. They

THE SELECTION OF MODERN AMERICANS 325

represent the matured judgment of a group of people from whose deliberations the question of religious affiliation was rigidly excluded. Yet here is the reply that I received when I asked a prominent Unitarian for a list of the persons of his denomination included in the Hall of Fame:

> It is always somewhat difficult to define a Unitarian. Ours is a Church of the Spirit rather than of the Letter. Our church membership is less distinctively marked than in better disciplined organizations. A man is sometimes justly claimed as a Unitarian whose habits of mind and principles of conduct obviously ally him with that fellowship, but who has no definite membership in any Unitarian church or congregation.
>
> In reviewing the list of names in the Hall of Fame it may safely be said that twenty are the names of men and women who had a fairly definite allegiance to the Unitarian fellowship, and there are in addition the names of several persons whose sympathies were evidently with that fellowship though they were probably never actually or actively connected with a Unitarian church. The names of the Unitarians are:
>
> John Adams.
> John Quincy Adams.
> Louis Agassiz.
> *George Bancroft.
> William Cullen Bryant.
> William Ellery Channing.
> Peter Cooper.
> *Ralph Waldo Emerson.
> Benjamin Franklin.
> Nathaniel Hawthorne.
> *Oliver Wendell Holmes.
> Henry Wadsworth Longfellow.
> *James Russell Lowell.
> Horace Mann.
> John Lothrop Motley.
> *Francis Parkman.
> Joseph Storey.
> Daniel Webster.
> Charlotte Cushman.
> Maria Mitchell.
>
> In addition the Unitarians may not unreasonably claim Thomas Jefferson (who lived where there was no Unitarian Church in which to worship but who wrote: "I trust there is not a young man now in the United States who will not die a Unitarian"), John Greenleaf Whittier and Lucretia Mott, who were born Quakers but who recognized their connection with the Unitarian habit of mind and often attended Unitarian churches.
>
> There is one name in the list whose denominational connections I cannot identify. It is that of George Peabody. His family connections in Salem were all in Unitarian churches but whether he maintained that connection or not I do not know.

This discussion of the Hall of Fame brings out two remarkable facts. First, out of the sixty-three people thus far adjudged worthy of a place among America's most useful leaders at least one-third are Unitarians. Second, among the twenty whose Unitarian connections are positive no less than five were the sons of ministers. Among the entire forty-three names mentioned in connection with our two lists of Unitarians seven were the sons of ministers. The remarkable quality of these facts can best be appreciated by realizing that there were no Unitarians as a distinct denomination until the early part of the nineteenth century. The denomination has never been large, and to-day numbers only about 100,000 adherents, or one in 1,000 of our population. At present the Unitarians have less than 600 ministers, or perhaps one for every forty adult men among their nominal parishioners. At the time when the people who are enrolled in the Hall of Fame were born, the relative number of Unitarians was probably larger than now, perhaps one in 300 of the total population, but the proportion of ministers was presumably not greatly different. In that case the productivity of the Unitarians in supplying leaders of the first rank has been about 150 times as great as that of the remainder of the population, while that of Unitarian ministers has been nearly 1,500 times as great.

The eminent sons of Unitarian ministers seem to represent the final result of a long process of natural selection. Many centuries ago there came to East Anglia some migrant Saxons, Danes, and Normans, selected groups of people, uncommonly intelligent and competent. At a later date many of their most thoughtful descendants became Puritans, a second selection. Then from among the Puritans a third selection separated those with special earnestness, determination, and adaptability, and led them to migrate to America. In the fourth place, among their descendants another group characterized by unusual intellectual activity and high moral purposes became Unitarians. Finally, among these intellectual people a fifth selection during the first half of the present century picked out many of the most earnest and thoughtful as clergymen. These Unitarian clergymen, more than any other group in proportion to their numbers, have been the fathers of the recent leaders of America. Their case resembles that of the Khmers of Cambodia, the Hakkas of China, the ancient Athenians, the Normans in Sicily, and the Vikings in Iceland.

If such a stock as this, or any other similar stock, should be isolated in a good environment and should become the sole ancestors of the inhabitants of some great state, what heights of attainment might be achieved.

I have devoted much space to the Puritans because they appear to be the most notable instance of natural selection in America. But there are many other cases which will well repay study. Let us select one as far removed from New England as possible. Up to 1849 California was a distant Mexican province. Then it suddenly began to grow into one of the greatest States in the Union. I have studied all sorts of statistics again and again in order to discover how the various States compare with one another in real progress. Time after time I have been struck by the fact that in a great variety of activities California ranks higher than the States immediately east of it and finds no rivals until the Northern States to the east of the Mississippi are reached. This is true in education, in literary achievement, in commerce, in manufacturing, in religious activity, and in other ways. Take the universities and scientific institutions of the country, for example. Among the universities California has two that are famous all over the country. The University of California is probably better known than any other west of the Mississippi, and there is perhaps none south of Baltimore to compare with it. Leland Stanford also is one of the most famous universities in America, while the new California Institute of Technology at Los Angeles seems to be achieving remarkable things in proportion to its age. Among scientific institutions California has several which are becoming famous all over the world, for example, the Lick Observatory, the Mount Wilson Observatory, and the Scripps Institute, and others are beginning to arise. Nowhere else to the west of the Mississippi River does one find anything equally well known of similar kind. Again in politics California has become one of the pivotal States not so much because of its size, but because its people vote independently, and send to Washington men who command attention. In many other ways the whole country is conscious that California is a State to be reckoned with. Neither its area nor the size of its population cause this, for few people have the same feeling about Montana, or New Mexico, which almost rival it in area, or about Georgia, which is almost as populous, or about Texas, which far exceeds California in both respects. Mere size counts for almost nothing, for Massachusetts, although only one-nineteenth as large as California, is one of the States that has to be reckoned with most fully. San Francisco, like Boston, is known all over the country as a place where things are done, a city of achievement. The way in which the city rebuilt itself after the earthquake and fire of 1906 is admired all over the world. And Los Angeles likewise has an extraordinarily wide reputation for wealth, culture, and progress, as well as for esoteric sects and moving pictures.

One reason why California stands out so pre-eminently is doubtless its good climate, but on the basis of climate alone it is probable that Oregon and Washington should stand equally high or higher. California, however, seems to have another reason for its eminence. Like New England it had a good start because of natural selection. The first important group of English-speaking people in California was the Forty-niners. They came rushing to the State when gold was discovered. In one sense they were a wild lot, for they were adventurers, but in another sense they were highly selected. It was a long, difficult journey to California in 1849. Overland it took three months or more of hard and dangerous travelling. By sea it took about as long, for even if one went by Panama instead of Cape Horn it was often necessary to wait for a ship on the Pacific side. And the ships in those days were sailing vessels. The trip to the gold mines took money as well as time. Only rarely did a genuinely poor man get there. Courage and the spirit of adventure were also highly necessary. Hence the Forty-niners were on the whole a group of young and vigorous men, well-born, fond of adventure, fairly well-to-do, and animated by the desire to see the world as well as make a fortune. An uncommonly large number were college graduates. In other words, the early Californians were selected for qualities somewhat like those of the Puritans, except

THE SELECTION OF MODERN AMERICANS 329

that the religious and moral element was lacking, but there was no lack of ability.

Many of the Forty-niners went back to their old homes, but a fair percentage stayed in California. Some sent for their wives or sweethearts; many went home, found wives, and returned to raise wheat, which in the long run was more profitable than mining gold. In this case, as among the Puritans, the women exercised the strongest selective influence. Only when a man with the spirit of daring, adventure, and initiative was mated to a woman of good physique and of the same spirit was a family likely to establish itself in California in those early days. For two decades the journey to California continued to be especially difficult and dangerous—more so than to almost any other part of the United States where settlement was then taking place at all rapidly. Until the railroad was completed in 1869 the majority of the settlers in California, especially those who brought up families and were not mere floaters, were rigidly selected. Even after that time the distance to California and the expense of settling there exerted a distinct selective effect. Thus San Francisco and the neighboring cities were first peopled by an unusual proportion of families who were selected because of the spirit of adventure coupled with comparative prosperity. Of course many people of other kinds came even in the earlier years, and still more have come later. But that does not alter the fact that because of its early inaccessibility coupled with the attractiveness of its gold, its climate, and its crops, the part of California around San Francisco attracted more than the normal proportion of able people.

In southern California the course of events was much the same except that the attraction of gold was replaced by that of climate. The early settlers were families who had the courage, the resources, and the adventurous spirit to make the long, hard journey across the deserts and mountains or to sail the difficult route by Cape Horn, with its storms, or by Panama with its malarial fevers. In later days the pleasant, equable climate has attracted a great number of people of the more prosperous classes.

Los Angeles has not been a poor man's retreat. It has been a place to which have gone many prosperous people in search of health, and many other prosperous people in search of pleasant homes. Those people carried an inheritance of ability—a physical inheritance—and they have passed it on to their children. Just as San Francisco is notable for its great activity, so Los Angeles is notable for its general air of high culture, its large number of scientific people, its artists and religious colonies, its love of pleasure, and its moving-picture colony. All these selected types have come largely because of certain qualities of climate. They have brought with them certain ideals and customs, certain *mores*, and they have modified these somewhat to fit their new environment and their new neighbors. But for our present purpose the outstanding fact is that a vast number of the settlers in California came because they inherited special aptitudes or mental tendencies differing somewhat from those of the average of their race. For that reason a detailed study of the type of migration and natural selection may be expected to throw a flood of light on the present and future state of society in California. The people of California, as well as the early New Englanders, have been unconscious agents in causing certain distinct types of racial character to be accentuated in definite geographical regions.

It seems to me that this chapter may well be concluded with a quotation from *Adventures in Journalism* by Sir Philip Gibbs. His summing up of American character brings out the kind of qualities which have apparently resulted from the various types of selection that have operated in giving the United States its present population. He does not, to be sure, dwell on the less commendable qualities which are now growing strong among us, but one can see that he has them in mind.

> I was and still remain convinced that the United States will shape, for good or ill—and I believe for good—the future destiny of the world, for these people, in the mass, have a dynamic energy, a clear-cut quality of character, and a power not only of material wealth, but of practical idealism, from which an enormous impetus may be given to human progress,

in the direction of the common well-being, international peace, liberty, decency, and average prosperity of individual life. I was not unconscious of a strong strain of intolerance; a dangerous gulf between the very rich and—not the very poor, there are few of those—but well-paid, speeded-up, ugly-living, dissatisfied labor; something rather hysterical in mass emotion when worked up by the wire-pullers and the spell-binders; and the noisy, blatant, loud-mouthed boasting vulgarity of the mob. I saw the unloveliness of "Main Street," I met "Babbitt" in his club, parlor car, and private house. But though I did not shut my eyes to all that, and much more than that—a good deal of it belongs to civilization as well as to the United States—I saw also the qualities that outweigh these defects, and, in my judgment, contain a great hope for the world. I met, everywhere, numbers of men and women who have what seems to me a clean, sane, level-headed outlook on life and its problems. They believe in peace, in a good chance for the individual, in a decent standard of life for all people, in honesty and truth. They are impatient of dirt, however picturesque, of ruin, however romantic, of hampering tradition, however ancient. They are, in the mass, common-sense, practical, and good-natured folk, who, in the business of life, cut formalities and get down to the job.

But behind all their common sense and their practicality they are deeply sentimental, simply and sincerely emotional, quick to respond to any call upon their pity or their charity, and when stirred that way enormously generous. I agree with General Swinton, the inventor of the "Tanks," who, after a tour in the United States, told me with a touch of exaggeration that he thought the Americans, as a nation, were the only idealists left in the world. Europe is cynical, remembering too much history and suffering too much disillusionment. The United States, looking always to the future, and not much backward to the past, is hopeful, confident of human progress, and strangely and wonderfully eager to find a philosopher's stone of human happiness, for which we, in Europe, have almost abandoned search.

I think that, as a people, they are more ready than any other to do some great work of rescue for humanity (I have told how they fed ten million people a day in Russia), and to adopt and carry out an ideal on behalf of humanity in the way of peace and reconstruction, at some personal sacrifice to themselves. That is possible at least in the United States, and it may almost be said that it is impossible in any other nation.

CHAPTER XXII

THE RACIAL TENDENCIES OF CIVILIZATION

We have now examined many of the main factors through which in past times man's physical and social environments have selected certain types of human character for preservation or destruction. We have seen in detail how these factors cause the people of some regions to differ markedly from others. We have likewise concluded that in spite of the extreme persistence of heredity, certain races have acquired new characteristics almost in our own day. Is there any reason to think that the mutations and selective processes through which racial character has been evolved are acting more slowly than formerly? We are sometimes told that although society may evolve indefinitely, man as an animal has become so highly specialized that the next step must be degeneration. The only hope of mankind, it is said, lies in creating a social system far more complex and perfect than that of to-day, a system in which each individual is fully utilized for the general good, and all are as dependent upon one another as are the ants and bees.

That man's future will depend largely upon his capacity for social organization and his ability to work for the common good regardless of himself seems certain. But this does not necessarily mean that either physical or especially mental evolution has come to an end, or that the next step will be degeneration. According to the general opinion of geologists a few thousand years of a severe climate, such as a glacial epoch, have more effect than millions of years of monotony in destroying some forms of life and bringing others to the front. The last glacial epoch appears to have been an essential factor in man's latest great stride in mental evolution whereby he reached the plane where civilization became possible. That such evolution would have taken place at a slow rate even without the glacial period is highly probable, but the stress of climate seems greatly to have hastened it.

Suppose another glacial epoch were to overwhelm the earth. Unless man is wholly different from the other animals, racial evolution would be decidedly accelerated. We have, to be sure, no reason for predicting another glacial epoch, but we are equally without reason for predicting the contrary. If it occurred, mankind would be subjected to tremendous stress. Most of the regions now in the forefront of progress would become uninhabitable. In North America agriculture would probably be out of the question except from the southern and southwestern United States equatorward. Parts of the United States and Canada where perhaps seventy million people now live would actually be covered with ice. In Europe a similar fate would overtake the homes of two or three times as many; only the Mediterranean and Caspian regions would be good for farming. In China the dry winter might be so long and cold and the summer rains so irregular and torrential that half the country would be untillable and much of the rest by no means favorable.

If all this happened, as well it may, half the world's population might be forced out of its present homes. Millions upon millions would either perish by slow degrees, or be forced to migrate into regions that now are deserts or tropical forests. Such a change in the location of population would almost inevitably cause tremendous stress and strife. Nation might be pitted against nation, race against race in a terrific combat in which victory would depend on many complex conditions. One race would have an advantage through the individuality and initiative of its common people, as in the United States; another through the submissiveness of a competent people to their leaders, as in Germany; a third through its ability and willingness not only to cajole its stronger rivals but ruthlessly to exterminate those that are weaker, as in Turkey. Other races, such as the English and Dutch, would have a great advantage through their commercial prowess. The tough fiber of the Chinese and their ability to live economically and with any kind of food would be a tremendous asset. In such a strife of races the victory might go to the race which had best learned to combat disease. But would such a race be able permanently to compete with a race which had acquired an inherent resistance to disease through a long process of selection? Which would be a greater asset, the ability of the common people to submit their wills to the public good, or the capacity to produce great leaders? It would be interesting, but futile, to speculate further as to what might happen if another glacial period should occur. The point is that there are endless possibilities for the development of races with powers far superior to any that the world has yet seen. It seems to me that we are nowhere near the limits of profitable specialization.

We have used a possible glacial epoch as an example of the kind of crisis which may overwhelm the world, but other important crises are much more imminent. The growing density of population bids fair within a generation or two to bring a state of stress almost as great as would glaciation, as East well shows in *Mankind at the Crossroads*. The exhaustion of metals and mineral fuels may have a similar effect. Unless radical measures are soon taken the exhaustion of the soil and of our stock of mineral fertilizers may produce distress and hence natural selection far beyond anything that now is dreamed. So, too, may the spread of new forms of disease, not only among men but among plants and animals. Civilization is still so young that these possibilities have not yet had time to manifest their importance. We do not appreciate what they may lead to because we think in terms of days, months, years, and decades. But races must think in generations, centuries, millenniums. Perhaps the ability to think in these large units, and act accordingly, may be a primary factor in enabling one race or another to survive and flourish.

It may still seem that we are dealing merely with remote possibilities, and not with any real problems of the immediate future. But there is another change to be considered. During the last few centuries the great racial evolution of hundreds of thousands of years has suddenly borne strange fruit. Throughout the successive glacial and interglacial epochs the winnowing of the future races of Europe by alternate pressure from expanding deserts in Asia and expanding ice in Europe was presumably producing a group of races with unusually keen active minds. Five to ten thousand years ago, in western Asia and north Africa, those races began to build up civilization. Babylonia, Egypt, Syria, and Palestine made their contributions; they were followed by Greece and Rome, and then by the countries of western Europe. In India and China a similar evolution of civilization went on, although not quite so rapidly. But suddenly, under the stimulus of a marvelously favorable combination of racial mixture, physical environment, new ideas, and mechanical inventions the European type of civilization has flared up almost incredibly. Modern transportation, manufacturing, and the growth of cities have altered man's environment and transformed life in a single century more than it was previously altered by ten thousand years of civilization and barbarism or half a million years of savagery.

Universal education, modern medicine, the democratic organization of society and of government, the adoption of new moral and religious standards, the emancipation of women, and the partial substitution of altruism for egoism are producing equally revolutionary results. These changes are substituting an indoor for an outdoor life; they are putting desks, pens, pushbuttons, and levers in place of the plow, hoe, axe, and scythe; the creamery and flour-mill in place of the churn and grinding stone that once occupied so large a share of women's time. They are preserving weak children as well as strong. They are sifting good minds from poor and dooming many of the good to die childless. They are leading to suicidal movements like modern war and feminism. The moral, social, political, and industrial revolution in Europe and North America during the last few centuries is so great that its effect on the evolution of racial character may well be compared to that of a glacial epoch.

In what direction is this epoch-making revolution carrying us? No one yet knows, but some hints may be obtained from a brief résumé of some of the present evolutionary tendencies in western Europe and especially the United States. In order to appreciate the magnitude of the cultural revolution, consider again the great contrast between the modern factory hand in the United States and of his farmer ancestor in Central Europe. In the old home an active outdoor life, shared by the women as well as the men; pure, fresh air for both sexes many hours out of the twenty-four; frequent exercise of kinds that employ all the muscles and especially the abdominal muscles so important in digestion and childbearing; an occupation varied enough to demand the exercise of at least a moderate amount of judgment and initiative; and with all this the ever-present necessity of planning far ahead for the winter's supply of food, clothing, and fuel. In the factory of America scarcely an hour of outdoor air each day for months at a time; unduly warm, dry, stagnant indoor air day and night for half the year, and dusty germ-laden city air the rest of the time; a cramped position involving the maximum harm to the vital abdominal muscles and the internal organs; an occupation which develops one limited set of muscles and of mental responses at the expense of all others, and which demands little exercise of the high faculties of judgment and initiative; and with all this an almost complete release from the necessity of forethought, for the grocery store is around the corner, the Union or the Associated Charities will feed the hungry and clothe the naked, and the Lord will provide a job.

In addition to all this, racial mixture is now proceeding in the United States at a probably unprecedented rate. In one place the white and the black are crossing on a large scale; in another the English and Italian; elsewhere the Jew and Gentile, for recent statistics show that Jews are marrying outside their own race and faith to an astonishing degree. The more diverse the mingling of races, the greater the number of extreme types of people, and the greater the opportunity for the selection of new qualities for preservation and the formation of new racial characteristics. If the presence of these new types and the stress of a new environment and of new occupations do not cause some change in racial character, the people of America would seem to be breaking an immutable law of nature. But they are not breaking this law, for it is a matter of common observation that new types of character—several of them—are actually developing in the United States.

Is this a good thing? In the course of centuries will this country and the world be stronger because of these new types? A conclusive answer is impossible. The whole matter is too complex, and there are no reliable statistics. The only thing we can do is to suggest certain probable tendencies. Take factory work in America. What kind of people can best stand it? Here are two men of equal ability otherwise, but one is phlegmatic and the other nervous. Neither has brains enough to be more than an unskilled worker. The phlegmatic man would seem to have an advantage, for he can endure the endlessly monotonous task of sticking a million bits of metal into the same machine, day after day for years. The other cannot stand it more than so long. Then he throws up his job, loafs a while, looks for work, lets his family be cared for by charity, finds a new job, moves on again, and finally drifts away and leaves his family. Such a man, it would seem, is likely to have fewer children than the other, and in spite of philanthropic assistance his children are more likely to die than are those of the steadier, more phlegmatic man. Thus one tendency of modern manufacturing may be to increase the proportion of dull, phlegmatic, unenterprising people at the bottom of the social ladder. Perhaps our mechanical inventions are helping to give part of our people a certain Chinese quality. But factory work does not put any such premium on industry, economy, and endurance as does the life of the Chinese farmer. In fact, the methods of many modern unions and the invention of automatic machinery go far toward removing all advantages from the laborer who is industrious, thrifty, and persistent. The union likewise joins with public and private charity in diminishing the selective value of economy. So it may be that our occidental factory system favors the preservation of a type of people less valuable than the common Chinese farmer.

Let us carry our investigation into other fields. From the standpoint of racial evolution which is better off, war-ridden Europe or peaceful America? The biological effect of war has been much discussed. The question is not so much the number as the kind of people who are killed, and not so much the kind who are killed as the kind of children who are born compared with those who might have been born had there been no war. The direst indictment of war has always been its effect on the babies. The great mistake of militarists is their failure to recognize this. All wars tend to take from home a large percentage of the strong, the brave, the adventurous; they leave at home a corresponding percentage of the weak, the cowardly, and those who lack that fine spirit of adventure which is one of the great roots of human progress. Of those who are taken away a considerable number die and leave no children, while among those who come back some are wounded and crippled, while many must wait before establishing new homes and having children. On the other hand, the undesired remnant who stay at home often enjoy unusual prosperity because of the scarcity of labor, and succeed in marrying sooner or in marrying better women than would normally be the case. Hence they bring to maturity more children than would otherwise be their share, while the number of high-grade children is lessened.

In war the mental selection is probably greater than the physical. In ancient days the man with brain as well as brawn wore armor and rode a horse. He doubtless fought more than the common soldier, but he was less likely to be killed. In battle he might hew down scores of unprotected peasants, and himself come through unscathed. Even when attacked by his fellow knights, his armor often saved him from death when wounded. On the other hand, he was far more likely to be killed than was the stupid yokel who was left at home on the farm because he was too dull and cowardly to make a good soldier. With the introduction of gunpowder the partial immunity of the upper classes largely disappeared, for armor was no longer a real protection. Then the bravest and most competent were in more danger of death than the more cowardly and the incompetent. The born leader, the young officer who led the charge was the one most likely to be shot. Thus war lost whatever beneficial selective action it may have had in more primitive ages.

Modern war has not changed all this. The officers, to be sure, are now protected as far as possible, but it is the higher and older officers who profit by this, the ones whose families in many cases are already complete. The younger ones, the lieutenants and the non-commissioned officers who are in line for commissions, are more likely than ever to be killed. They lead the charge; they go out between the firing lines to obtain information; from their ranks are recruited the men who take their lives in their hands for special missions. Moreover, we have lately seen the development of a new type of warfare more deadly than any that ever preceded it and at the same time demanding higher qualities. In the whole realm of human ingenuity aerial warfare is one of the most effective methods ever devised for killing the unusually able young men. In the Air Service the common soldier stays safely on the ground far behind the lines; the highly selected, highly trained, and exceptionally skilled aviator flies high in the air, assails the enemy, and dies.

Of a piece with this biological iniquity is the modern labor battalion. What is the biological meaning of such a battalion? Among a hundred recruits perhaps twenty are of unusually poor mental caliber. Let us pick these out for the labor battalion; that will permit them to stay behind the lines; it will prevent most of them from being killed; and it will let them go home to become the fathers of the next generation. But here are five youths of exceptional ability. Let us make aviators of them and kill them.

In addition to all this, modern war calls from home a large share of the most able of the men above thirty years of age, the ones who are most likely to be the fathers of valuable children. These men take up the burden of supplying arms and munitions, of running the transportation system, of gathering information, dealing with other countries, controlling the supplies of food and clothing, and a hundred other occupations which formerly were allowed to take care of themselves. The more able a man is, the more likely he is to be taken away from home. He may not be killed, but he is not likely to have children while the war lasts. On the other hand, his less competent neighbor stays at home and brings up a family. Whatever may have been the case in the past, it is hard to see any important way in which modern war tends to improve the inheritance of the succeeding generations. It is easy to see scores of ways in which it lowers that quality. A few decades of war like the World War might injure the inheritance of the western world almost irretrievably. If war is followed by wholesale execution and expulsion of the most competent parts of the population, as in Russia, it is almost impossible to estimate the degree to which the general caliber of the people may be lowered.

Sad as are the effects of war on the racial inheritance of the western world, the effect of peace seems to be almost worse. Religion and its modern sister philanthropy have perhaps rivaled war in lowering the biological caliber of the human race. We have here one of the strange contradictions which are so common in this whole subject. An agency which works almost incalculable good to the individual and to the social organism appears to work equally great harm to the race. In this I do not refer mainly to the fact that religion and philanthropy with the aid of modern medicine are preserving a larger and larger share of the weak and incompetent. This is probably a minor matter compared with the diminution of the competent. What have religion and philanthropy to answer for in this respect? As far back as the days of the early Christians, and indeed in still earlier times groups of unusually religious men separated themselves from the world. They assumed vows of celibacy; they planned to honor God by breaking away from what they called the world, the flesh, and the devil. Even so wise a man as Saint Paul was himself unmarried and urged the Christian priest and minister to

THE RACIAL TENDENCIES OF CIVILIZATION 341

follow his example: "I would that ye also were as I am." And hundreds, thousands, millions, have followed that example, women as well as men. How many monks, nuns, and celibate priests there have been since the beginning of the Christian era cannot be estimated even approximately, but there have almost certainly been millions. To-day the Roman Catholic Church is estimated to have somewhere in the neighborhood of 1,500,000 persons vowed to celibacy. Galton expresses himself so forcefully on this subject that I am impelled to quote him at considerable length.

... The long period of the dark ages under which Europe has lain is due, I believe, in a very considerable degree to the celibacy enjoined by religious orders on their votaries. Whenever a man or woman was possessed of a gentle nature that fitted him or her to deeds of charity, to meditation, to literature, or to art, the social condition of the time was such that they had no refuge elsewhere than in the bosom of the Church. But the Church chose to preach and exact celibacy. The consequence was that these gentle natures had no continuance, and thus, by a policy so singularly unwise and suicidal that I am hardly able to speak of it without impatience, the Church brutalized the breed of our forefathers. She acted precisely as if she had aimed at selecting the rudest portion of the community to be alone the parents of future generations. She practised the arts which breeders would use, who aimed at creating ferocious, churlish, and stupid natures. No wonder that club-law prevailed for centuries over Europe; the wonder rather is that enough good remained in the veins of Europeans to enable their race to rise to its present, very moderate level of natural morality....

The policy of the religious world in Europe was exerted in another direction, with hardly less cruel effect on the nature of future generations, by means of persecutions which brought thousands of the foremost thinkers and men of political aptitudes to the scaffold, or imprisoned them during a large part of their manhood or drove them as emigrants into other lands. In every one of these cases, the check upon their leaving issue was very considerable. Hence the Church, having first captured all the gentle natures and condemned them to celibacy, made another sweep of her huge nets, this time fishing in stirring waters, to catch those who were the most fearless, truth-seeking, and intelligent in their modes of thought, and therefore the most suitable parents of a high civilization, and put a strong check, if not a direct stop, to their progeny.

... The extent to which persecution must have affected European races is easily measured by a few well-known statistical facts. Thus, as regards martyrdom and imprisonment, the Spanish nation was drained of free-thinkers at the rate of 1,000 persons annually, for the three centuries between 1471 and 1781; an average of 100 persons having been executed and 900 imprisoned every year during that period. The actual data during those three hundred years are 32,000 burnt, 17,000 persons burnt in effigy (I presume they mostly died in prison or escaped from Spain), and 291,000 condemned to various terms of imprisonment and other penalties. It is impossible that any nation could stand a policy like this without paying a heavy penalty in the deterioration of its breed, as has notably been the result in the formation of the superstitious, unintelligent Spanish race of the present day.

Italy was also frightfully persecuted at an earlier date. In the diocese of Como, alone, more than 1,000 were tried annually by the inquisitors for many years, and 300 were burnt in the single year 1416.

The French persecutions, by which the English have been large gainers, through receiving their industrial refugees, were on a nearly similar scale. Three or four hundred thousand Protestants perished in prison, at the galleys, in their attempts to escape, or on the scaffold, and an equal number emigrated. Mr. Smiles in his admirable book on the Huguenots, has traced the influence of these and of the Flemish emigrants on England, and shows clearly that she owes to them almost all her industrial arts and very much of the most valuable life-blood of her modern race. . . . (Galton.)

The Protestant church is better off than the Roman Catholic because it permits its ministers to marry. How important this is we have already seen in our study of the eminent sons of clergymen. We are led to a similar conclusion by the fact that according to Cattell's data in *American Men of Science*, 1921, each thousand of the members of the various professions in 1850 produced the following numbers of sons who rank among the 885 leading American men of science for whom the facts are available:

Clergymen	3.3	Teachers	2.4
Lawyers	2.5	Physicians	1.6

The relative preponderance of clergymen's sons among scientific men is the more remarkable when it is remembered that when these scientists were born there was supposed to be a bitter conflict between science and religion. In other lines of eminence the sons of ministers seem to stand equally high, as we have already seen, and this is especially true of the ministers of early New England stock who preserve the biological as well as the social and religious heritage of the Puritans more fully than almost any other group in America.

Of late, however, the Protestant church has fallen into bad ways so far as heredity is concerned. Religious orders with vows of celibacy are indeed rare among Protestants, but there has grown up in many quarters the idea that both men and women can do better service to God and their fellows by remaining unmarried than by having homes. I listened to a missionary about forty years of age. She was sweet, strong, finely built, and most attractive; she had the subtle quality of charm and no small share of mental ability. That fine thing called character showed itself in her earnestness, her clearness, and above all in her determination. I know nothing of her personal experiences. Yet I feel quite sure that if she were asked why she was not married, she would say that she was serving God better by engaging in Christian service than she could by bringing up four or five children of her own. Hundreds of our finest young women say that and believe it, little recking that by cutting off the supply of strong personalities they may hurt future generations infinitely more than they help the present generation.

Consider the meaning of the age-long procession of monks, nuns, celibate priests, unmarried clergy, and religious and philanthropic workers who give up marriage and children. It means that for hundreds of years there has been a steady tendency to eradicate from the human race the qualities that make people religious and philanthropic. There are some weaklings in every group, but the vast majority of this great body of religious and philanthropic workers stand far above the average not only in reverence, love of man, humility, and devotion to duty, but in strength of will, capacity for leadership, and power of initiative. Remember too that so far as such people marry they tend to marry their own kind, and to have children who by both heredity and training resemble themselves. The number of missionaries' children who become missionaries is extraordinary. It does not seem possible that this great process of exterminating the religious stock of Europe can have persisted one or two thousand years without lessening the innate religious tendencies of the people as a whole. Perhaps the fact that this has gone on so much more in countries like Italy and Spain than in Scotland and Switzerland has something to do with the fact that the Latin countries where priests and monks have been most numerous are by no means the countries where the people display the strongest tendencies toward personal and effective religion. On the contrary, religion is most vital and influential where the ministers have had large families. That the monasteries kept alive the spark of learning; that organized religion at all times, and organized philanthropy in our own day have been among the greatest benefactors of mankind, I do not question for a moment. But how about the price?

The whole argument which applies to religion applies also to education. From time immemorial there has been a tendency for the scholar to abstract himself from the world, and to rid himself of the ties of family. This has been partly because of his absorption in his work, but quite as much because the world has not recognized the value of original thought. Many a thinker, many an inventor has remained impoverished, unmarried, and childless because the world has not been able to see the value of his ideas and hence has not been willing to pay even a living wage. It has laughed at such seekers-for-truth, and has permitted the source of such personalities to dry up.

According to the census of 1920 some 635,000 women are teaching school in the United States. They may not be so highly selected as they were before other occupations were open to women, but nevertheless they are far above the average. Otherwise they could not have completed their education in high school, normal school, or college. The mere fact that they are teachers makes it certain that the percentage of marriages among them will be low and that those who marry will on the average marry late and have few children. They have chosen an occupation which more or less isolates them from men, and

THE RACIAL TENDENCIES OF CIVILIZATION

which in other ways withdraws them from the ranks of motherhood. Theirs is a noble and indispensable occupation, but in the process of building up the social structure of education, they are cutting off the supply of the kind of children who are best worth educating. Their occupation adds another to the many selective agencies which in the last few generations have begun rapidly to lower the innate character of the races that are most advanced.

CHAPTER XXIII

A RACIAL TEST OF CITIES, DEMOCRACY, AND FEMINISM

In the last chapter we discussed some of the evolutionary agencies which are especially active in highly civilized modern communities. We did not examine many statistics, however, and hence our results were merely suggestive rather than conclusive. In the present chapter we shall look at certain other agencies statistically, and shall see that there is little question as to their highly selective activity.

One of the greatest of these agencies is the attraction which draws the young people from the country to the city. On an average it is the more enterprising and versatile who go to the cities; whereas the less competent, generally although by no means always, stay in the country. It is almost universally agreed that among people of the same race and social position the birth-rate is lower in cities than in the country. Thus there is to-day a strong tendency, first, to remove the brighter young people from the country districts, and then to reduce the number of their descendants so drastically that they by no means maintain their due proportion among the population as a whole. This general idea has become familiar through the writings of men like Stoddard, Humphreys, and others, but it is by no means universally accepted. For example, President Eliot voices the contrary view: "The country breeding gives a vigor and an endurance which in the long run outweigh all city advantages, and enable the well-endowed country boys to outstrip their city-bred competitors."

Let us examine the actual facts and draw our own conclusions. From a study of 885 scientists whom the votes of their fellows had picked out as leaders, Cattell puts the cities far ahead of the country districts. His method was to ascertain how many of the scientists were born among families where the fathers pursued various occupations, and then to compare these numbers with the number of men in each occupation. He makes his comparisons on the basis of the census of 1850, since the best-known leaders in science, as in most professions, are fairly advanced in years. His results show that if the number of families were the same in all occupations, the relative numbers of eminent scientists in each of three great groups would have been as follows:

> Farmers.................................... 1.0
> Manufacturers and merchants.................. 2.2
> Professional men............................ 29.0

These figures mean that where a given number of American farmers' families in the middle of the last century produced only a single scientific leader, the same number of families of persons engaged in manufacturing and trade produced 2.2, while an equal number of professional families gave rise to 29.0. This seems to be due partly to greater opportunities and better training in the professional families than in the others, and partly to greater inherent capacity. Whatever may be its cause, it seems to indicate a great difference between rural and urban communities, for the farmers are mainly rural, while the manufacturers, merchants, and professional men live largely in towns and cities. Moreover, Cattell states specifically that in proportion to their population, regardless of occupation, the cities have produced twice as many scientific men as the rural districts, and the tendency in this direction is growing.

The recent investigation by Visher, already referred to, supplies even more conclusive evidence to this same effect. By comparing 18,400 persons in *Who's Who* for 1922–1923 with the census for 1870 he finds that among the Americans born in different types of communities the relative numbers who were sufficiently notable to be included in *Who's Who* vary approximately as shown in the last column of the following table.

Type of Birthplace	Total Number Reporting	Percentage of Population in 1870	Relative Value
Rural districts............................	4,750	69.9	1.0
Large cities of over 50,000 inhabitants....	3,789	10.0	5.6
Small cities of 8,000 to 50,000 inhabitants	4,571	10.9	6.1
Villages and small towns up to 8,000 inhabitants............................	4,488	8.2	8.9
Suburbs of large cities...................	758	1.0*	10.9*

*Approximate.

The rural parts of America, in proportion to their population, have supplied only about one-sixth or even one-tenth as many of the present leaders as have the portions where people congregate more closely. Part of this deficiency is undoubtedly due to lack of opportunity and training, and part to the fact that most of the negro population is rural, but the depletion of the rural stock by migration to the villages and cities is almost certainly another cause of the rural deficiency in the production of leaders. A second significant fact in Visher's table is that although the cities, both large and small, stand far ahead of the rural districts in the production of leaders, their relative values (5.6 and 6.1) are decidedly below those of the villages and small towns (8.9), and likewise of the suburbs of the large cities (10.9). The cities are generally considered to be the places of greatest opportunity; but they present unusual opportunities for evil as well as good; the suburbs and small towns are better in this respect. Those are the places whence come the leaders in greatest numbers, and they are also the places where able, home-loving, sober-minded people tend most to congregate, and to be least intermixed with the duller rural types, on the one hand, and the pleasure-loving as well as the degraded city types, on the other hand.

The effect of cities upon the distribution of persons of eminence has been much discussed. Ward, in his *Applied Sociology* and Cattell in his work on *American Men of Science* have emphasized the importance of the urban environment in providing opportunities and incentives and thus in bringing out latent genius. Davies, in *A Statistical Study of the Influence of Environ-*

ment, has presented a large number of correlation coefficients which seem at first sight to show convincingly that density of population is the main factor in bringing men into prominence. That there is much truth in the position of these authorities I do not question for a moment. But Ward takes no account of natural selection, Cattell is so engrossed in the admirable task of improving the educational system that he sees almost everything in terms of education, while Davies overlooks the important fact that great cities as a rule are numerous in stimulating climates. Such climates, as we saw in the case of Europe and as I have shown for the United States in *Civilization and Climate,* have the same general effect as education; they tend to bring out whatever ability a man may have. The activity which they stimulate also tends to cause people to engage in manufacturing and commerce to a high degree. Hence great cities and an abundance of eminent men would often occur in the same States even if climate were the only factor to be considered. In the same way education alone would cause the number of eminent people in some places to be large and in others small, provided all other factors remained constant. And the same is true of natural selection. In other words, the proportion of eminent people produced in a given community depends first upon the presence of good material, which means inheritance; then upon the presence of a stimulating physical environment with good conditions of climate, food, and health; and third upon the presence of a good social environment which includes the right conditions of education, religion, and social habits.

The interplay of these three main factors in causing people to achieve something unusual is well illustrated by Clarke in his study of the thousand most eminent literary men born in America prior to 1851. Here is one of his tables which I have recalculated in such a way that it shows the number of eminent literary persons per 10,000 of the average white population from 1781 to 1850. I have included all places whose average population was over 5,000.

	Average Population 1781-1850	Eminent Literary Persons Born 1781-1850	Eminent Literary Persons per 10,000 Inhabitants
*Cambridge	5,000	9	18.0
Portland	7,500	13	17.3
*Newburyport	6,000	9	15.0
New Haven	7,900	9	11.4
Portsmouth	6,250	7	11.2
*Salem	11,200	12	10.7
Hartford	6,600	7	10.6
*Boston	51,500	53	10.3
Charleston	10,000	7	7.0
*Charlestown	6,750	4	5.9
Washington	9,000	5	5.6
Providence	14,600	8	5.5
*Roxbury and Dorchester	9,250	5	5.4
Albany	17,400	9	5.2
New York	168,000	80	4.8
Utica	5,250	2	3.8
Philadelphia	130,000	49	3.8
Buffalo	8,900	3	3.4
Baltimore	52,500	16	3.4
St. Louis	13,900	4	2.9
Richmond	7,500	2	2.7
Rochester	8,400	2	2.4
Troy	9,250	2	2.2
*New Bedford	6,600	1	1.5
Brooklyn	34,250	5	1.5
*Lowell	7,500	1	1.3
Louisville	10,000	1	1.0
Pittsburgh	11,900	1	0.8
New Orleans	26,800	2	0.8
Cincinnati	24,900	0	0.0
Newark	9,100	0	0.0
*Lynn	5,400	0	0.0

Look first at the nine cities that are starred. All of them lie in eastern Massachusetts, and most are suburbs of Boston. Cambridge (18.0), Boston proper (10.3), Charlestown (5.9), and Roxbury and Dorchester (5.4) are now parts of one continuous city, and Lynn (0.0) lies only fifteen miles away. It is clear that climate can have nothing to do with these differences. It is equally clear that food and mode of life can account for only a small percentage of them. Opportunities for public education and other benefits which come from life in a city can scarcely play more than a minor rôle. The only things that can account for the differences seem to be inheritance and the immediate environment of the home. But the home environment depends so intimately upon the innate temperament and mentality of the parents that the two can scarcely be separated. The importance of inheritance becomes still more clear when we go a little farther afield. Newburyport, with 15.0 eminent literary persons per 10,000 inhabitants, and Lynn with none are near neighbors north of Boston; but Lynn has an advantage over Newburyport in nearness to Boston, Harvard, and all the influences that radiate from a great city. Nevertheless, up to 1850 Lynn had not given birth to a single person among the one thousand most eminent literary people of America, whereas Newburyport, with an average population only about 12 per cent greater than that of Lynn, produced nine. This seems to be explicable only on the supposition that during the first half of the nineteenth century there was greater inherent ability in Newburyport than in Lynn. Newburyport at that time was an especially pleasant place of residence and attracted many fine families. Lynn has long been a relatively unattractive manufacturing town from which the most successful people are likely to move to Salem, Boston, Cambridge, Newburyport, and similar places.

Turning again to our table, it may be that New Orleans (0.8) stands almost at the bottom, because its climate is relatively debilitating. I feel quite certain that one of the important factors in causing the South in general to be backward in producing persons of eminence is the climate and the diseases which it fosters. Nevertheless Charleston, South Carolina (7.0), stands well toward the top of our list, coming next to Boston, in spite of having a climate which is by no means so stimulating as that of Brooklyn (1.5), Pittsburgh (0.8), and Cincinnati (0.0). The low position of Pittsburgh and Cincinnati may have been due to their newness and to the fact that their able men were engrossed in building up a new country. But this explanation can apply only in small measure to Brooklyn, Lowell, New Orleans, and Newark. The type of inherent ability that expresses itself in literature seems simply to have been deficient in those cities; for one reason or another they have not attracted and have not retained people of that particular type. They may have produced other types of ability; Cincinnati, for example, has produced an unusual number of eminent scientists, according to Visher. In the cities at the head of our list (Cambridge, Portland, Newburyport, New Haven, Portsmouth, Salem, Hartford, and Boston), and likewise at Charleston, which probably would stand well above Boston if allowance were made for the adverse effect of climate, we seem to have instances of what Thompson, the author of *Outline of Science*, has called "the virtuous circle." The migration of able people to attractive centers creates a good environment; unusually able children are born there, and the good environment stimulates them to achievement; such achievement attracts other able people, and thus still further improves both the inheritance and the environment.

The sum and substance of the whole matter seems to be that in the long run a good inheritance, on the one hand, and a good social environment and good training, on the other hand, are inseparable. Which is more important in causing a community to send out able men no one can say, for both are essential. Because the able young people press into the centers of population, and especially because now even more than in 1870 those of them who have families tend to take up their residence in the suburban parts of metropolitan areas, those are the places where the proportion of children with a high inheritance is largest, and where the social environment and the training are best. Hence they give rise to the largest proportion of eminent people.

The depletion of the innate capacity of the rural population by migration to the cities may help to explain one of the most serious economic problems which now confronts the United States. That condition is the chronic depression and dissatisfaction among the farmers. The early farmers of the United States, quite unlike the peasants of Europe, were the backbone of the country and produced a large share of leaders. This condition continued more or less vigorously so long as there was new land to be taken up in the West, so that there was an incentive to men of ability and initiative to remain on the farms. But now, even in our Western States, the more able sons of the farmers tend to flock to the cities. They are full of ambition to go to college or to work their way up in business. And thousands of them succeed. If boys of that kind stayed on the farms, while those with less ability and initiative went to the cities, would the farmers now be at the mercy of the city people? May it not be that what the farmers need is constructive leaders and organizers rather than laws, loans, and tariffs?

Having satisfied ourselves that the more able people drift from the farms to the cities, and that they transmit at least a portion of their ability to their children, let us inquire how numerous are the children in the cities compared with the rural districts. It might be supposed that for such information we would turn directly to the reports of the census on birth-rates in our two types of regions. That, however, would be misleading, for the census does not show how many children are born per family in the cities versus the country. It does, however, give exact data as to the size of families among persons of various occupations. Farming, fishing, lumbering, and mining are good representatives of rural occupations, while manufacturing and trade are largely urban. Among the families in the United States where children were born in 1920 the following conditions prevailed in these great groups of occupations:

	Total Number of Children Born in 1920	Average Number of Children, Both Living and Dead, per Family	Average Number of Living Children
Farming, fishing, and lumbering	365,275	3.7	3.3
Mining and quarrying	51,693	4.1	3.5
Manufacturing and industry	686,062	2.9	2.4
Trade	117,922	2.5	2.2

In large groups like these there is no reason to think that the average age of one set of parents differs materially from that of the others. It is well known that the proportion of foreign-born immigrants engaged in manufacturing is very large. For example, in 1920 about 49 per cent of all the foreign-born men engaged in gainful occupations were employed in manufacturing and mechanical industries, whereas only 36 per cent of those of mixed native and foreign parentage and 28 per cent of those of native white parentage were so engaged. On the farms, on the other hand, the corresponding percentages were: foreign-born 14, mixed 21, native white 36. Among the foreign-born population of America it is well known that the birth-rate is high. For example, among the mothers to whom children were born in 1920, the Poles and Italians had had an average of 4.5 children, the Germans 4.4, Austrians 4.3, and Hungarians 4.2, compared with only 3.0 for the white women born in the United States. Moreover, if we take account only of surviving children, no large foreign-born group of women had so few children as the American-born women, 2.7, while the German-born women had 3.9 and the Italians 3.8. Such conditions are normal.

On this basis we should expect that the families in the manufacturing industry where the foreign-born are numerous would be larger than among farmers where native whites predominate. Quite the contrary is actually the case. The average farmer's family into which a child was born in 1920 had up to that time produced 3.7 children, including the newcomer, while 3.3 of the children were still alive. But among the people engaged in manufacturing and industry, the average family into which a child was born had produced only 2.9 children, of whom 2.4 survived. In the same way the mining population had an average of 4.1 children per family, with 3.5 survivors, while the average family engaged in trade had given birth to only 2.5 children, of whom only 2.2 still lived in 1920. Race and social position, as well as the contrast between rural and city life, probably have something to do with causing the miners' families to be sixty per cent larger than those of the merchants and clerks. A large percentage of the miners are poor foreigners, while the merchants comprise a great many middle-class native Americans. Nevertheless, the great contrast between the two groups is presumably in part due to the difference between city and country. On the other hand, the contrast between the rural farmer and the urban manufacturing population would presumably be still greater if the racial and social composition of the two groups were similar and if both contained the same proportion of recent immigrants. In other words, the mere fact of city life, with its mercantile and industrial conditions, seems to reduce the size of families. Those who stay in the country, although they are less competent than those who migrate to the cities, are selected by nature to be the parents of many children, while those who go to the city are selected to be the parents of few. If this process continues for several generations, our farmers may deteriorate into dull peasants, while our cities become largely the home of the descendants of people from southern and eastern Europe.

It seems almost ridiculous to say that modern democracy as well as migration to the city may be exerting a harmful biological effect upon the human race, but that seems to be the truth. By democracy I mean not merely a political system, but a social system which allows the utmost freedom of movement from one social class to another. We pride ourselves on the facility with which an enterprising boy or girl can break away from old surroundings and make good anywhere and in any station, no matter how high. Is this an advantage biologically? The common answer is that it is the best possible thing, because it utilizes human ability to the fullest extent and also brings new blood into the effete upper classes. The first part of this answer is probably true, but what is the biological effect of our modern democratic form of society? When a bright boy is born among the lower classes, he is helped to go to college, separated from his early environment, and probably marries some one from a social group above that of his origin. Almost inevitably he passes on to his children a mixed inheritance. He bequeaths to them something of his own ability, but likewise something of the low degree of ability which belongs to the general class from which he sprung.

If an unusually able man of the lower classes is held in his own social group by a rigid social system he fails to achieve the self-development and usefulness which are within his capacity, but his children raise the general biological level of his group.

If he is raised to a higher social plane his children lower the biological level of the new group, unless by some happy chance he and his children marry persons whose inheritance happens to overcome their weaknesses. That, alas, is rare. But being raised to a higher group, the man from the lower classes tends to have fewer children than would have been the case had he remained in his own class. Thus, as a rule, although by no means always, the man's inheritance is applied in such a way that it takes away relatively strong elements from the lower class and tends to increase the numbers of that part of the middle class which though temporarily efficient may at any time revert to the "shirt-sleeve" stage. Moreover, because of the lack of scientific control of marriage, the building up of a numerous group of pure-bred and highly competent leaders is checked.

What the final effect of thus depleting the ability of the lower classes and at the same time diluting the more capable classes will be, no one can yet tell. This much is certain, the animal breeder is very slow to permit his blooded stock to become mixed with mongrel stock, no matter how good a few individual animals from the mongrel herd may be. Yet on the other hand, the blooded stock originally was produced by selecting just such animals from the mongrel herd. Hence we are left in doubt except as to one thing: by raising the promising young man and young woman from the lower ranks of society we almost inevitably reduce the number of their children, and their children are likely to be above the average of the group from which the parents were derived.

Putting aside these somewhat speculative considerations, let us look at certain evidence which shows how rapidly the lower classes of society are increasing compared with the upper classes. Here are some figures from *Birth Statistics for the Birth Registration Area of the United States for* 1920. They show the number of children in the families of men in various pursuits to whom children were born in 1920. The figures take no account of differences between the professions as to the age of the fathers, the number of men who are not married, and the number who, although married, had no children. They simply show the actual size of the families in which children were born in 1920.

OCCUPATIONS OF FATHERS AND SIZE OF FAMILIES IN WHICH CHILDREN WERE BORN IN 1920 IN THE REGISTRATION AREA OF THE UNITED STATES

Occupations	Number of Children Born in 1920	Average Number of Children to End of 1920 per Pair of Parents, Including Children Who Have Died	Average Number of Living Children per Pair of Parents in 1920
Mining			
Mine operatives (young)	45,544	4.3	3.6
Mine foremen, overseers, and inspectors	1,022	4.6	3.9
Mine operators, officials, and managers	584	3.2	2.9
Chemists, assayers, and metallurgists	1,692	1.9	1.8
Manufacturing			
Factory laborers (young)	244,365	3.7	3.1
Factory foremen and overseers	9,534	3.3	2.9
Factory managers, superintendents, owners, and officials	14,142	2.5	2.3
Technical engineers (civil, electrical, mechanical, mining)	5,971	2.1	1.9
Trade and Inventions			
Laborers in coal and lumber yards, and warehouses (young)	1,165	3.4	2.9
Retail dealers	63,840	3.1	2.8
Bankers, brokers, and money-lenders	5,045	2.3	2.1
Designers, draftsmen, and inventors	2,886	2.0	1.8
Personal Service			
Bootblacks (young)	454	3.9	3.3
Janitors and sextons	2,228	4.1	3.4
Barbers	9,000	3.2	2.8
Lawyers, judges, and justices	4,338	2.4	2.2
Public Service			
Garbage men, scavengers, and other laborers (young)	1,744	3.4	2.9
Guards, watchmen, doorkeepers	1,502	4.1	3.5
Policemen	3,752	3.2	2.8
Officials and inspectors (city and county)	3,752	3.2	2.8
Officials and inspectors (State and U. S.)	2,690	2.3	2.1

In each section of the table the occupations which require least skill are placed at the top and those requiring most skill at the bottom. In most cases there is a regular gradation from families averaging three or four among the unskilled to families averaging not far from two among the highly skilled. In each group of occupations I have added the word "young" to the upper line because the men in that particular occupation average younger than the others, and therefore their families are not so near completion as are those of older men, such as foremen, bankers, officials, and the other higher categories. It seems safe to say that at least six children, and perhaps more, are born to the average mine operative, and that not less than five of these live well beyond infancy. On the other hand, it is very doubtful whether the chemists, assayers, and metallurgists in our table will have an average of as many as three children, so that not much over two and a half will survive far into childhood. After infancy is past the chemists' children may have nearly as high a death-rate as the children of the miners. Snow, in a paper on *The Intensity of Natural Selection in Man,* has confirmed Pearson's conclusion that if the death-rate is high in the first year of life, as it is among the children of mine operatives, it is relatively low in succeeding years. On the contrary, a low death-rate in infancy means that the later death-rate is higher than it would have been under the same conditions of environment if the weaker infants had perished.

The case of the chemists appears still worse when we remember that even though the proportion of married men (91 per cent) among a picked body of leading scientists has been found by Cattell to be as large as in the population as a whole, the proportion among Harvard graduates over fifty years of age is only 72 per cent. Moreover, in both groups the number of childless marriages is large because of the relatively advanced age of the wives. For example, among leading men of science whose families were presumably complete, 22 per cent of those who were married were childless. Again among these scientific men the average completed family amounts to only 2.23 children. As the death-rate among the children up to the usual age of marriage is about 120 per thousand, the average number of children reaching that age is 1.96. This, however, takes no account of the scientists who are not married. If they are included,

A RACIAL TEST 359

the average number of children who live to maturity is approximately 1.75 per scientist among people who have passed the age when they are likely to have more children.

Among graduates of Harvard the case is even worse. In 1902 the average Harvard graduate, both married and unmarried, who had been out of college twenty-five to thirty years had only 1.43 children, as reported by President Eliot. The figures for other men's colleges are similar, while for women's colleges they drop as low as an average of only about one child per graduate. In the same way Kuczynski is reported by Cattell to have found that among the native white population of Massachusetts the number of children who survive to maturity was only 1.92 per family as long ago as 1900. The conditions among all these groups are probably still worse to-day. In other words, these competent portions of the community are simply dying out.

The net result of all this is that when allowance is made for people who are not married, for childless marriages, and for deaths before maturity, the mine operatives appear to be at least doubling their number in each generation. On the other hand, it is extremely doubtful whether the chemists and the kind of people whom they represent are reproducing themselves, for they appear to bring to maturity less than one child for each potential parent. Among persons engaged in manufacturing, trade, personal service, and public service the general conditions are the same as in the mining group, as is obvious from the preceding table. The inefficient people are reproducing themselves very rapidly, while the more efficient groups are certainly not increasing in proportion to the population and are probably dying out. Bootblacks, janitors, garbage men, and watchmen are not in the least danger of becoming scarce, but scientific men, lawyers, and high officials are producing so few children that their descendants will probably be less numerous than the present generation. If our conclusions as to natural selection and the evolution of racial character are anywhere near correct, it would appear that America is suffering a very rapid and harmful type of selection. It is a selection similar to that which appears to have made northern

China so inert, so lacking in public spirit, and so averse to progress.

The position of women and the status of the family appear to represent another important selective factor. Here, as in most cases, sociological advantages must be set over against biological disadvantages. In fact, throughout our comparisons we find the same contrast almost everywhere. Even in war there are some social advantages through the stimulus to patriotism and to the spirit of bravery and devotion. In religion, education, and industrial life; in our modern democratic organization of society; and in the cultural progress represented by cities the world reaps great advantages through the development of the individual and the building up of a splendid body of social, political, religious, scientific, industrial, and other usages for the benefit of future generations. But at the same time there seems to be a terrible biological disadvantage through the withdrawal of many of the finest people from parenthood. Feminism of the right sort, like religion, education, and democracy, is an admirable thing in itself, but to an almost greater extent than the others it harms our racial inheritance and thereby defeats its own end. It is truly pathetic to see how rapidly the feminists are killing their own movement. Of course, there are all kinds of feminists, ranging from the confirmed man-hater to the mother of six children. But the mothers are rarely strong in the faith. They are willing to help their sisters to gain a freedom of which they themselves feel little need, but very few would sacrifice a single year of their children's love for all the women's rights in the world.

I sympathize with the desire of women for full freedom to have their own careers. Until I realized the biological significance of the movement, I was willing to aid it. Now I am still willing to do so, for true feminism, like religion, philanthropy, education, and democracy is of fundamental value. But it needs modification to eliminate its harmful biological aspects. The truer and nobler feminists believe strongly in marriage, motherhood, and home. They merely contend that a woman as well as a man should have her own career, and should not be bound to give it up because of children. That sounds reasonable, and perhaps it is. In actual practice, however, the woman who has a career rarely has more than one child even if she is married, and generally she is not married. She puts her career first, and it usually becomes both the first thing and the last in her life. The practical result is that whether they wish it or not, the women who stand strongly for women's rights are rarely the mothers of the next generation. The mothers are women of a less independent spirit, those whose love of home, husband, children is their strongest motive. They are the ones who are passing on their qualities to the next generation, and who have in their hands the early training of that generation. Let the present rapid process of selecting the strong-minded women for destruction and the gentler, less aggressive ones for motherhood continue a few generations, and the banner of feminism will have to be hung on the wall—an interesting trophy. No type can persist long if it is rigorously picked out for destruction whenever it appears.

In this connection divorce has an interesting bearing on the evolution of racial character. For a long time I viewed with deep alarm the rapid increase of divorce. One divorce for every 14.3 marriages in the United States in 1896, 1 for every 11.8 in 1906, and 1 for 9.3 in 1916. This is a terrible record of incompatibility and unhappiness. But look deeper and see what it means. In former days when divorce was not so easy, married couples inevitably made more effort than now to adjust their differences. The presence of children helped greatly in this. So too did the disfavor in which divorce was then held. Now all this is changed; divorce is common in many circles, although fortunately not in all. What is the result? Unhappiness? Yes. The breaking down of the home? Not at all. The contrary will probably be the result in the long run. Among people who are divorced, the number of children is inevitably smaller than among those who are happily married. Even when divorce was rare, the number of children per family among happily married couples averaged larger than among the unhappily married. The present freedom of divorce greatly increases this disproportion. In other words it tends to reduce the proportion of children born of the type of parents who through jealousy, bad disposition, hasty marriage, infidelity, incompetence, or other causes of discord are not able to maintain happy homes. Thus here, as in the case of feminism, we probably have a rapid process of selection whereby the home-loving, faithful, self-sacrificing type of people are more and more allowed to be the parents of most of the children. Thus in each new generation these qualities presumably tend to become more wide-spread.

Look now at the future of the countries which to-day are leaders of civilization. The great outstanding fact is that the industrial revolution, the growth of facilities for transportation, the development of science, the spread of good government, and many other new conditions have permitted an enormous increase in population. So long as that increase does not exceed the increase in man's capacity for making a living it is not dangerous. Already, however, we have practically exhausted the easily available new land which was one of the chief reasons why such rapid growth of population was possible. We have also quite surely passed the stage of most rapid increase in our capacity for transportation, for the development of new and more productive types of plants and animals, for improvements in other agricultural methods, and for the exploitation of undeveloped resources. Hence, as is well shown by East in *Mankind at the Crossroads* and by Carr-Saunders in *The Population Problem*, the present rate of increase in population must speedily be checked.

Nature's blind way of checking increase at the present stage of human progress happens to be through a natural selection which begins at the top, where it works painlessly through voluntary restriction of families, late marriages, celibacy, and the like. Then it works down into the less competent classes through increasing poverty, agricultural distress, industrial stagnation, famine, war, pestilence, and other disasters. The particular phase of nature's method which is operating among us resembles in many ways the selection which has produced such disastrous effects in the famine-stricken portions of China. In China the best stock has largely been lost through migration, while the poorer stock has remained and multiplied until dire famines were almost the only means of getting rid of the surplus population. With us the best material is being selected for destruction by war, by religious zeal, by education, by a democratic social system with free opportunities for all, by industrial development, by feminism, by celibacy, by late marriages, by birth-control, by unhealthful occupations, and in numerous other ways. Moreover, the upper classes are being constantly diluted by mixture with less competent stock from every other class, so that there is no chance to build up anything at all similar to the pure strains which among animals are always the ones that can be steadily relied on for great achievement. Thus our upper classes decline both in numbers and efficiency. At the same time we are constantly culling out the best material from the lower classes, and partially sterilizing it by making it part of the upper classes. Hence the lower classes grow lower while the upper receive little permanent benefit. Joined with this steady adverse selection is the ominous growth in the numbers of the lower classes. Unless a change occurs promptly there is genuine danger that within the lifetime of our children this country may be so densely populated that it will be face to face with problems as serious as those of Europe, while Europe may be confronted by problems almost like those of China.

If this book carries any lesson to the people of to-day it is this: A race or a nation can apparently be made or marred by natural selection. Mere numbers count for nothing: in many cases a dense population is the greatest of curses, as it has been in Ireland, China, Japan, and Germany. Quality is what counts, and quality can be obtained only by diminishing the number of people who inherit low moral and mental capacities and increasing the proportion who inherit the high qualities which lead to racial dominance. In the past, without man's conscious intervention, natural selection has been actively at work, sometimes for good and sometimes for ill. The only question is whether a race or nation will control such selection so that it will always act beneficially as in the first and greatest days of ancient Greece and Iceland, or will permit it to continue to work haphazard and perhaps toward great unhappiness, as in China or in the Greece of later days.

America has gone far on the way toward racial impoverishment, so far that the country already feels that it is slipping. Historical examples show that the danger is real and great. Many thoughtful people are already considering how we may get back to safety; history points out the road. We must set ourselves diligently to limit and control the future growth of population. Such conscious control of natural selection may perhaps be nature's own next method. One aim must be to insure that all newcomers to our land are of a quality *above the average* of our own people; then they will do us good, not harm. Another aim must be a low birth-rate and a low death-rate, so that our population may increase less rapidly than does our command over natural resources. Thus we shall be able not only to maintain, but raise our standard of living. But a low birth-rate is of little advantage unless it is *differential;* and the differences must be of the right sort. To-day, as we have seen, the people of highest ability largely fail to reproduce themselves; the garbage men double themselves in every generation. Unless such differences are reversed, the future is dark. The ideal condition would be large families among the tenth of the population having the soundest combination of physical health, good intellect, strong wills, and fine temperaments. Among persons standing lower in the scale there should be progressively fewer children until the lower tenth has none at all. In the Middle Ages, according to Carr-Saunders, an imperfect approximation to such a system actually prevailed in western Europe. In the future a better approximation is entirely possible. We may reasonably hope that some day the average family among the most competent people will consist of perhaps four to six children, while the numbers among those of less ability will diminish to such a point that the population as a whole will increase only at the slow rate which accords with the growth of man's control over nature.

Of course there are innumerable objections to any such system. What incentive will ever induce the lower classes to limit their families? Is it possible to check the present feminist tendencies so that *all* of the finest women will unhesitatingly choose motherhood? And how can we make it economically possible for persons engaged in intellectual, religious, and philanthropic pursuits to support and educate half a dozen children? These questions must be answered by future generations, but they are no more difficult than a hundred others which mankind is successfully answering.

Meanwhile we must ascertain with more certainty the exact biological effect of a differential birth-rate. Here is a sample of the kind of problem that must be faced. It is generally agreed that the interval between one child and another ought to be two or three years. Hence if the more competent people, as a rule, are to have families of five or six children, marriage should take place while the men are under thirty years of age and the women in the middle twenties. But according to Redfield these ages are much too young; the human race would be greatly benefited if the birth of children were delayed till the parents are not far from forty years of age. He bases this conclusion on many interesting facts as to the relatively high competence of the progeny of old parents among dogs, cows, horses, and men. Here is a little table which sums up his findings as to the average age of fathers at the birth of their sons. The degree of capacity increases as one goes down the table. According to Redfield, "the average quality of the men in the different groups is well represented by the attached figures."

AVERAGE AGE OF FATHERS WHEN CHILDREN WERE BORN

Chicago, all births registered in 1913..........................31.2 years
Persons born in New England in the 18th and 19th centuries....33.8 years
  865 living American men of science..........................35.0 years
  100 eminent English men of science, per Galton...............36.0 years
  299 eminent British men of genius of all sorts, per Ellis........37.1 years

39 eminent men of the 18th and 19th centuries in whole world,
Yoder...37.8 years
1,028 cases in pedigrees of 571 great men of all times and countries,
Redfield..40.7 years

The consistent increase in the ages of the fathers as the degree of eminence increases evidently means something. Redfield thinks that it means that parents transmit to their offspring the ability which they acquire through their own efforts. Hence the older the parents the more they are likely to have children of unusual ability.

But what does the table really mean? If Redfield is right, the youngest son or daughter in any group of brothers and sisters ought on an average to show more ability than the eldest. Accordingly I asked a dozen friends to classify their acquaintances who belong to large families, basing the classification on general competence and influence. Brothers and sisters were classified separately in groups of three; a few cases were taken from reference books. If a person's achievements have no relation to his birth-rank, the averages given below ought to be exactly 2.00. If the first-born is always the most competent and the youngest least competent the averages should be 1.00, 2.00, 3.00. Here is what we actually find:

AVERAGE RANK IN GENERAL COMPETENCE AND INFLUENCE

	Oldest	Second	Third
First hundred cases......................	1.93	2.02	2.05
Remaining 87 cases.....................	1.94	2.10	1.98
Average of 187 cases...................	1.93	2.05	2.02

This suggests that the oldest son or daughter is a trifle more influential than those born later, but the differences are so slight that they are probably accidental. The youngest person who made this classification, a man of about twenty-four years, placed the oldest brother or sister relatively higher than did any other of my collaborators. This is natural, for among his young acquaintances the oldest in any group has had most chance to show what is in him. On the other hand, among twenty cases that I estimated from books, the rank of all three sons was identical. The people classified in the preceding table belonged largely to professional families or to those of successful business men—just the ones among whom Redfield's supposed inheritance of acquired characteristics ought to be most prominent. The average difference in age between oldest and youngest was at least seven years and probably more, or approximately as much as the difference between the highest and lowest groups in Redfield's table. But the results show no hint of superiority in the youngest child, and the faint hint of superiority in the eldest is probably mere accident.

In order to test the matter in another way I have selected from the genealogical memoirs of the Adams, Churchill, Lincoln, Lyman, and Huntington families a group of persons whose biographical sketches show that they were leaders and possessed unusually high moral and intellectual qualities. Of course it is impossible to pick out all such leaders, but the persons here chosen certainly average far above the general level of those who have borne these five family names, and likewise above their brothers and sisters. To each person thus chosen two birth-ranks were assigned, one counting from the eldest child and one from the youngest. If there were five children in a family, for example, the second would have a birth-rank of two, counting from the eldest, and of four counting from the youngest. The average of each rank for all five children would be three. Now if the age of the parents has any appreciable effect upon the ability of the children, the three hundred unusually competent and valuable persons here chosen ought to show at least a slight tendency to fall in the younger rather than the elder halves of their respective families. In other words, the birth-rank reckoned from the oldest child ought to be greater than from the youngest, but here is what we find:

BIRTH-RANK OF LEADERS

	Reckoned from	
	Oldest child	Youngest child
88 Churchills..................................	3.9	4.6
92 Adamses, Lincolns, Lymans..................	4.2	4.4
112 Huntingtons................................	4.2	4.4

I do not feel sure that there is any biological significance in the uniform, though slight, tendency for the leaders here and in our previous comparison to fall among the older rather than the younger members of their families. The important point is that all the evidence is adverse to the idea that the age of the parents is in itself a factor in determining the ability of the children.

As a final test we may investigate the age of the seventy fathers of the members of the Adams family who were used in the preceding comparison. All these fathers were born previous to 1780, for in this particular case I wished to avoid complications due to the decline in the size of families during the past century. We will compare their age when their notable sons were born with that of the 420 fathers whose 2,850 children form the group from which the 70 leaders were chosen. Families with one child, two children, three children, and so forth, were treated separately in order that there might be no error because large families produce an absolutely greater number of leaders than small families. The net result is that at the time of the birth of their first child the fathers of the leaders were 0.2 year younger than the average father at the birth of the first child; at the birth of the last child they were correspondingly older than other fathers of equally large families by 0.4 year. At the birth of their notable children these same seventy fathers averaged 33.7 years, which is 0.1 year older than the estimated age of the other fathers at the birth of corresponding children. Here again there is no evidence that the age of the parents makes any difference in the ability of the children. Eminent men undoubtedly tend to be the sons of relatively old fathers, but that presumably is not because of the fathers' age. The reason may be that the type of parents who produce eminent children tend to marry late; or perhaps unusual vigor in the parents causes them to have children late in life and helps to give the children the physical vigor which is one of the greatest aids to eminence.

Although this study of the age of parents lends no support to the hypothesis of the inheritance of acquired ability, it re-enforces a conclusion which Galton emphasizes, but which is only half appreciated. The conclusion is that when one member of a family displays unusual ability there is a strong tendency for his near relatives, especially his brothers and sons, to do likewise. Thus the important thing is not to be born of old parents or young parents, but to be born in the right family. The great biological danger is not from early marriages, but from the fact that in our day families which might produce eight or ten unusually talented children usually produce only one or two, while those which cannot produce a single competent child produce eight or ten incompetents.

This brings us back to our main problem of selection. What do the preceding facts indicate as to how our modern system of late marriages and birth-control influences natural selection? The answer seems to be implied in the following fact: If the Anglo-Saxon families of the United States up to about 1850 had been limited to three children, about half of the leaders in the various family groups discussed above would not have been born. In other words, one of the main things done by the modern system of small families is to prevent the birth of more than about three children to parents who are strong physically, mentally, and morally, and who under a wiser social system might have produced numerous children with more than the usual capacity for leadership. Although we have no statistics, it is highly probable that among people who would have had only two or three children under the old system the number of children has not diminished in anything like so large a ratio as among the people who might have had a dozen. Among the incompetents, no matter whether they are physically strong or weak, the number of children has probably not yet declined much, if at all, in America, while the proportion of survivals has increased.

This whole line of reasoning adds another to the many types of evidence which show that our present system of marriage and birth-control is acting as a strong adverse selective factor. There seems to be no remedy except a gradual, but deep-seated social revolution whereby society shall impose new, drastic, and thoroughly scientific criteria as to what kind of people shall have many children and what kind few. It seems imperative that we devise social rather than legal incentives to persuade one kind of people to refrain from having large families and another type to have such families. We must teach the physically weak and the mentally and morally incompetent to say: "Why should we be burdened by having many children?" But those who have an unusually good inheritance must be persuaded to say: "To us has been given strength of body and mind. Our children are almost sure to be above the average. Society stands in imperative need of such children; they are its salvation. Therefore we must have children, even although it means sacrifice."

Long before society as a whole has decided what kind of people shall increase in numbers and what kind shall diminish, there will probably be significant progress in one of those great processes of selection which have been the theme of this book. Already we see not only a beginning of the restriction of the birth-rate among the lower as well as the upper classes, especially in France, but in England and America the tide of restriction among the upper classes is beginning to turn. A compact group of thoughtful, conscientious, progressive, and strong-minded people are acquiring a new sense of the value of the home as the center of progress and a new sense of responsibility for the production of leaders for the next generation. They are the kind of people who leave the conveniences of city apartments, and move to the suburbs for the sake of the children. They put up with discomforts and annoyances in order to have homes where the children will thrive. They gladly curtail parties, travel, the theater, and the club because they feel that the home and their children are of infinitely greater value. They likewise watch over their children as never before, not only to train them aright, but to insure marriages free from taint and full of promise of children who will be bold and noble leaders.

This ends our interpretation of the parable of races. The full interpretation is so complex and vast that perhaps the greatest of libraries will not suffice to contain all the books that will some day be written about it. Those books will doubtless explain the great mystery which now underlies all our studies, the mystery of the origin of new characteristics and of the curiously divergent types which we call mutants and variants. A knowledge of these will vastly help in understanding the process of natural selection. Yet even now, although we are so ignorant as to how new types arise, we understand something of the way in which the survival of certain types is fostered by various factors of physical environment such as the relief of the earth's surface, climate, mineral resources, vegetation, parasites, disease, food, and many other conditions. We see that human inventions, social customs, religious practises, forms of government, the progress of science, modern industry and transportation, and a host of other cultural conditions play an enormous part in picking out one type for slow destruction and another for preservation. The more we study this process of selection, the more we realize why one race differs from another in temperament and mentality as well as in physique, and why the spirit of one age is diverse from that of the next.

Throughout the course of human history the pioneer, cityman, plowman, and seeker-for-truth have always been important types. In the earliest days the anthropoid ape who first essayed to walk across a broad clearing was a pioneer; the one that lost his life because in eager search for fruit he failed to scent a tiger was a primeval cityman; the plowman breed of apes clung closely to the trees; while the seeker-for-truth picked up a crooked stick, handled it curiously, swung it aimlessly, and discovered that with it he could strike a blow that would hurt his enemy. Thus it has been all through the ages. At frequent intervals some crisis has sent out pioneers to found an empire in India or China, to mingle with other pioneers and form new races in the belt surrounding the Mediterranean, or to people new continents in post-glacial Europe, America, Australia. With equal frequency men of a certain type have flocked to the market-places and cities in hope of gain. Those cities have been the graveyards of racial progress, for luxury, ease, excitement, vice, and lack of good air and exercise have sapped the strength of their inhabitants. The plowmen who have remained in the old homes have not perished as have the city men, but often they have been scarcely more than the dregs from which the better portions have been drained, as in parts of north China. All through the ages the seekers-for-truth have been ruthlessly exterminated because they sacrificed themselves for the good of others, because they would not conform to strict archaic laws, and because out of sheer love of knowledge, or dissatisfaction with things as they are, they ruined their health by overwork, put off marriage until it was too late, or sought asylum in monasteries.

From many points of view this eternal drama of nature's choice of some types for destruction and others for preservation seems unutterably sad. It has been a powerful factor in the decay of almost every fallen nation from ancient Egypt to modern Spain. Yet the drama has also its encouraging aspects. The process which left certain Chinese villages almost subnormal has helped to give the Chinese community of northern Manchuria an almost American progressiveness. The persistence of a chosen inheritance in Iceland and the high quality of that inheritance after a thousand years are veritable messages of glad tidings. It is inspiring to realize that prolonged selection could cull out pioneers from the rough peoples that invaded England one or two thousand years ago, and later could cull from the descendants of those same pioneers a group of seekers-for-truth such as the famous New Englanders in the Hall of Fame. If such things can happen under Nature's seemingly haphazard regime, why should they not happen ten times more often under the sane direction of the science that is gradually evolving?

Ever since the discovery of America the world has been in the throes of a colossal revolution whereby man has increased his dominion over Nature almost incredibly. This era of material progress seems to be nearing its climax. What more can we expect now that we can travel in the air, the ocean, and the earth as well as on the ocean and the earth? Can we not see and hear that which is happening thousands of miles away? Shall we not soon send power wheresoever we will without pipe, belt, or wire? We are learning to transmute metals, to contend on victorious terms with minute parasites, and to break the atom into its electrical components. Much more that is wonderful is doubtless yet to come, but we may well question whether any material invention for many a long century, not even the harnessing of the power of the sun, will change human life so much as have modern transportation, communication, and manufacturing, modern medicine and chemistry, and the discovery of vast continents untouched by the main streams of progress. So great has been the change that mankind is dazed, perplexed, and seems to be hurrying toward self-destruction.

But while the material revolution thus whirls madly to its climax, we faintly see the beginnings of another and greater revolution. It is the biological revolution. Thus far modern science has scarcely touched man's own self. It has, to be sure, made marvelous strides in healing the sick and preventing disease, and in improving the breeds of domestic animals and plants, but man as a biological species is not appreciably different from what he was at the beginning of civilization. Our aim for ten thousand years has been the mastery of *things* rather than the mastery of *ourselves*. Yet already the revolution is at hand. Whether we will or no, it seems probable that the advanced nations will attempt to control their own physical evolution, and that the rest of the world will follow suit or perish. This attempt is sure to create terrific rivalries, dissensions, revolutions, and even wars; no man can foresee all that will happen. To timid souls the prospect is appalling. To the true pioneer and still more to the seeker-for-truth it is as the cry of battle to the warrior. Behold what natural selection seems to have done in the past, in Greece, in Iceland, in colonial America! Dare we attempt to do the same once more in a grander, better way? And can we succeed? These are the questions which centuries of struggle will answer during the biological revolution whose dawn even now reddens the eastern sky.

LIST OF REFERENCES

Abbott, Evelyn	History of Greece. New York, 1900.
Adams, Brooks	The Emancipation of Massachusetts. Boston, 1919.
Andrews, Charles M.	Preface, The Colonising Activities of the English Puritans. By A. P. Newton. New Haven, 1914.
Backhouse, E., and Bland, J. O. P.	Annals and Memoirs of the Court of Peking. Boston and New York, 1914.
Barrell, Joseph	Probable Relations of Climatic Changes to the Origin of the Tertiary Ape-Man. Scientific Monthly, January, 1917.
Bashford, J. W.	China, an Interpretation. New York, 1919.
Bates, H. W.	The Naturalist on the River Amazon. London, 1892.
Bishop, C. W.	The Geographical Factor in the Development of Chinese Civilization. Geographical Review, vol. 7, 1922.
Bland, J. O. P.	See E. Backhouse.
Boulger, D. C.	A Short History of China. New York, 1891.
Bryce, James	Preface, Denmark and Sweden, with Iceland and Finland. By Jon Stefansson. New York, 1917.
"	Race Sentiment as a Factor in History. London, 1915.
"	Letter quoted in Huntington's Civilization and Climate, pp. 161 ff. New Haven, 1915.
"	Studies in History and Jurisprudence. New York, 1901.
Burton, Richard R.	Ultima Thule, or a Summer in Iceland. London, 1875.
Campbell, George	Origin and Migrations of the Hakkas. Swatow, 1912.
Carr-Saunders, A. M.	The Population Problem, a study of human evolution. Oxford, 1922.
Cattell, J. McK.	Families of American Men of Science, in American Men of Science, third edition, 1921.
Clarke, Edwin L.	American Men of Letters: Their Nature and Nurture. Studies in History, Economics, and Public Law, Columbia Univ., vol. 72, 1916.
Clutton-Brock	Religion Now. Atlantic Monthly, vol. 124, 1919.
"	Problem of Martha. Atlantic Monthly, vol. 126, 1920.
"	Pursuit of Happiness. Atlantic Monthly, vol. 124, 1919.
Conybeare, C. A. V.	The Place of Iceland in the History of European Institutions. London, 1877.
Davies, Geo. R.	A Statistical Study of the Influence of Environment. Quart. Journ., Univ. N. Dak., vol. 4, 1914.
Dixon, R. B.	The Racial History of Man. New York, 1923.
Doughty, C. M.	Travels in Arabia Deserta. London, 1888 (also 1921).

LIST OF REFERENCES

East, Edward M.	Mankind at the Crossroads. New York, 1923.
Edwards, H. J.	Article on Colonies, in A Companion to Greek Studies. Cambridge, 1905.
Eliot, Charles W.	Family Stocks in a Democracy, in American Contributions to Civilization. 1898.
Elliot, G. F. Scott	Prehistoric Man and His Story. London, 1915.
Elliot (Odysseus)	Turkey in Europe. London, 1900.
Ellis, Havelock	A Study of British Genius. London, 1904.
Forbes, Charles S.	Iceland; Its Volcanoes, Geysers, and Glaciers. London, 1860.
Freeman, Edward A.	Articles on the Normans and Sicily, in Encyclopædia Britannica, tenth and eleventh editions.
Galton, Francis	Hereditary Genius, an Inquiry into Its Laws and Consequences. New York, 1870.
Gautier, E. F.	Nomad and Sedentary Folk of Northern Africa. Geographical Review, vol. 6, January, 1921.
Gibbs, Philip H.	Now It Can Be Told. New York and London, 1920.
"	Adventures in Journalism. London, 1923.
Gilfillan, C. S.	The Coldward Course of Progress. Political Science Quarterly, 1920.
Gobineau, Arthur de	The Inequality of Human Races. London, 1915.
Goodwin, John A.	The Pilgrim Republic. Boston, 1895.
Grant, Madison	The Passing of the Great Race. 4th edition. New York, 1923.
Gray, Dr.	See article by Lennox in China Medical Journal, July, 1919.
Grinnell, Geo. B.	The Cheyenne Indians, Their History and Ways of Life. New Haven, 1923.
Hodge, F. W.	Handbook of American Indians North of Mexico. Washington, 1911.
Humphrey, Seth K.	The Racial Prospect. New York, 1920.
Huntington, Ellsworth	See list of books opposite title page.
Icelandic Books of the Ninth Century.	Islandica, vol. 14, 1922.
Keary, C. F.	The Vikings in Western Christendom. 1891.
"	Article on Vikings in Encyclopædia Britannica.
Keith, Arthur	The Antiquity of Man. London, 1915.
Kendrew, W. G.	The Climates of the Continents. Oxford, 1922.
Kneeland, Samuel	An American in Iceland. Boston, 1876.
Kroeber, A. L.	Articles on Indians of California in F. W. Hodge: Handbook of American Indians.
Lambert, R. A.	Post-war Work Among the Armenian Refugees by the Red Cross. Military Surgeon, vol. 49, pp. 314–332. Washington, 1921.
Latourette, K. S.	The Development of China. Boston, 1917.
Laughlin, H. L.	Racial Stocks in a Democracy. New York, 1923.
Leaf, Walter	Homer and History. London, 1915.
Le Bon, Gustave	The Psychology of Peoples, Its Influence on Their Evolution. London, 1899.
Lee, Mabel P.	The Economic History of China, with Special Reference to Agriculture. New York, 1921.

Lennox, Wm. G.	Chinese Vital Statistics. China Medical Journal, July, 1919.
Lowie, Robert H.	Primitive Society. New York, 1920.
Lull, R. S.	Organic Evolution. New York, 1917.
MacDougall, Wm.	Is America Safe for Democracy? New York, 1921.
Matthew, W. D.	Climate and Evolution. New York Academy of Science, 1915.
"	The Arboreal Ancestry of the Mammalia, American Nat., vol. 38, 1904.
Mayor, R. J. G.	Article on Population in A Companion to Greek Studies. Cambridge University Press, 1905.
McCurdy, G. G.	Human Origins, a Manual of Prehistory. New York, 1924.
Mimir Icelandic Institutions, with addresses, 1903.	
Myers, John L.	Greek Lands and the Greek People.
Newton, Arthur P.	The Colonising Activities of the English Puritans. New Haven, 1914.
Nicoll, James	An Historical and Descriptive Account of Iceland, Greenland, and the Faröe Islands. Edinburgh, 1841.
Novakovsky, Stanislaus	Unpublished manuscript on Siberian Hysteria as a Reflex of the Geographic Environment, published in part in Ecology, vol. 5, 1924.
Odin, Alfred	Genèse des Grands Hommes. Gens de Lettres Français Modernes. Paris, 1895.
Osborn, H. F.	Men of the Old Stone Age. New York, 1918.
"	Preface to The Passing of the Great Race, by Madison Grant. New York, 1916.
"	The Pliocene Man of Foxhall in East Anglia. Natural History, November and December, 1921.
"	The Dawn Man of Piltdown, Sussex.
"	Old and New Standards of Pleistocene Division in Relation to the Prehistory of Man in Europe. 1922.
Parker, E. H.	China, Her History, Diplomacy, and Commerce, from the Earliest Times to the Present Day. New York, 1917.
Pettersson, O.	Climatic Variations in Historic and Prehistoric Times. Svenska Hydrogrefisk-Biologiska Kommissioners Skrifter, Heft V. Goteborg, 1914.
Pfeiffer, Ida	A Journey to Iceland and Travels in Sweden and Norway. New York, 1852.
Powell, F. Y.	Article on Iceland, Encyclopædia Britannica, eleventh edition.
Redfield, C. L.	Control of Heredity. Chicago, 1903.
"	Dynamic Evolution. New York, 1914.
"	Great Men and How They Are Produced. Chicago, 1915.
"	Human Heredity. Chicago, 1921.
Ridgeway, Wm.	The Early Age of Greece. Cambridge (Eng.), 1901.

LIST OF REFERENCES

Russell, Bertrand	The Problem of China. London, 1922.
Russell, W. S. C.	Iceland, 1914.
Semple, Ellen C.	Influence of Geographic Environment. New York, 1911.
Sergei, G.	The Mediterranean Race. A Study of the Origin of European Peoples. New York, 1901.
Smith, A. H.	Chinese Characteristics. New York, 1894.
"	Village Life in China. New York, 1899.
Snow, E. C.	The Intensity of Natural Selection in Man. Drapers' Company Research Memoirs. London, 1911.
Stefansson, Jon	Denmark and Sweden, with Iceland and Finland. Preface by James Bryce. London, 1916.
Stefansson, V.	The Friendly Arctic. New York, 1921.
"	The Northward Course of Empire. New York, 1922.
Stoddard, T. Lothrop	The Revolt Against Civilization: the Menace of the Under Man. New York, 1922.
Taylor, Griffith	Climatic Cycles and Evolution. The Geographical Review, December, 1919.
"	The Evolution and Distribution of Race, Culture, and Language. New York, 1921.
Thompson, J. Arthur	Lectures on Evolution at Yale University, 1924.
Tyler, John M.	The New Stone Age in Northern Europe. New York, 1921.
U. S. Census	A Century of Population Growth in the United States, 1790–1900. Washington, 1909.
Vanden Bergh, L. J.	On the Trail of the Pigmies. New York, 1921.
Venn, John	Cambridge Anthropometry. Journal Anthropol. Inst., vol. 18, 1889.
Visher, S. S.	A Study of the Type of Place of Birth and Occupations of Fathers of Persons in Who's Who in America. 1924–1925.
Waller, S. E.	Six Weeks in the Saddle. London, 1874.
Ward, Lester F.	Applied Sociology. A Treatise on the Conscious Improvement of Society by Society. Boston, 1906.
Wetzel	Quoted by Elliott, G. F.
Whitney, Harry	Hunting with the Eskimo. New York, 1910.
Williams, F. W.	A History of China. New York, 1901.
"	The Manchu Conquest of China. The Journal of Race Development, vol. 4, no. 2, 1913.
"	Unpublished manuscript on China.
Willis, J. C.	Age and Area. A Study in Geographical Distribution and Origin of Species. Cambridge (Eng.), 1922
Wissler, Clarke	Man and Culture. New York, 1923.

INDEX

A

Abbott, E., cited, 245, 374
Abdul Hamid, 145
Aboriginal America, anomalies of, 156 ff.
Achæans, 236, 244; migrations of, 240 ff.
Acquired characteristics, inheritance of, 365 ff.
Acquisitiveness, racial trait, 215
Adams, B., cited, 324, 374
Adams family, 324, 367 ff.
Adams, Henry, 324
Adams, John, 325
Adams, J. F., 324
Adams, J. Q., 325
Adaptability, of Greeks, 240 f.; 267 f.; of Norse, 267
Adaptation to environment, 34
Aegeans, 236, 238
Aerial warfare, 339
Afghans, raids among, 120
Africa, aborigines, 73; migrations in, 21, 59; Norse in, 254; pastoralism in, 112; Proto-Australoids in, 89; Proto-Negroids in, 89
Agassiz, L., 325
Age, and area, 80; of fathers, 365 ff.
Agriculture, in Arizona, 108; Icelandic, 274 f., 283; origin of, 60; primitive American, 92
Aigun, 200
Air Service, 329
Alaska, 62; climate of, 63; Indians of, 53; insanity in, 67
Alcæus, 241
Alcott, L. M., 324
Aleppo, 140
Aleutian Islands, 102
Algeria, nomads of, 123; pirates of, 297
Algonkians, 90, 91 ff.
Alikaluf, 89 f., 92
Alpine race, 83 f., 207, 210 ff.; achievements of, 231; in Crete, 237 f.; defined, 5; genius of, 228; racial position of, 77
Althing, 291 f.
Amazon River, 71
America, causes of greatness, 308 ff.; contrast with Europe, 161; early culture of, 61; early immigrants, 61; eminent men in, 349; migrations in, 77; natural selection in, 301 ff.; political leaders in, 227; racial character in, 90; racial impoverishment in, 364; scientists in, 342; suppression of, 61 ff.
American Indians, origin of, 63 ff.
American Museum, 33, 35
American Revolution, 306
Americans, character of, 330 f.; selection of modern, 301 ff.
Amoy, vehicles in, 164
Anadyr Peninsula, glaciation in, 55
Andrews, C. M., 302, 374
Angkor Tom, 11
Angkor Wat, 10 ff.
Angles, 248, 265, 303
Anhwei, famines in, 177, 182
Animals, during famines, 180; in glacial Europe, 51; origin of domestic, 21
Anomalies of aboriginal America, 88 ff.
Antarctica, winds of, 48
Anthropoids, as human ancestors, 27
Anthropology, generalizations of, 74
Ape-man, 31; of Java, 24
Arabia, migrations in, 59; nomadism in, 114 ff.; deserts of, 57
Arabs, character of, 115 f.; leadership among, 153; migrations of, 208; in Sicily, 266
Aral, Sea of, 57
Arcadia, 243
Architecture, Norman, 265 ff.; relation to selection, 18
Arctic hysteria, 68 ff.
Arctic problem, 67
Area, and age, 81
Aridity, in Central Asia, 27; and Chinese migrations, 149; and human dispersal, 24; and natural selection, 44
Aristocracy, in England, 250; in Greece, 245
Aristotle, 246
Arizona, 92; agriculture in, 108
Arkansas, eminent men in, 317
Armenians, character of, 137 ff., 146; deportations of, 139 ff.; physique of, 140; migrations of, 208

Armor, effect on natural selection, 338
Art, in Belgium, 225 f.; Bohemia, 225 f.; Cambodia, 11; Europe, 223; Holland, 244 f.; Italy, 226; among nomads, 125 f. of Normans, 265 f.
Artists in Europe, map, 223
Aryans, 5
Asia, aridity in, 42; centre of evolution, 23 f.; contrast with Europe, 53; deserts of, 63; glacial period in, 52 ff.; migrations from, 77; original human home, 20 ff.; nomads in, 112 ff.; original home of man, 20 ff., 81
Asia Minor, 54; Greeks in, 241 ff.; migrations in, 59
Asiatic languages, in America, 62
Associated charities, and natural selection, 336
Athelred II, 292
Athens, natural selection in, 244 ff. See Attica
Atlantic Drift, 274
Atlantis, 61
Attica, serfs in, 237; settlement of, 242 ff.
Australia, aborigines, 73; early migrants in, 21; Norse type in, 266
Australoids, in South Africa, 85
Austria, achievements of, 231; political leaders in, 227; birth rate in, 354
Aural index, 77
Aviators, 329
Aztecs, 88, 92
Azubah, 129

B

"Babbitt," 331
Babylonia, deportations to, 134 ff.
Bacon, Sir Francis, 301
Backhouse, E., 156, 374
Balkans, 227
Baltic Sea, 233
Bancroft, G., 325
Banditry, in China, 175, 193
Bantus, 85 f.
Baptism, in Iceland, 289
Baptists, eminence among, 319, 321 ff.; natural selection among, 322
Barrell, J., 27, 29, 374
Bashford, J. W., 374
Bates, H. W., 24, 71, 374
Bathing, among Chinese, 168; relation to climate, 289
Beans, 92
Beduin, 114 ff.
Beggars, in China, 162, 164, 175

Belgium, 254; achievements of, 231; art in, 225; political leaders in, 227; scientists from, 312
Beni Sakr, 115
Benjamin, tribe of, 131
Berbers, 123
Bering Island, 102
Bering Strait, 62 f.; glaciation near, 55
Bermaz, 137
Bey Shehr Lake, 142 f.
Bible, 129 ff.
Biographies, 235
Birthplaces of eminent persons (maps), 223, 225, 227, 229, 286
Birthrate, American, 310 f., 354, 356; urban, 194, 346, 353; restriction of, 365, 370
Bishop, C. W., 374
Bland, J. O. P., 156, 374
Block, cited, 281
Blondness, in Greece, 238, 246
Boas, F., 74
Bœotia, 243
Bohemia, achievements of, 231; art in, 225; political leaders in, 227
Boomerang (map), 79
Boulger, cited, 196, 374
Brahmaputra River, 14, 26
Brain, evolution of, 30, 50; size of, 31
Breidifjord, 291
Brides, in China, 190
Brigandage, in China, 175, 193
Brinton, D. G., 71
British, character of, 214 f.; contact with China, 200; as settlers in U. S., 65; See also England and English
British Charitable Hospital, 194
British Isles, in Britannica, 230
Broad-heads, location, 77
Brothers, relative ability of, 365 ff.
Brownists, 322
Bryant, W. C., 325
Bryce, James, 2 ff., 269, 278, 291, 374
Brythons, 303
Burmah, 15
Burton, R. R., 292 ff., 374
Bushmen, 85 f.
Byzantine Emperors, 354

C

Caleb, 129
Calendar, Maya, 63
California, 88; agriculture in, 107 f.; genius in, 316; gold in, 328; human relics in, 61; Indians of, 100, 103 ff.;

INDEX

leadership in, 327; natural selection in, 329
California Institute of Technology, 327
Cambodia, ruins in, 10 ff.
Cambridge University, cephalic index at, 31
Campbell, G., 168, 195 f., 197 f., 374
Canada, ice in, 63; scientists from, 312
Canals, in Shantung, 171
Cannibalism, in China, 177 ff., 188 f.
Canton, 199, 202; first Chinese in, 149; Hakkas in, 167
Cantonese, character of, 160
Cape Horn, 91
Caravan roads, importance of, 131
Caribbean Sea, colonization schemes in, 302
Carr-Saunders, 362, 374
Caspian race, Dixon's, 83 f., 89, 207, 209, 257
Caspian-Mediterranean blend, 90, 92
Cathay, 152
Catholics, eminence among, 320. *See* Roman Catholic Church
Cattell, J. McK., 342, 346, 358 f., 374
Caucasus, migrations in, 59
Celibacy, 340 ff.
Central America, civilization in, 61; ruins in, 11
Central Asia, decline of glacial period in, 58; migrations in, 57; upheaval of, 27
Central Europe, migrations in, 77
Cephalic Index, 77 (map); 79 (map). *See* Head form, Broad-heads, Long-heads, and Round-heads
Chang La, 26
Channing, W. E., 325
Chaplicka, quoted, 70
Character, of Americans, 333; of Dutch, 333; of English, 333; and environment, 286 ff.; geographical distribution of, 215 ff.; of Germans, 333; of Indians, 65; of Turks, 333
Charity, among Chinese, 192; and natural selection, 336
Charlemagne, 258 f., 261
Chehsi, 182
Chekiang, 180, 182; famines in, 196; Hakkas in, 196; officials from, 161
Chemists, children of, 358
Chi, 178
Chicago, age of fathers in, 365
Chihli, famines in, 177 f.; migrations from, 200 ff.; officials from, 161
Children, 357; effect of migrations on, 13, 304; per family, 194, 353 ff.; mortality of, in Crete, 238; sale of, in China, 177, 187 f.
Ch'in, 148
China, early migrations in, 21, 58; European contact with, 164 f.; famines in, 170 ff.; history of, 148, 179, 203 ff.; leaders in, 155, 160 f.; progress in, 60; rainfall in, 170 ff.; selection in, 362; social origins in, 60
Chinese, vs. Americans, 337; brains of, 78; head form of, 77; as invaders, 148; in Manchuria, 200; natural selection among, 184 ff.; officials (map), 161; physique of, 185, 333 f.; warlike quality of, 155
Chinese Bannermen, 161
Chinook Indians, 103
Chios, 242
Chosen, 58
Christianity, in Iceland, 269, 289, 292 f.; origin of, 133 f.
Chronicles, book of, 130
Chukchees, 53; hysteria among, 70; suicide among, 69
Churchill family, 367
Cities, birth rate in, 346; effect on selection, 193 f.; leadership in, 346 ff., 348, 352; Mediterranean, 213; test of, 346 ff.
Cityman, 7, 371
Civilization, defined, 228; European, 335; and glaciation, 332 ff.; localization of, 59; map, 225; origin, 60; racial tendencies of, 332 ff.
Clapham, 78
Clarke, E. L., 309 ff., 349, 374
Clarke, J. F., 324
Class distinctions, in America, 307
Clay, as food, 179
Cleanliness, relation to climate, 289
Cleomenes, 246
Clergymen, children of, 318 ff., 321, 342 ff.
Climate, of California, 328 f.; China, 164, 170; Greece, 247; Iceland, 273; Ireland, 248 ff.; North Sea region, 252 f.; Norway, 219, 268; Sicily, 219, 268
Climate, effects of, on civilization, 88; in cold regions, 65; evolution, 34; famines, 183; health, 289 f.; human character, 92, 288 ff.; human origins, 22; leadership, 329; migrations, 349
Climatic changes, 55, 206; in Iceland, 293 ff.; and migrations, 205 ff.; in Norway, 260 ff.
Climatic cycles, 47
Climatic energy, 88; map, 232
Clothing, 21

Clutton-Brock, 3, 374
Coffee, among Arabs, 116
Coleman, R. V., 174
Colonists, qualities of, 215, 217
Colorado, leaders in, 316
Colorado River, Indians on, 107 f.
Como, persecutions at, 342
Communism, and natural selection, 215
Complexion, 78
Coneybeare, C. A. V., 259, 268, 374
Confucius, temple of, 164
Congregationalists, eminence among, 319; natural selection among, 321 ff.
Connecticut, eminence in, 315 f.
Conservatism, in China, 177
Constantinople, 254
Constitution, of Iceland, 291; of U. S., 306
Continents, distribution of headform in, 75; evolution in, 34
Cooper, Peter, 325
Copper Island, 102
Coprolite jaw, 42
Corn, 92
Cossacks, 69
Couvade (map), 79
Cranial capacity, 77
Cranial types, among women, 84
Crete, civilization of, 239 ff.; settlers in, 237 ff.
Cross, sign of, 137 f.
Cultural conditions, 79 (map)
Curiosity, racial trait, 211 f.
Cushman, C., 325
Customs, racial, 79 (map)
Cycles, in history, 148, 203
Cyclonic storms, 47
Czechs, migrations of, 77

D

Dakotas, eminence in, 315 ff.
Dana, R. H., 324
Danes, 248, 326; in England, 255, 265; among Vikings, 303
Daniel, Book of, 134
Davies, G. R., 348 f., 374
Dead Sea, 57
Death rate, in cities, 194; in Europe, 279 f.; in famines, 184; in Iceland, 279 f., 286 (map); among infants, 358
Deformation, of heads, 79 (map)
Delhi, Prince of, 12
Demetrius Phalerius, 246
Democracy, 346 ff.; among Arabs 125, biological effect of, 355 ff.

Demosthenes, 245
Denmark, 55; political leaders in, 227
Denominations, natural selection among, 321 ff.
Deportation, of Armenians, 139 ff.
Deserts, effect of, on evolution, 58, 73; map, 48, migration, 206; racial character, 56
Deterioration, of races, 18
Diagrams. *See* Maps
Diet, in Iceland, 300. *See* Food
District of Columbia, eminent persons in, 315
Divorce, in Europe, 213; effect on racial inheritance, 361
Dix, Dorothy, 324
Dixon, R. B., 75 f., 77 f., 81, 85, 88 ff., 102, 166, 209, 212, 234, 256 f., 374
Dnieper River, 354
Docility, racial trait, 214
Domestic animals, and civilization, 112; in Iceland, 275
Domestic plants, origin of, 21
Door court, 292
Dorians, 240 ff., 244
Doughty, C. M., 121, 374
Drake, 301
Drought, in Arabia, 121 ff.; in China, 154, 170 ff.; as cause of famines, 176 ff.; effect on raids, 117 f.; in Sicily, 218
Dublin, 254
Duckworth, 78
Dutch, character of, 333
Dyradomr, 292

E

Earthquakes, in Iceland, 271, 294, 298
East, E. M., 334, 362, 375
East Anglia, 326; home of Puritans, 302
Eastern Tsin, 195
Economy, relation to famines, 185 ff.
Eddas, 256 ff.
Education, in America, 306; in Europe, 223; effect on inheritance, 314, 320; effect on leaders, 349; effect on selection, 344; leaders in, 223 (map)
Edwards, H. J., 244, 375
Egypt, 54; domestic animals in, 113; relation to America, 61; migrations to, 59
Eirik Eymundsson, 259
Elbe, Norse at, 254
Eliot, C. W., 324, 346, 359, 375
Elizabeth, Queen, 301
Elliot, G. F. L., 27, 375
Elliot (Odysseus), 144, 375

INDEX

Ellis, H., 302 *f.*, 365, 375
Emerson family, origin of, 303
Emerson, R. W., 325
Eminence, distribution of, 228, 303, 309, 312, 314, 347 *ff.*
Emotional selection, 322 *f.*
Empress Dowager, 151
Encyclopædia Britannica, 220 *ff.*, 230, 235, 277, 279
Energy, and genius, 233
Engineers, eminent children of, 319
England, character of, 222, 333; genius in, 220 *ff.*, 302 *f.*; immigration to, 248, 342; versus Ireland, 248 *f.*; migrations from, 250; Normans in, 265 *f.*; Norse in, 261 *f.*, 264; persecutions in, 342; scientists from in U. S., 312; in sixteenth century, 301 *f.* See British, and English
English, in Iceland, 295 *ff.*; migrations of, 77. See British, and England.
Environment, adaptation to, 34; and character, 286 *ff.*; and genius, 228; in Iceland, 299; and inheritance, 309 *ff.*, 320; and leadership, 348 *ff.*; potency of, 92; and social position, 78
Ephesus, 242
Epidemics, in Iceland, 294
Episcopalians, 321 *ff.*; children of, 319
Erna, 257
Esau, 129
Eskimos, 53, 67, 89 *f.*, 92; cleanliness among, 289
Esther, Book of, 134
Ethelbert, 259
Euphrates River, 120
Europe, celibacy in, 341; character of, 220 *ff.*; in China, 165; center of civilization, 60; contrast with America, 88; contrast with Asia, 53; glacial period in, 49, 51 *f.*, 56, 58; evolution in, 336 *ff.*; land connections of, 42; leadership in, 234; migrations in, 206 *ff.*; races of (map), 76, 77; type of civilization, 325
Europeans, classification of, 235; in tropics, 51
Evangelists Island, 91
Evolution, centers of, 33; human, 36 *ff.*; mental, 332 *ff.*
Examinations, Chinese, 152, 161
Extroverts, 213

F

Factories, effect on character, 337
Family, in China, 188 *f.*; rural *vs.* urban, 355; size of, 310, 354, 357, 366 *ff.*; size according to occupation of fathers, 357; status of, 360
Famine, in Chekiang, 196; in China, 154, 156, 170 *ff.*, 176 *ff.*; deaths in, 184; in England, 295, 297; in Iceland, 270, 274, 295, 297; in Ireland, 248 *f.*; and migration, 199; relief of, 189
Farmers, birthrate among, 353 *ff.*; chronic depression of, 352; leaders among, 347, 352
Fathers, of eminent men, 357; age of, 365 *ff.* See also Eminence
Feet, of Chinese women, 162 *f.*; of Hakka women, 168
Feminism, 346 *ff.*, 360 *ff.*
Fidelity, among raiders, 127
Filial piety, 197
Finland, health in, 23; literature in, 226; Norse in, 255
Fishermen, birth rate among, 353 *ff.*; death rate among, 286 *f.*; among Haidas, 101; in Iceland, 298
Flemings, as migrants, 303, 342
Floods, in China, 170 *ff.*, 176 *ff.*; in Iceland, 271
Florida, eminent persons in, 316, 318; human relics in, 61
Food, among Arabs, 116, 120; in China, 179, 184; in Iceland, 300; of primitive man, 28; of shepherds, 116
Forbes, C. S., 375
Foreign-born, in U. S., 353
Forests, Miocene, 28
Formosa, Chinese in, 150
Forty-niners, 328 *f.*
Foxhall man, 42
France, migrants in, 77, 303; Norse in, 254, 261, 264; persecutions in, 342
Franklin, Benjamin, 325
Freeman, E. A., 265 *ff.*, 375
French, in Iceland, 297
French Indo-China, ruins in, 10 *ff.*
Friends, eminent people among, 319
Fró á, 291 *f.*
Fuchow, 202; migrants from, 196; vehicles in, 164
Fukien, 198; Hakkas in, 196
Furness, H. H., 324
Furniture, among Arabs, 124

G

Gaelic belt, map, 79
Galilee, 131 *f.*
Galton, F., 341 *f.*, 365, 368, 375
Ganges River, 14

INDEX

Garonne, Norse in, 254
Gautier, cited, 123, 375
Genius, distribution in England, 303; in Europe, 233; maps of, 223, 229, 286, 316. *See* also Eminence and Leaders
Geographical distribution, of character, 215 *ff*
Germans, birth rate of, 354; character of, 214, 222, 333; head form of, 77; migrations of, 77; political leaders in, 227; relation to Turkey, 142
Gibbs, Philip, 266, 330, 375
Gilfillan, S. C., 375
Girls, in China, 162 *f.*, 187 *f.*
Glacial period, 38, 47 *ff.*, 58, 63; causes, 47; effect on evolution, 49, 332 *ff.*; in Europe, 53, 205 *f.*
Glaciation, map of, 48
Gobineau, A. de, 1, 375
Goodwin, J. A., 375
Gorm, 259
Government, relation to famines, 177 *f.*, 180 *ff.*; among nomads, 115 *f.*
Grand Canal, 171
Grant, M., 1, 375
Grass, distribution of, 37; as food, 176
Gray, Dr., cited, 194, 375
Greece, achievements of, 231; advantages of, 60; population of, 246
Greeks, character of, 236 *ff.*; migrations of, 208; racial feeling of, 2; science among, 212; in Sicily, 266
Greenland, hysteria Arctica in, 69; winds of, 48
Grimsey Island, 280
Grinnell, G. B., 375
Gulf Stream, 274
Gunnlaug Ormstunga, 292

H

Haidas, character of, 109; civilization of, 99 *f.*; origin of, 102
Hainan, settlement in, 150
Hair, as racial trait, 77 *f.*; loss of, 21
Hakkas, 167, 194 *ff.*
Hale, E. E., 324
Hall of Fame, 324 *f.*
Han dynasty, 177
Hands, evolution of, 29 *ff.*
Hangchow, 182
Harbin, 165, 200
Harems, 140, 145
Harold Fairhair, 258 *ff.*, 262
Harvard graduates, children of, 358 *f.*
Hawes, cited, 237

Hawthorne, N., 325
Head, deformation of, 79 (map); evolution of, 29 *ff.*, 75; form of, 62, 74, *ff.*, 77 (map), 228
Head-form, criterion of race, 75; inheritance of, 85; relation to genius, 228
Health, in Iceland, 300; map of, 225, 231 *ff.*; and progress, 88
Heber, 130
Hebrews. *See* Jews
Hecate, 101
Hecatæus, 237, 244
Helga, the Fair, 292
Hellas, settlement of, 242 *ff.*
Herd instinct, 212, 214
Herding, 112. *See* Nomads, Nomadism.
Herodotus, 246
Hezron, 129
Higginson, T. W., 324
Himalayas, effect on migrations, 59; in Miocene period, 25
Hing-Ning, 168
Historians, in Europe, 223 (map)
Hiung-nu, 148, 202
Hoang-Ho, 174, 181
Hodge, F. W., 107, 375
Hoklos, 196 *f.*, 203
Holland, 255; art in, 225; migrations into, 303; Norse in, 261; political leaders in, 227; Puritans in, 304; scientists from, 312
Holmes, O. W., 325
Homer, 237, 239, 241
Homicide, in Europe, 213 *f.*
Honan, famines in, 176 *f.*; migration from, 196–198
Hongkong, 200
Hopeh, famines in, 177
Horses, among Arabs, 124; evolution of, 35 *f.*
Hospitality, among nomads, 126
Hottentots, 85 *f.*
Houses, in China, 172
Howeitat Arabs, 117
Hrafn, 292
Hsu, 178
Hu, 180
Huguenots, 222, 248, 342; in America, 305
Human character, first steps in, 20 *ff.*
Human evolution, abnormal environment of, 36 *ff.*
Human remains, 39, 61, 64
Humidity, 23
Humphreys, S. K., 346, 375
Hungarians, birth-rate of, 354

INDEX

Hungary, achievements of, 231; literature in, 226
Huns, invasions of, 195
Huntington family, 367 ff.
Hupa Indians, 103
Hupeh, famines in, 177
Hutchinson, Mrs. Anne, 322
Hysteria Arctica, 68

I

Ibn Khaldun, 123
Ice Age, 63. *See* Glacial Period
Iceland, 264 ff.; agriculture in, 274 f., 283, 294; architecture in, 295 f.; books in, 284; changes in character in, 284; Christianity in, 289; cleanliness in, 289; climate of, 273 ff., 289 ff., 300; compared with New Hampshire, 280; constitution of, 291; death rate in, 279 f., 286; diet in, 300; domestic animals in, 275; earthquakes in, 294, 298; education in, 280 f.; eminent men in, 279; environment of, 270 ff., 300; epidemics in, 274, 295; fisheries in, 275 f., 284, 298; foreigners in, 272; health in, 300; illegitimacy in, 281; isolation of, 272; library references to, 277; literature of, 226, 278 ff.; migrations to, 211 f.; mode of life, 282 f.; natural selection in, 284, 299; Norse in, 254 ff., 262; Norwegian relations, 294 ff.; pirates in, 297; printing in, 279; population of, 270; racial character in, 299; Reformation in, 296 f.; religion of, 272; ship-building in, 296; selection of settlers, 268; smallpox in, 297; summary of history, 298 ff.; syphilis in, 281 f.; trade of, 295 ff.; trees in, 274, 294; temperature of, 91; volcanic eruptions in, 294, 297 f.; witches in, 297
Iceland moss, 283
Iceland spar, 272
Illegitimacy, in Iceland, 281
Illinois, eminent men in, 316
Immigrants in U. S., character of, 309 ff.; factory work of, 353; head-form of, 74
Imperial clan, 152
Implements, racial relations of, 79 (map)
Incas, 88, 92, 273
India, 205; migrations in, 12, 59; mountains of, 26; racial origins in, 60
Indians (American), character of, 66, 71, 88; natural selection among, 68; of South America, 24
Individualism, racial trait, 211 ff.

Indo-China, Chinese in, 150; ruins in, 10 ff.
Indo-European race, 5
Indus River, 26
Industry, and birth rate, 353 ff.
Inheritance, and achievement, 352; of acquired characteristics, 365 ff.; Athenian laws of, 245; *vs.* education, 314, 320; *vs.* environment, 309 ff., 320; of headform, 85; persistence of, 277 f.
Intellectual awakening, in England, 301 ff.
Intellectual selection, 322 f.
Interest, rate of, in China, 180
Introversion, racial trait, 213 f.
Invasions, of China, 183, 195; and migration, 199
Inventions, and birth rate, 357
Ionian migrations, 225
Ireland, eminent persons in, 225; land tenure in, 181; library references to, 277; migrations from, 77; Norse in, 254 f., 264, 261; political leaders in, 227
Irish, character of, 215, 236, 247 ff.; in Iceland, 254
Iroquois, 90 ff., 109
Irrawadi River, 15
Irrigation, primitive, 107 f.; in Turkey, 142
Ishmael, 130
Isolation, of Iceland, 271; factor in religion, 133 f.
Israelites. *See* Jews
Italy, birth-rate in, 354; celibacy in, 344; land tenure in, 177, 181; Normans in, 265 f.; Norse in, 254 f.; persecutions in, 342

J

Jackson, H. H., 324
Jacob, 129
Jael, 130
Japan, 205; effect of ice age in, 58
Japanese, adaptability of, 267; scientists in U. S., 312
Jarls, among Norse, 256 f.
Java, apeman of, 39; Chinese in, 150
Jaw, evolution of, 30
Jebel Druze, 118
Jerahmeel, 129
Jerioth, 129
Jesus, 132 f.
Jewett, S. O., 324
Jews, character of, 129 ff.; head-form of, 74; natural selection among, 135 ff.;

386 INDEX

racial feeling of, 2 f.; scientists among, in U. S., 312
Job, 129
Jochelson, 69
Jordan, D. S., 324
Joseph, 129
Judah, Tribe of, 131
Judaism, 133 f.
Judea, isolation of, 130, 132
Judges, Book of, 130
Junkers, 214

K

Kaffirs, origin of, 86
Kalahari, 87
Kansas River, floods of, 174
Kansu, officials from, 161
Karakorum Pass, 26
Karls, among Norse, 256 f.
Katla, volcano, 271
Kautso, 177
Keary, C. F., 267, 375
Keith, A., 375
Kelts, 248
Kendrew, W. C., 91, 375
Kent, migrants in, 303
Khirghiz, 127; agriculture of, 109; cleanliness among, 289
Kiang Ling, 177
Kiang River, 177
Kiansi, 198; famines in, 177; Hakkas in, 196
Kiangsu, 180; famines in, 177
Kiang Wei, 177, 180
Kia-Yian, 168
Kin words, 79 (map)
Kings, Book of, 130
Kingship, in Europe, 259
Kitans, 152
Kjartan, 291
Kneeland, S., 375
Konia, 142
Koryaks, hysteria among, 70
Kroeber, A. L., 103, 375
Kublai Khan, 151
Kuczynski, 359
Kurds, Armenian origin of, 137 f., 140; raids among, 120
Kwan Chung, 178
Kwang Tung, 177
Kwantung, 198

L

Labor battalion, 329
Laki, volcano, 270
Lambert, R. S., 140, 375

Lamuts, hysteria among, 70
Land, Norse ownership of, 256, 258 f.; tenure in China, 177, 181 f.; value in China, 171
Landnamabok, 294
Language, relation to race, 79
Laos, 196
Lapps, 53; cleanliness among, 289
Latin countries, celibacy in, 344
Latitude, and progress, 158
Latourette, K. S., 375
Law, enforcement in Iceland, 291 f.
Lawyers, eminent children of, 319, 342
Laziness, among nomads, 128; among Turks, 144
Leaders, among Jews, 135 f.; among nomads, 126 ff.; among Turks vs. Armenians, 143, 145
Leaf, W., 236, 239 f., 375
Le Bon, G., 3, 375
Lee, Miss M. P., 175 ff., 195, 375
Leland Stanford University, 327
Lennox, W. G., 194, 376
Levirate, 79 (map)
Lick Observatory, 327
Limbs, of Pygmies, 78
Lincoln family, 367 ff.
Linguistic stocks, in America, 62
Literacy, among Hakkas, 168
Literary men, map, 223
Literature, pursuit of, in America, 309 f., 349 ff.; in Europe, 223; in Iceland, 226; in Portugal, 225 f.; in Spain, 226
Li Tsuchung, 154
Lodge, H. C., 324
Loire River, Norse at, 254
Longfellow, H. W., 325
Long-heads, location, 76
Los Angeles, 328 f.; skull at, 61, 64
Lop Nor, 57
Lot, 130
Low Countries, 264
Lowell, J. R., 325
Lower California, Indians of, 104 f.
Lowie, R. H., 376
Lumbermen, birth-rate among, 353 ff.
Lull, R. S., 376
Lyman family, 367

M

MacDougall, W., 209, 211, 267, 376
Maichan, 168
Maidu Indians, 104
"Main St." 331
Malaria, 219, 246 f.; in Crete, 238; in Sicily, 268

INDEX

Malay Peninsula, Chinese in, 150
Malays, in China, 166; cleanliness of, 289
Malnutrition, in China, 185
Mamelukes, and Mongols, 148
Mammals, origin of, 33
Man, origin in Asia, 20
Manchu Dynasty, 150 *ff.*
Manchuria, Chinese in, 200; invasion of, 148; progress in, 165, 169; rainfall of, 170 *ff.*
Manchus, 183, 203; in China, 148, 166; leadership among, 153; literary degrees of, 161
Mann, Horace, 325
Manufacturing, and birth-rate, 353 *ff.*, 357; and character, 272; and leadership, 347
Maps and diagrams: artists in Europe, 223; birthplaces, of Chinese officials, 161; of eminent Americans, 286; of eminent Europeans, 223, 229; boomerang, 79; cephalic index, 76, 77, 79; civilization, 229; climatic energy, 229, 232; couvade, 79; cultural conditions, 79; customs, 79; death-rates in Iceland, Norway, and Switzerland, 286; deserts, 48; educational leaders, 223; European races, 76, 77; Gaelic belt, 79; genius, in America, 286, 314, 316, 317; in Europe, 223, 229; glaciation, 48; head deformation, 79; health, 231; historians, 223; kin words, 79; levirate, 79; literary men, 223; migrations, 76, 77; military leaders, 229; monoliths, 79; philosophers, 223; political leaders, 229; pressure zone, 48; rainfall, 231; races of Europe, 76, 77; religious leaders, 223; residence of eminent Americans, 316; scientists, 223; snake and sun cults, 79; swan-maiden tales, 79; tattooing, 79; tree growth, 231
Marco Polo, 152
Marriage, among leaders, 358; among nomads, 127; postponement of, 343, 365, 368 *f.*; among teachers, 344
Massachusetts, children in, 359; eminent persons in, 515 *ff.*
Massacres, of Armenians, 137 *ff.*; of Jews, 135 *f.*
Matthew, W. D., 21, 23, 33, 376
Mayas, 92; cleanliness of, 289; hieroglyphics of, 63; ruins of, 11
Mayflower, 304
Mayor, R. J. G., 286, 376
McCurdy, G. G., 376

Meat, among nomads, 121
Mechanical inventions, and human character, 337
Mediterranean-Caspian blend, 234
Mediterranean race, 5, 83, 89, 207 *f.*, 237 *f.*; character of, 209 *ff.*; genius among, 228; in Ireland, 247
Mediterranean region, 60, 205; history of, 237
Mei-chau, 198
Menam River, 15
Mental evolution, 332 *ff.*
Mental selection, in war, 338
Mentality, of apeman, 31
Merchants, birth-rate among, 354; leaders among, 347
Mesopotamia, 54; civilization in, 59; Jews in, 134; migrations in, 59
Methodists, eminent children among, 319; natural selection among, 321 *ff.*
Mexico, civilization in, 61; land tenure in, 182; library references to, 277
Middle Kingdom, 153
Middle States, birth-rate in, 310
Migrations, to America, 62 *ff.*; in China, 148 *ff.*; concentration in West Asia, 60; in Europe, 206 *ff.*; and glaciation, 58; hardships of, 54; maps of, 76 *ff.*; and natural selection, 12, 18, 151, 203, 303 *f.*; nature of, 65, 241; primitive, 20, 33; tropical, 58
Miletos, 242
Military leaders, 223 (map), 227
Milk, among Arabs, 121
Mimir, 376
Miners, birth-rate among, 353 *ff.*, 357
Ming dynasty, 150
Ministers, eminent sons of, 326, 342 *ff.*
Minoans, in Greece, 239 *ff.*, 244
Minotaur, 239
Miocene period, 25, 37
Misgovernment, causes of, 154
Missionaries, 192, 199; children of, 343; in China, 187 *ff.*, 189; wide knowledge of, 168, 175
Mississippi, eminent persons in, 318
Mitchell, Maria, 325
Mixture of races, 19
Moab, 114; raids in, 131
Mohammedanism, converts to, 137 *f.*
Mohave Indians, 107 *f.*
Mongol dynasty, 150 *f.*
Mongolia, after glacial epoch, 58
Mongoloid race, 83 *f.*, 209; in South Africa, 85

Mongols, in China, 148, 166, 197; leadership among, 153; literary degrees among, 161; social position of, 77
Monoliths, 79 (map)
Montana, eminent persons in, 317
Moquelumne Indians, 104
Moral code, among Arabs, 122; in Iceland, 211 f.
Morocco, 54
Mortality, among children of Crete, 238; in Europe, 232
Motley, J. L., 325
Mount Wilson Observatory, 327
Mousterian people, 53
Mukden, 165, 200
Mutations, 6, 78
Mycenæ, 236; art of, 240
Myers, J. L., 237 f., 246, 376

N

Nanking, 182, 195
Napoleon, 214
Natural selection, 1 ff.; in America, 301 ff.; and acquisitiveness, 215; in California, 328; among Chinese, 184 ff.; in Europe, 253; in Greece, 246 f.; in Iceland, 299; among Indians, 68; in Ireland, 249 f.; among Jews, 133 f.; among Manchus, 152; among nomads, 123; among Pilgrims, 303; and racial position, 78; among religious bodies 321 ff.; in tropics, 50. See Selection
Nature vs. nurture, 309 ff.
Negrito, racial position of, 77
Negro, 24, 218; optimum temperature for, 23; racial position of, 77
Negroids, in South Africa, 85
Nehemiah, Book of, 130
Nervous diseases, in Europe, 213
Nevada, eminent persons in, 316
New England, age of fathers in, 365; birth-rate in, 310; colonization of, 302 ff.; eminent persons in, 315 ff.; racial stock of, 343
New Hampshire, compared with Iceland, 276
New Mexico, agriculture in, 92, 108; eminent persons in, 315; people of, 92
Newton, A. P., 302, 376
New York City, eminent persons in, 316 f.; head-form of immigrants, 74
New Zealand, distribution of plants in, 81
Nicoll, J., 298, 376
Ning-hua, 198
Nomads, and agriculture, 109; Asiatic, 112 ff. See also Arabs, Khirghiz.

Nomadism, effect on character, 144
Nordics, 5, 77, 90, 92; brains of, 78; character of, 207, 210 ff.; genius among, 228; hair of, 78; leadership among, 153; racial composition of, 83
Norman Conquest, 255, 264 f.
Normandy, Norse in, 254, 264
Normans, 248, 264 ff., 326; character of, 265 f.; in England, 265; as migrants, 303
Norse, 253 ff.; character of, 216, 260, 267 f.; in England, 265; migrations of, 258 ff.
North Africa, early civilization in, 59; glacial period in, 58; racial origins in, 60
North America, early migrations to, 21; ice in, 63
Northmen. See Norse and Vikings
North Sea, advantages of, 60; cultural center, 233, 252 ff.
North vs. South, in China, 158, 185 f., 203
Norway, death-rates in, 287; environment of, 216; and Iceland, 294 ff.; political leaders in, 227
Norwegians, in England, 255
Nose, measurement of, 82
Novakovsky, S., 66, 68, 70, 376
Novgorod, 254
Nuchins, 152
Nurhatchu, 153, 155
Nurture vs. nature, 309 ff.

O

Occupations, of fathers of notables, 318, 357
Odin, A., 320, 376
Odinism, in Iceland, 293
Oeræfa Jäkul, volcano, 270
Officials in China, 186; birth-places of, 161; degrees of, 152
Officers, protection of, 339
Oklahoma, eminent persons in, 316
Olafr Tryggvason, 292
Optimum climate, 22
Orbital indices, 77
Oregon, 328; eminent persons in, 317; Indians, 103
Osborn, H. F., 1, 33, 376
Ottoman Turks, 153
Over-population, 181 f., 203, 219; in China, 156, 171, 174 ff.; in Scandinavia, 260

INDEX

P

Pah Sauh, 177
Pale-Alpine race, 83 f., 209
Palestine, 129 ff.
Palmyra, 119
Parable, of races, 7, 370 f.
Parker, E. H., 148 ff., 376
Parker, T., 324
Parkman, F., 325
Parsis, 245
Parthians, relation to Mongols, 148
Pastoral migrations, of Nordics, 217
Pastoralism. See Nomads, Nomadism, Arabs, etc.
Peace, effect on racial inheritance, 340
Pearson, cited, 358
Pekin, birth-place of officials, 161; modernness of, 164; vital statistics of, 194
Pekin Union Medical College, 194
Pelasgians, 243
Periodicity, of famines, 183
Persecution, of Armenians, 137 ff.; effect on character, 341; among religious denominations, 321; effect on selection, 135
Persia, 54, 59
Persians, racial feeling of, 2 f.
Personal service, and birth-rate, 357
Pertag, 120
Peruvians, 92
Petersson, O., 294, 376
Pfeiffer, Ida, 376
Philanthropy, and racial inheritance, 340
Philistines, 133
Philosophers, 223 (map)
Philosophy, and curiosity, 212; distribution in Europe, 223
Phœnicians, racial feeling of, 2 f.
Physical environment, and racial character, 233
Physicians, eminent children of, 319, 342
Pilgrims, origin of, 303 ff. See Puritans
Pioneer, in parable, 7, 371
Piracy, in ancient Greece, 243; in Iceland, 297
Pithecanthropus erectus, 39
Plague, 178; in Greece, 246
Plains, of China, 171 ff.
Plants, age of, 80
Pliocene man, 31, 61
Plowman, in parable, 8 ff., 371
Pogroms, 135
Poland, birth-rate in, 354; movement from, 77
Pole of cold, 70
Political causes of migration, 258 ff.
Political leaders, in America, 227; in Europe, 224, 227; map of, 225
Population, of Europe in 1800, 235; of Iceland, 270; of Ireland, 248 f.; of Greece, 246; increase in, 362 f.; stress of, 334
Portugal, political leaders in, 227; literature in, 225 f.
Potatoes, in Ireland, 228
Powell, F. V., 294 f., 376
Preglacial man, 42
Presbyterians, eminent children of, 319; racial selection among, 321 ff.
Pressure zone, 48 (map)
Price, Mr., 161
Primogeniture, in England, 250
Printing, in Iceland, 279
Professional men, children of, 342 f., 347
Progress, relation to health, 88
Protestants, leaders among, 342 ff.; in France, 320, 342; origin of, 214 f.
Proto-Australoid race, 83 f., 90, 92, 166, 209; in Africa, 89
Proto-Negroid race, 83 f., 90, 92, 166, 209; in Africa, 89; among Norse, 256
Public service, relation to birth-rate, 357
Pueblo Indians, 92, 108
Puget Sound Indians, 103
Puritans, descendants of, 318, 326, 343; migrations of, 302 ff.; selection among, 303 ff.
Pygmies, 40, 78

Q

Quakers, 305, 322
Quarrymen, birth-rate among, 353 ff.
Queen Aud, 269
Queen Charlotte Islands, 99 ff.
Queue in China, 162 f.

R

Races, birth-rate among, 354 ff.; classification of, 73; defined, 3; Dixon's 83 f.; genius among, 228; kinship of, 74; mingling of, 54 f.; plasticity of, 20; relation to environment, 92
Racial character, 1 ff.; in early America, 90; in California, 330; distribution of, 227; in Iceland, 299; and physical environment, 233; and natural selection, 287
Racial feeling, 2 f.
Racial inheritance, 234
Racial mixture, 7, 78; in China, 166; in Europe, 253; and glaciation, 59; in U. S., 336

Racial superiority, 77
Racial tendencies of civilization, 332
Racial test of cities, 346
Rags, in China, 162, 164
Raids, 115, 117 ff.; in China, 175; season of, 123; of Vikings, 256 ff.
Rain, in Arabia, 121; in China, 170 ff.
Rainfall, in California, 227 (diagram)
Raleigh, Sir Walter, 301
Ran, goddess, 292
Randolph, John, 324
Red Crag, 42
Red Cross, 140, 188
Redfield, C. L., 365 ff., 376
"Red land," 182
Reindeer, 66
Relief, in famines, 179, 189
Religion, distribution in Europe, 223; and environment, 132 f.; and natural selection, 341; and persecution, 222; and racial inheritance, 340
Religious denominations, eminence among, 321 ff.
Religious leaders, 223 (map)
Residence, of American leaders, 316 (map)
Revenue and famine, 179, 181
Reykir hot springs, 289
Rhine, Norse at, 254
Rhode Island, eminent persons in, 317
Rhodesia, 86
Rhone, Norse at, 254
Rich vs. poor in China, 177 ff., 182
Richards, L. E., 324
Ridgeway, W., 236, 376
Rigsmal Edda, 256
Roaring Forties, 91
Rocky Mountain region, Indians of, 100; eminent persons in, 317
Roman Catholic Church, 322 f.; celibates in, 341; eminent persons in, 319 f.
Romans, as migrants, 248, 303; science among, 212
Rome, advantages of, 60
Roundheads, in U. S., 77
Rumanians, broad heads of, 77
Rural districts, birth-rate in, 353; leaders in, 348; natural selection in, 194
Rurik, 254
Rus, 254 f.
Russell, B., 377
Russell, W. S. C., 377
Russia, 227; contact with China, 165, 200; Jewish persecution in, 135; migrations in, 77; Norse in, 254 ff., 261; scientists from, 312; selection in, 340
Russians, hysteria among, 70

S

Sacramento Indians, 104
Sagas, 293
Sahara, 57, 206; migrations in, 59
Saigon, ruins near, 10 ff.
Saint Paul, 340
Salwin, R., 15
Samaria, accessibility of, 132; Israelites in, 131
Samoyedes, 53
Samson, 133
San Francisco, 328 f.
Sand, in Chinese fields, 174
Sappho, 241
Saracens, in Sicily, 266
Savanna belt, 42
Saxons, 248, 326; conquest of, 258 f.; in England, 265; relation to Normans, 264 f.
Scandinavia, center of culture, 252 ff.
School teachers in U. S., 344
Science, and curiosity, 212; among missionaries, 195
Scientists, age of fathers, 365; distribution in Europe, 223 (map); origin, in America, 312 ff., 342, 346
Scotch, character of, 215, 222; migrations of, 77
Scotch Irish, 247
Scotland, celibacy in, 344; eminent persons in, 220 ff., 225; Norse in, 254, 264; political leaders in, 227; Presbyterians in, 322; scientists from, in U. S., 312
Scripps Institute, 327
Sea captains, children of, 319
Sea route, to America, 102
Sea, selective power of, 217
Seeck, O., 212
Seeker-for-Truth, in parable, 8, 371
Seine, Norse at, 254
Selection, in Crete, 238; among denominations, 321; postglacial, 59. See Natural selection
Self-assertion, racial trait, 214 f.
Self-determination, 308
Selfishness, in China, 171, 186 ff.
Self-reliance, among nomads, 128
Semites, racial feeling of, 3 f.
Semple, E. C., 269, 377
Sequoia trees, 260
Serfs, in Greece, 244
Sergei, G., 377
Shakespeare, 301
Shanghai, character of people, 160
Shansi, famines in, 177 f.

Shantung, character of people, 160, 163 *f.*; famines in, 177; migrations from, 200 *ff.*; official greed in, 171; physique in, 185
Sheep herders, in U. S., 113
Shen, 180
Shensi, famines in, 176 *ff.*; migrations from 149
Show, 180
Shyok River, 26
Siam, 15
Siberia, 66; changes of climate in, 56, 58; climate of, 55, 63; environment of, 217; hysteria in, 70
Sicily, 84, 216; character in, 218; environment of, 217; health and seasons, 23; head form of immigrants from, in U. S., 74; Normans in, 265 *ff.*; Norse in, 255
"Sick Man of Europe," 145
Sissera, 130
Size of families, 357
Skin, of man *vs.* animals, 22
Skull, evolution of, 30, 74; measurements of, 81
Small-pox, in Iceland, 297 *f.*
Smiles, Mr., 342
Smith, A. H., 190, 377
Snake and sun cults, 79 (map)
Snorri, 292
Snow, E. C., 358, 377
Snow-fields, effect on winds, 48
Sociability, racial trait, 212
Social environment, and achievement, 352; and birth-rate, 354 *ff.*
Socialism, racial effect of, 215
Soil, and Chinese floods, 173 *f.*
Solar activity, 47
Somme River, Norse at, 254
South Africa, races in, 85 *f.*
South America, Caspian type in, 89; early migrants to, 21
South Carolina, eminent persons in, 315, 318; immigrants to, 305
South *vs.* North, in China, 158, 185 *f.*, 203
Spain, celibacy in, 344; literature in, 226; migrations in, 77; Norse in, 254; persecutions in, 341 *f.*
Sparta, helots in, 237; natural selection in, 244 *ff.*; population of, 246
Speaker-of-the-law, 289, 291
Specialization, and environment, 34
Species, origin of, 80
Spiker, Mr., 168, 199
Spitzbergen, 273
Squashes, 92
"Squeeze," in China, 186

S'rutavarman, 16
Stature, 78
Stefansson, J., 377
Stefansson, V., 67, 279, 288, 377
Stoddard, T. L., 346, 377
Storey, J., 325
Storminess, 23
Storms, 47
Strandloopers, 85 *f.*
Su, 180
Suburbs, leadership in, 348, 352
Suicide, in China, 187 *f.*, 190, 192; among Chukchees, 69; in Europe, 213
Sultans, capacity of, 145
Summer Palace, Pekin, 151 *f.*
Summer rain, value of, 108
Sun cult, 79 (map)
Sunspots, 47
Sun Yat Sen, 167
Swan maiden tales, 79 (map)
Swatow, 199; Hakkas in, 167; vehicles in, 164
Sweden, political leaders in, 227
Swedes, character of, 198
Swimming, in Iceland, 289
Swinton, General, 331
Swiss, character of, 220 *f.*; cleanliness among, 289
Switzerland, achievements of, 231; celibacy in, 344; death-rate in, 287; political leaders in, 227; scientists from, in U. S., 312
Syphilis, in Iceland, 281 *f.*
Syria, civilization in, 59
Syrian Desert, 114 *ff.*; deportations to, 139
Szechuan, 177, 203

T

Taft, W. H., 324
Taimir Peninsula, glaciation of, 55
Taiping rebellion, 167 *f.*
Taishan, 164
Talifer, 265
Tartars, 183, 227; invasions of China, 148, 202; leadership among, 153; racial feeling of, 3 *f.*
Tattooing, 79 (map)
Taxation, in Norway, 258 *ff.* See Revenue
Taylow, Griffith, 75 *f.*, 77 *f.*, 79, 81, 88, 206, 231, 377
Teachers, eminent children of, 342; and marriage, 344
Temperature, in Iceland, 91; and natural

INDEX

selection, 44; optimum, 22 f.; of Siberia, 55, 90
Tents, dwellers in, 112 ff.
Texas, eminent persons in, 316
Thais, 17
Thaxter, R., 91
Theft, in China, 175
Theseus, 239
Thessaly, 243
Thing court, 292
Thirteenth century, disasters of, 17
Thompson, J. A., 352, 377
Thorodd, 291
Thorwaldsen, 279
Thralls, among Norse, 256 f.
Thrift, and farming, 185 ff.
Thucydides, 242, 246
Thurio, 291
Tibet, uncleanliness in, 289; migrations in, 59, 149; plateaus of, 25 f.
Tientsin, modernness of, 164
Tierra del Fuego, 53, 91
Tlingits, 100
Tools, first use of, 28
Torrey, R. A., 184
Totem poles, of Haidas, 99
Totemism, 79
Town meeting, 306
Trade, relation to birth-rate, 353 ff., 357
Trade unions and character, 336
Training, relation to achievement, 352
Transcaspia, races of, 59
Trees, diagram of growth, 227; in glacial Europe, 52; in Miocene era, 37
Trojan War, 240 f.
Tropical environment, effect on human evolution, 23, 40 ff., 49, 51, 58
Tsi, 178
Tsinan, compared with Canton, 163
Tsing, 178
Tso Hou, 177
Tsok, 198
Tuaregs, 123
Tungus, uncleanliness among, 289; hysteria among, 70; migrations of, 149, 195
Turkestan, invasions from, 148
Turkey, character of races, 333
Turkomans, agriculture of, 109; raids of, 120, 126
Turks, 208, 227; and Arabs, 115; character of, 142 ff.; head form of, 77; leadership among, 153; massacres by, 137; and Mongols, 148; political leaders among, 227
Tyler, J. M., 377

U

Ulflgot, 291
Ulster, 247
Unitarians, eminent persons among, 319, 321 ff., 324 ff.; origin of, 324
United States, British settlers in, 65; census of, 378; character of people, 330 ff., 333 f.; divorce in, 361; evolutionary tendencies in, 336 ff.; Indians of, 88; racial mixtures in, 336; roundheads in, 77; school teachers in, 344; size of families in, 357
Unionism, effect on character, 336
Universalists, eminent persons among, 319, 321 ff.; origin of, 322
University of California, 327
University of New York, 324
Upsala, 259
Ural Gulf, 54
Ural race, 83 f.
Urban environment, 348 ff.
Urban vs. rural birth-rate, 353
Utah, eminent persons in, 315 f.

V

Vanden Bergh, L. J., 377
Varangian Guard, 254
Variability, 23
Venereal diseases, 194
Venice, colonies of, 237 ff.
Venn, J., 31, 377
Verkhoyansk, 70
Vermont, eminent persons in, 317
Vienna, 227
Vikings, 212, 254 ff. See Norse
Villages, leaders from, 348
Virginia, eminent persons in, 318; immigrants to, 305
Visher, S. S., 312 ff., 318 ff., 347, 352, 377
Volcanoes, in Iceland, 270 f., 294, 297 f.

W

Wai River, 177
Waller, L. F., 377
Walloons, 303
Wanderers, in China, 173, 184 ff., 190 ff.
Wang Chau, 197
War, biological effect of, 338 ff.; European leaders in, 224
Ward, L. F., 377
Washington, city, eminent persons in, 315
Washington Family, origin of, 303
Washington State, climate of, 328; Indians of, 103

INDEX

Webster, D., 325
Wei, 182
Wei River, 177
Wesley, John, 321
Wetzel, cited, 30, 377
Whites, children of, in Massachusetts, 359; racial position of, 77
Whitney, H., 69, 377
Who's Who in America, 314 *ff.*
Wilder, Dr., 174 *f.*
William the Conqueror, 255, 264 *f.*, 267
Williams, F. W., 151, 154 *f.*, 377
Williams, Roger, 322
Willis, J. C., 80 *f.*, 377
Wilson, W., 308
Winthrop family, origin of, 303
Wissler, Clark, 377
Witches, in Iceland, 297
Wives, purchase of, 190 *f.*
Women, among Arabs, 117 *f.*; in China, 162, 175; cranial types of, 84; "of the Gentiles," 133; among Hakkas, 167 *f.*; hysteria among, 69; and marriage, 343; and migrations, 13, 304; number of children of, 354; Norse, 218; at Plymouth, 305 *f.*; position of, 307, 360; quarters of, in tents, 130; sale of, 187, 189 *ff.*; self-reliance among nomads, 128; 128; in South China, 166; as teachers, 344

Wrangel, cited, 66
Wurm ice age, 53
Wyoming, eminent persons in, 315 *f.*

Y

Yaghans, 91
Yakuts, hysteria among, 70
Yakutsk, in glacial period, 55
Yale University Library, 277
Yang, 180
Yangtse River, 181 *f.*, 195, 202; settlement of valley of, 150
Yellow River, floods of, 177. *See* Hoang Ho
Yen, 178
Ying, 180
Yoder, cited, 365
Yucatan, cleanliness in, 289
Yukagir, 66
Yuman Indians, 104 *f.*
Yunnan, settlement of, 150

X

Xenophon, 246

Z

Zambesi River, 86
Zenobia, 118
Zulus, origin of, 86